Islet Biology and Metabolism

Islet Biology and Metabolism

Editors

Melkam A. Kebede
Belinda Yau

MDPI • Basel • Beijing • Wuhan • Barcelona • Belgrade • Manchester • Tokyo • Cluj • Tianjin

Editors
Melkam A. Kebede
Faculty of Medicine and Health
University of Sydney
Camperdown
Australia

Belinda Yau
Faculty of Medicine and Health
University of Sydney
Camperdown
Australia

Editorial Office
MDPI
St. Alban-Anlage 66
4052 Basel, Switzerland

This is a reprint of articles from the Special Issue published online in the open access journal *Metabolites* (ISSN 2218-1989) (available at: www.mdpi.com/journal/metabolites/special_issues/Islet_Biology_Metabolism).

For citation purposes, cite each article independently as indicated on the article page online and as indicated below:

LastName, A.A.; LastName, B.B.; LastName, C.C. Article Title. *Journal Name* **Year**, *Volume Number*, Page Range.

ISBN 978-3-0365-2737-6 (Hbk)
ISBN 978-3-0365-2736-9 (PDF)

Contents

metabolites

Editorial

Islet Biology and Metabolism

Belinda Yau and Melkam A. Kebede *

School of Medical Sciences, Faculty of Medicine and Health, Charles Perkins Centre, University of Sydney, Camperdown, Sydney 2006, Australia; belinda.yau@sydney.edu.au
* Correspondence: melkam.kebede@sydney.edu.au

This Special Issue, *Islet Biology and Metabolism*, was intended as a collection of studies highlighting the importance of the pancreatic islet—in both form and function—to our growing understanding of metabolic physiology and disease.

The pancreatic islets of Langerhans are composed of five distinct secretory cell types that influence metabolism via the secretion of carefully balanced mixtures of islet hormones into the circulation. By mass, the beta-cells contribute to 60–70% of the islet and are responsible for the secretion of the only glucose-lowering hormone insulin [1]. Alpha-cells, which secrete glucagon, make up another 20–30% of the islet, with delta-cells, epsilon-cells, and PP-cells, which secrete somatostatin, ghrelin, and pancreatic polypeptide, respectively, representing the approximate 10% remaining [1] endocrine cell types. As beta-cells are the primary cell type of the islet, islet function is often studied correlatively with beta-cell function, which in turn is often studied synonymously with insulin secretion.

In most species, including humans, insulin secretion is bi-phasic, with a rapid first phase followed by a slower but sustained second phase of secretion. These distinct phases of secretion are mediated by two pathways. The first, triggered by nutrient or glucose-stimulated closure of the ATP-sensitive K^+-channels and subsequent depolarisation-induced influx of calcium, is termed the "triggering pathway" [2]. The second, termed the "amplifying pathway", which relies on metabolic amplification of the initial stimulus originating at the mitochondria, is reviewed in this Special Issue by Rustenbeck et al. [3]. Gerber et al. also presently reported on dose-dependent responses within mouse and human islets with extremely high glucose concentrations driven by the amplifying pathway—outcomes that have been unappreciated despite comparable clinical observations [4]. Relatedly, the role of mitochondrial metabolism is further highlighted in a study by Kabra et al., who utilised a diet-induced obesity model to correlate and classify properties of islet mitochondrial respiration with respect to glucose-stimulated insulin secretion in the islet [5].

Zooming in deeper within the beta-cell, we approach the subcellular compartments inside which insulin is synthesised, processed, and stored. These vesicles, termed insulin secretory granules, and the proteins involved in their formation, maturation, and secretion, are comprehensively discussed by Germanos et al. [6] within this Special Issue. This article is further complemented by a review on the technical advances and limitations in the isolation of insulin secretory granules for analysis by Norri et al. [7], which reflects particularly on knowledge gaps in the field with respect to insulin granule proteomics.

As we continue to advance our characterisation of the beta-cell and the islet, new techniques and technologies are becoming available. The use of machine learning to augment our analyses is showcased in a study by Cottle et al. [8], which demonstrated the utility of deep learning to model 3D pancreatic islets and measure subcellular proteins of beta-cells within pancreatic slices. As these analyses show their capability to identify cellular polarity within islets—a phenomenon closely associated to beta-cell regulated secretory behaviour [9]—they further highlight their potential to make functional assessments of whole islets in situ.

Indeed, a better understanding of the functional islet in its native environment is especially critical to our understanding of the progression of disease. For example, it

Citation: Yau, B.; Kebede, M.A. Islet Biology and Metabolism. *Metabolites* **2021**, *11*, 786. https://doi.org/10.3390/metabo11110786

Received: 2 November 2021
Accepted: 16 November 2021
Published: 18 November 2021

is well known that maternal nutrition plays an important role in programming beta-cell development and the function of her offspring. In a detailed review, O'Hara et al. discussed our current understanding of fetal exposure to either maternal caloric excess or nutrient restriction. Significantly, they discussed the effects of fetal malnutrition with respect to multiple outcomes, from mitochondrial metabolism to islet morphology and beta-cell function and the consequences for Type 2 Diabetes development [10].

In Type 1 Diabetes (T1D), autoimmune-destruction of the pancreatic beta-cells results in insulin insufficiency [11], and islet transplantation is an established approach to beta-cell replacement therapy for patients with T1D. This involved isolating islets from the pancreas of a deceased donor and implanting them into a T1D patient. Although recent advances in islet isolation and culture techniques have improved the quality of transplant islets and thus the outcomes of patients, there are still several experimental and logistical issues that could be optimised. Presently, Hawthorne et al. discussed the implications and outcomes for islet transplant across large distances with respect to the national islet transplant network in Australia [12]. Furthermore, in an experimental mouse transplant model, Leibiger et al. demonstrated a proof-of-concept technique that allows the expression and functional action of a non-native hormone in intraocular transplanted pseudo-islets [13].

T1D pathology is further reviewed in this Special Issue in the context of the gut microbiome by Priyadarshini et al., who particularly implicated short-chain fatty acid receptors as potential targets for therapy [14]. Additionally, two research articles focused on the prevention of T1D incidence and progression using the spontaneous diabetic NOD mouse model. The first, by Waters et al., investigated the role of the SLC6A19 amino acid transporter in the development of T1D using SLC6A19-deficient female NOD mice [15], and the second, by Borg et al., assessed the benefit of the anti-advanced glycation end products drug Alagebrium Chloride as a pre-diabetic therapy and its subsequent effects on pancreatic function [16].

Altogether, these articles present a high-quality perspective of both innovative and established islet biology research. As guest editors, we would like to thank all the authors for their noteworthy studies, the peer reviewers for their assessments and comments for the refinement of these articles, and the Metabolites Editorial Office for their support and contributions to this *Islet Biology and Metabolism* Special Issue.

Funding: This research received no external funding.

Conflicts of Interest: The authors declare no conflict of interest.

References

1. Da Silva Xavier, G. The Cells of the Islets of Langerhans. *J. Clin. Med.* **2018**, *7*, 54. [CrossRef] [PubMed]
2. Henquin, J.-C. Triggering and amplifying pathways of regulation of insulin secretion by glucose. *Diabetes* **2000**, *49*, 1751–1760. [CrossRef] [PubMed]
3. Rustenbeck, I.; Schulze, T.; Morsi, M.; Alshafei, M.; Panten, U. What Is the Metabolic Amplification of Insulin Secretion and Is It (Still) Relevant? *Metabolites* **2021**, *11*, 355. [CrossRef] [PubMed]
4. Gerber, K.; Whitticar, N.; Rochester, D.; Corbin, K.; Koch, W.; Nunemaker, C. The Capacity to Secrete Insulin Is Dose-Dependent to Extremely High Glucose Concentrations: A Key Role for Adenylyl Cyclase. *Metabolites* **2021**, *11*, 401. [CrossRef] [PubMed]
5. Kabra, U.; Affourtit, C.; Jastroch, M. Respiratory Parameters for the Classification of Dysfunctional Insulin Secretion by Pancreatic Islets. *Metabolites* **2021**, *11*, 405. [CrossRef] [PubMed]
6. Germanos, M.; Gao, A.; Taper, M.; Yau, B.; Kebede, M. Inside the Insulin Secretory Granule. *Metabolites* **2021**, *11*, 515. [CrossRef] [PubMed]
7. Norris, N.; Yau, B.; Kebede, M. Isolation and Proteomics of the Insulin Secretory Granule. *Metabolites* **2021**, *11*, 288. [CrossRef] [PubMed]
8. Cottle, L.; Gilroy, I.; Deng, K.; Loudovaris, T.; Thomas, H.; Gill, A.; Samra, J.; Kebede, M.; Kim, J.; Thorn, P. Machine Learning Algorithms, Applied to Intact Islets of Langerhans, Demonstrate Significantly Enhanced Insulin Staining at the Capillary Interface of Human Pancreatic β Cells. *Metabolites* **2021**, *11*, 363. [CrossRef] [PubMed]
9. Gan, W.J.; Do, O.H.; Cottle, L.; Ma, W.; Kosobrodova, E.; Cooper-White, J.; Bilek, M.; Thorn, P. Local Integrin Activation in Pancreatic β Cells Tar-gets Insulin Secretion to the Vasculature. *Cell Rep.* **2018**, *24*, 2819–2826.e3. [CrossRef] [PubMed]
10. O'Hara, S.; Gembus, K.; Nicholas, L. Understanding the Long-Lasting Effects of Fetal Nutrient Restriction versus Exposure to an Obesogenic Diet on Islet-Cell Mass and Function. *Metabolites* **2021**, *11*, 514. [CrossRef] [PubMed]

11. Bellin, M.D.; Dunn, T.B. Transplant strategies for type 1 diabetes: Whole pancreas, islet and porcine beta cell therapies. *Diabetologia* **2020**, *63*, 2049–2056. [CrossRef] [PubMed]
12. Hawthorne, W.; Davies, S.; Mun, H.-C.; Chew, Y.; Williams, L.; Anderson, P.; Rogers, N.; O'Connell, P. Successful Islet Outcomes Using Australia-Wide Donors: A National Centre Experience. *Metabolites* **2021**, *11*, 360. [CrossRef] [PubMed]
13. Leibiger, B.; Moede, T.; Valladolid-Acebes, I.; Paschen, M.; Visa, M.; Leibiger, I.; Berggren, P.-O. Ectopic Leptin Production by Intraocular Pancreatic Islet Organoids Ameliorates the Metabolic Phenotype of *ob/ob* Mice. *Metabolites* **2021**, *11*, 387. [CrossRef] [PubMed]
14. Priyadarshini, M.; Lednovich, K.; Xu, K.; Gough, S.; Wicksteed, B.; Layden, B. FFAR from the Gut Microbiome Crowd: SCFA Receptors in T1D Pathology. *Metabolites* **2021**, *11*, 302. [CrossRef] [PubMed]
15. Waters, M.F.; Delghingaro-Augusto, V.; Javed, K.; Dahlstrom, J.E.; Burgio, G.; Bröer, S.; Nolan, C.J. Knockout of the Amino Acid Trans-porter SLC6A19 and Autoimmune Diabetes Incidence in Female Non-Obese Diabetic (NOD) Mice. *Metabolites* **2021**, *11*, 665. [CrossRef] [PubMed]
16. Borg, D.; Faridi, P.; Giam, K.; Reeves, P.; Fotheringham, A.; McCarthy, D.; Leung, S.; Ward, M.; Harcourt, B.; Ayala, R.; et al. Short Duration Alagebrium Chloride Therapy Prediabetes Does Not Inhibit Progression to Autoimmune Diabetes in an Experimental Model. *Metabolites* **2021**, *11*, 426. [CrossRef] [PubMed]

metabolites

MDPI

Review

Inside the Insulin Secretory Granule

Mark Germanos, Andy Gao, Matthew Taper, Belinda Yau and Melkam A. Kebede *

School of Medical Sciences, Faculty of Medicine and Health, Charles Perkins Centre, University of Sydney, Camperdown, Sydney 2006, Australia; mark.germanos@sydney.edu.au (M.G.); agao9724@uni.sydney.edu.au (A.G.); mtap8540@uni.sydney.edu.au (M.T.); belinda.yau@sydney.edu.au (B.Y.)
* Correspondence: melkam.kebede@sydney.edu.au

Abstract: The pancreatic β-cell is purpose-built for the production and secretion of insulin, the only hormone that can remove glucose from the bloodstream. Insulin is kept inside miniature membrane-bound storage compartments known as secretory granules (SGs), and these specialized organelles can readily fuse with the plasma membrane upon cellular stimulation to release insulin. Insulin is synthesized in the endoplasmic reticulum (ER) as a biologically inactive precursor, proinsulin, along with several other proteins that will also become members of the insulin SG. Their coordinated synthesis enables synchronized transit through the ER and Golgi apparatus for congregation at the *trans*-Golgi network, the initiating site of SG biogenesis. Here, proinsulin and its constituents enter the SG where conditions are optimized for proinsulin processing into insulin and subsequent insulin storage. A healthy β-cell is continually generating SGs to supply insulin in vast excess to what is secreted. Conversely, in type 2 diabetes (T2D), the inability of failing β-cells to secrete may be due to the limited biosynthesis of new insulin. Factors that drive the formation and maturation of SGs and thus the production of insulin are therefore critical for systemic glucose control. Here, we detail the formative hours of the insulin SG from the luminal perspective. We do this by mapping the journey of individual members of the SG as they contribute to its genesis.

Keywords: insulin; islet amyloid polypeptide (IAPP); granin; secretory pathway; trans-Golgi network (TGN); granule; pancreatic β-cell

Citation: Germanos, M.; Gao, A.; Taper, M.; Yau, B.; Kebede, M.A. Inside the Insulin Secretory Granule. *Metabolites* **2021**, *11*, 515. https://doi.org/10.3390/metabo11080515

Academic Editor: Amedeo Lonardo

Received: 8 July 2021
Accepted: 3 August 2021
Published: 5 August 2021

1. Introduction

The insulin secretory granule (SG) in the pancreatic β-cell is essential for glucose homeostasis in the body. It is both the site of proinsulin conversion into insulin and C-peptide [1], as well as the storage compartment for mature insulin to be readily available for secretion upon nutrient stimuli. Insulin is first synthesized as pre-proinsulin at the endoplasmic reticulum (ER), immediately converted to proinsulin, and transported through the Golgi to the trans-Golgi network (TGN). Here, proinsulin, along with other cargo proteins, is partitioned and sorted into its destination compartment, the immature SG (ISG) [1]. In the ISG, at least 99% of proinsulin is ultimately converted to insulin and C-peptide in a 1:1 molar ratio via proteolytic cleavages by the proprotein convertases PC1/3 and PC2 [2–5]. This coincides with several processes that facilitate SG maturation, including luminal acidification [6], selective removal of certain soluble components [7], and Zn^{2+}-mediated insulin crystallization [8]. Finally, in response to nutrient stimuli, these mature SGs (MSGs) are mobilized to fuse with the plasma membrane and deliver insulin to the bloodstream.

Importantly, ISGs can also undergo regulated secretion [9], which can be heightened in situations of increased β-cell demand [10–12] and may explain the higher circulating proinsulin to insulin ratio observed in both pre-diabetic and diabetic patients [13–19]. The mechanism behind increased proinsulin secretion is unknown; although it has been suggested to result from defective proinsulin trafficking or processing, and/or the premature release of ISGs [20,21]. Interestingly, β-cells from animal models of type 2 diabetes (T2D) display a compensatory expansion of the secretory pathway, characterized by increased proinsulin biogenesis but exhibit a thorough depletion of MSGs, pointing to the existence of a bottleneck in the secretory pathway resulting in an MSG replenishment defect during β-cell failure [12]. Therefore, there is a diversion away from SG maturation in favor of ISG secretion, limiting the compensatory capacity of the β-cell during metabolic stress.

Alongside insulin, the β-cell SG contains a cocktail of cargo proteins. These proteins drive trafficking through the regulated secretory pathway and are also released to affect systemic function [22–25]. Luminal enzymes accompany the cargo from synthesis in the ER through to storage in the MSG but are under tight regulation to restrict their activity to the correct site [4]. The ionic composition of the lumen controls protein behavior and is generated by a range of transmembrane channels and transporters that are stationed throughout the secretory pathway [26–31]. Finally, sorting receptors can escort unwanted components away from the maturing SG to refine its contents after formation [7,32]. In this review, we will explore the major luminal components of the β-cell SG. These components will be discussed in relation to secretory pathway dysfunction, providing context to critical aspects of β-cell failure. However, first, we will start with a historical overview of the process of insulin SG formation.

Historical Overview of Insulin SG Formation

Pioneering efforts in the 1980s elucidated the main concepts surrounding β-cell granule biogenesis. Orci first used immunogold labelling of total insulin, with an antibody that recognizes both proinsulin and insulin, to show that it is closely associated with membranes of the Golgi apparatus until the TGN, where it dissociates and concentrates into a mildly condensing core [33]. This core buds from the TGN into clathrin-coated ISGs, which develop into non-clathrin coated MSGs [33]. At the time, Halban was using pulse-chase methods to incorporate radiolabeled arginine and lysine analogues into newly synthesized proinsulin to inhibit its post-translational processing into insulin [9]. In collaboration, they inhibited proinsulin conversion and utilized autoradiography with clathrin-immunolabeling to provide the first direct evidence that proinsulin traffics from the TGN into clathrin-coated ISGs before its conversion into insulin and C-peptide [1]. Moreover, proinsulin conversion was shown to be required for complete SG maturation, as these analogue-treated cells could not form an electron-dense core which is characteristic of the MSG [1]. Indeed, the development of proinsulin and insulin-specific monoclonal antibodies later confirmed that proinsulin localization is most concentrated in the ISG compartment while insulin dominates the MSG compartment [6].

In 1987, Rhodes and Halban released a landmark study using radiolabeled proinsulin to follow the efficiency of its trafficking and conversion and the events of β-cell SG exocytosis [2]. The study found that 99% of proinsulin entered ISGs to lend itself for conversion, and that the resulting newly synthesized SGs were preferentially secreted over older SGs when exposed to glucose. Importantly, Halban had already shown that radiolabeled conversion-resistant proinsulin is released from the β-cell at the same rate as the non-resistant radiolabeled insulin product, therefore demonstrating that the ISGs housing proinsulin are also secretion-competent [9]. Collectively, foundational work from the 1980s suggested that the SG is the minimal functional unit for exocytosis, is formed through stringent processes, and is endowed with factors required for its regulated release early after formation. It would follow that delayed MSG production could result in the increased release of ISGs and hyperproinsulinemia, and thus a failure of the β-cell to respond to glucose with the secretion of insulin [12,34,35]. Ensuing efforts centering on answering how these carriers are formed have found that an ordered system of ionic and molecular factors underlie how SG proteins are sorted, packaged, and processed [36]. Likewise, efforts centered around understanding the preferential nature of exocytosis have facilitated the characterization of a vast network of components which confer mobility and fusion-competence to prepare the SG for release [37].

The MSG holds at least 50 unique soluble and transmembrane proteins [38], and the biosynthesis of many are thought to be commonly regulated at the translational level following exposure of the β-cell to glucose [39]. This enables their synchronized transit and congregation at the TGN, but from here, several proteins will traverse the ISG compartment on their way to other destinations. Due to this, the ISG intermediary was once the centerpiece of debate concerning the mode of transport that proinsulin and other regulated secretory proteins take en route to the MSG [40]. In the early 1990s, Arvan and colleagues found that C-peptide (the fragment generated from proinsulin after its complete conversion) could be released from the β-cell in molar excess to that of insulin during non-stimulatory conditions [41]. Follow up studies characterized the kinetics of this '*constitutive-like*' secretion, specifically showing that this pathway emanates from the SG compartment and temporally coincides with the maturation of ISGs into MSGs [42]. Subsequent demonstration that insulin, but not proinsulin, is capable of forming insoluble hexamers, led to the idea that insulin condensation within the core of the SG permits the excursion of C-peptide out of the maturing granule as the soluble fraction is removed [43]. Moreover, by analyzing the regulated secretion of lysosomal hydrolase cathepsin B at different time points following pulse-chase radiolabeling, it was revealed that pro-cathepsin B entered the ISG only to be removed from the ISG shortly after entry [43]. Taken together, these studies established the presence of post-Golgi sorting mechanisms that serve to facilitate SG maturation by refining its composition.

Arvan thus proposed that members of the SG were not exclusively trafficked into the regulated secretory pathway from the TGN, but rather that an assortment of proteins were delivered into ISGs through means of unregulated '*bulk-flow*'—largely due to the stoichiometric infeasibility of sorting receptors existing for each cargo [44]. Subsequent post-Golgi mechanisms served to remove and traffic non-regulated secretory proteins to other destinations and drive the maturation of the SG. The term '*sorting by retention*' was used to describe the selective condensation of proteins within the maturing SG, and '*sorting by exit*' was used to describe budding from the vesicle that sequesters parts of the soluble fraction to remove other proteins [44] (Figure 1). This proposal sparked a debate; in particular, proponents against *bulk-flow* asserted that entry of proinsulin and other key granular components into the ISG could not be through a passive, unregulated mechanism [45]. In the end, the field came to the consensus on a tripartite process where luminal TGN protein sorting was also involved in segregating proteins prior to ISG formation, termed '*sorting by entry*' [40]. Moreover, technological advances utilized by recent studies have revealed increasing levels of complexity, showing that some transmembrane components are in fact added to the SG after formation through retrograde plasma membrane/endosomal trafficking [46]. As we explore the luminal components of the insulin SG, we will come to appreciate that SG formation is difficult to lay out as a step-by-step mechanism. Individual components will contribute to multiple steps along the pathway, collaborating through a sequence of events to generate a functional entity that can be released upon stimulus. It is becoming more apparent that correctly forming this entity is crucial for systemic glucose homeostasis.

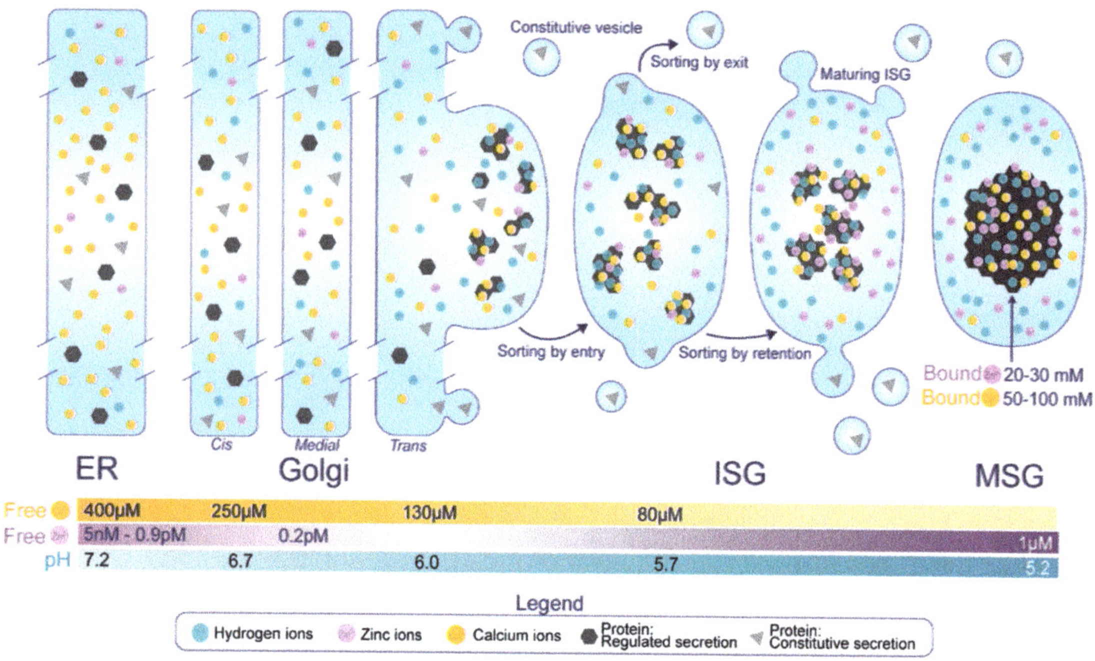

Figure 1. Overview of the β-cell secretory pathway. Following synthesis in the ER, proteins transit the Golgi apparatus to the TGN, and those destined for the regulated secretory pathway are sorted by entry into ISGs. This event relies on soluble protein aggregation which is under the control of Ca^{2+} and H^{+}, and several proteins may also interact with membrane components of the TGN that are enriched in β-cell SGs. Some non-SG proteins can also slip into ISGs but are removed as a byproduct of the sorting by exit mechanism, which specifically escorts proteins from the maturing SG via receptor mediated recognition and vesicle budding. Concurrently, Zn^{2+}, Ca^{2+}, and H^{+} taken up by the maturing SG will bind to certain proteins to enhance their condensation and prevent their exit, in a process termed sorting by retention. While the free concentration of Ca^{2+} is in the micromolar range and decreases proximal to distal along the secretory pathway, the β-cell SG holds 50–100 mM Ca^{2+} bound to luminal proteins. Similarly, the total amount of Zn^{2+} bound to luminal proteins in the SG is in the range of 20–30 mM, although its free concentration is elevated in the distal secretory pathway relative to the proximal secretory pathway. Finally, the pH of the newly formed ISG can be estimated as similar to that of a constitutive vesicle (~5.7), but this will drop to 5.2 in the MSG. Notably, these values represent the free H^{+} concentration, but there exists no indication of the amount of H^{+} that is bound to luminal proteins.

2. Luminal Components of the Insulin Secretory Granule

The luminal components of the insulin SG can be functionally segregated into four groups. These are cargo molecules, luminal enzymes and chaperones, ions (and their transporters and channels), and sorting receptors.

2.1. Cargo Molecules

The primary cargoes of SGs in the pancreatic β-cell are insulin, islet amyloid polypeptide (IAPP), the granins [chromogranin A (CgA), chromogranin B (CgB), secretogranin II (SgII), secretogranin III (SgIII), and VGF (non-acronymic)], and each of their precursors and derivatives. In addition to those covered in this review, the insulin SG also contains amines such as dopamine and serotonin [47–49], as well as nucleotides like ATP [50], which can be taken up by SG-localized pumps but as of yet, have ill-defined intragranular and post-exocytotic roles [51]. In this section, we will demonstrate what is known about the trafficking and processing of each individual cargo protein. These events are heavily dependent on the differential ionic composition of each compartment, where Ca^{2+}, H^{+} and

Zn^{2+} supplied by localized uptake pumps exist in an ascending concentration gradient proximal to distal (Figure 1).

Insulin. Insulin is synthesized as pre-proinsulin on the rough ER, and upon translocation has its N-terminal 24-residue signal sequence cleaved to form proinsulin [52]. Proinsulin undergoes folding in the ER where it acquires three disulfide bonds and dimerizes prior to ER exit [53,54]. En route to the TGN, proinsulin forms hexamers in the presence of Zn^{2+} [53,55], and importantly, proinsulin hexamers remain soluble [43]. Zn^{2+} binds to a histidine corresponding to residue 10 on the B chain of mature insulin (His-B10), and while the precise cisternal location of this event is undetermined, there is evidence of a Zn^{2+}-dependent rate limiting step for proinsulin trafficking around the TGN/ISG compartment [54].

After entry into the ISG, proinsulin is converted to insulin and C-peptide via ordered cleavage at two sites of dibasic amino acid residues by the subtilisin-related proprotein convertases, first by PC1/3 and then by PC2 (Figure 2A). The 31–32 Arg-Arg site is located at the C-peptide/B-chain junction and the 64–65 Lys-Arg site is located between the C-peptide/A-chain junction. Molecular modelling suggests that the co-ordination of Zn^{2+} by His-B10 works to position these sites along the exposed radial surface of the proinsulin hexamer [56], enabling accessibility for the two processing enzymes. PC1/3 preferentially cleaves the B-chain junction on the carboxyl side of Arg32, generating a proinsulin intermediate split between residues 32 and 33 (*split 32,33* proinsulin) [4,5,57]. PC2 preferentially cleaves the A-chain junction on the carboxyl side of Arg65 to generate the *split 65,66* proinsulin intermediate [4,5,57]. Following conversion by each of the subtilisin-related prohormone convertases, the exoprotease carboxypeptidase H/E (CPE) acts to trim the revealed dibasic residues to create the *'des'* intermediates, *des 31, 32*, or *des 64,65* proinsulin, with numbers denoting the excised residues [58]. A second round of endoprotease and CPE activity will generate insulin and C-peptide in a 1:1 molar ratio [3,4]. The insulin molecule consists of an A-chain and a B-chain, linked together by two disulfide bridges and maintained in hexameric oligomers through the co-ordination of two-Zn^{2+} by three of the six His-B10s [8,59]. Continual uptake of H^+ and Zn^{2+} into the developing SG affects the charge state of hexameric insulin and facilitates its packing into extremely insoluble crystals [60]. The low percentage of unprocessed/incompletely processed proinsulin can pack with crystalline insulin to some extent [61] and C-peptide can co-precipitate with insulin in pH conditions mimicking the MSG [62]. C-peptide can also undergo further exoproteolytic cleavage to generate *des 27–31* C-peptide, accounting for roughly 10% of the total C-peptide content [63]. Upon exocytosis, exposure to the neutral extracellular pH is likely to dissipate the insulin crystal rapidly [64], allowing monomeric insulin to circulate and signal via the insulin receptor expressed on target tissues.

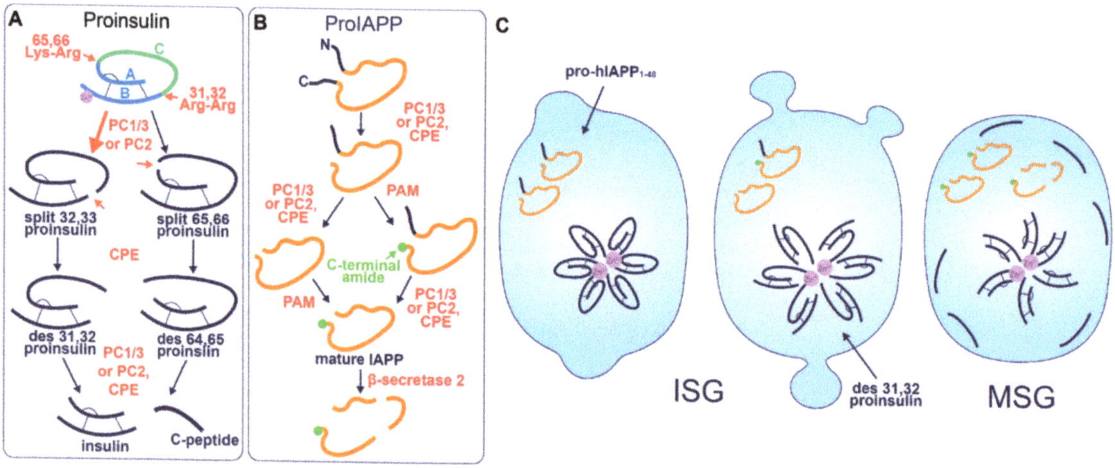

Figure 2. Prohormone Processing in the β-cell. (**A**) Sequence of proinsulin processing. After entry into the ISG, proinsulin is converted to insulin and C-peptide via cleavage at two sites of dibasic amino acid residues. The 31–32 Arg-Arg site is located at the C-peptide/B-chain junction and the 64–65 Lys-Arg site is located between the C-peptide/A-chain junction. Cleavage at one dibasic site by endoprotease PC1/3 or PC2 produces the split proinsulin molecules, which precedes C-terminal trimming of exposed residues by exoprotease CPE to produce the des proinsulin molecules. One round of endo/exoprotease activity is followed by the same action at the other dibasic site. (**B**) Sequence of proIAPP processing. The C-terminal proregion of proIAPP is cleaved in the TGN prior to ISG entry. Next, in no particular order, within the maturing ISG the N-terminal proregion of proIAPP is removed and the exposed C-terminal glycine residue is amidated to produce IAPP. IAPP may then be further processed into smaller fragments by β-secretase 2. (**C**) Processing events and products during secretory granule maturation in the human β-cell SG. des 31,32 proinsulin is the major proinsulin intermediate in human β-cells and is elevated in the circulation of those with T2D along with proIAPP$_{1-48}$.

A human mutation of His-B10 to aspartate (mAsp-B10) underlies familial hyperproinsulinemia [65] and represents a condition where mutant proinsulin is presumed to be excluded from wild-type proinsulin hexamers. While expression of this mutant in mice does not affect its intracellular conversion to insulin, there is an enrichment of non-crystallized SGs, and the constitutive release of proinsulin is increased by ~15% [66]. These phenotypes could indicate that mAsp-B10 proinsulin is correctly targeted into the ISG, but there is an increased constitutive-like release in the absence of Zn^{2+}-facilitated hexamarization prior to its conversion into insulin. Indeed, while contributing to the maturation of the SG, constitutive-like secretion is estimated to account for only 0.6% of the release of non-converted proinsulin [67]. This situation could represent an extreme example of protein exit out of the ISG, displaying the secretory capacity of the constitutive-like pathway. An alternative (and not mutually exclusive) explanation is that mAsp-B10 proinsulin leaks directly into the constitutive pathway from the TGN, however mAsp-B10 proinsulin degradation is also enhanced [66] suggesting that its transit to the PM occurs through the constitutive-like pathway (a route that travels via the endo-lysosomal system [43]). Nonetheless, these studies have highlighted that Zn^{2+}-facilitated hexamarization is a primary mechanism of proinsulin sorting and consequent SG maturation.

Early studies investigating human proinsulin and rat proinsulin isomers I and II in primary islets revealed that they are differentially processed. Human proinsulin tends to be cleaved first at the B-chain junction to produce *des 31,32* proinsulin [68] (Figure 2C), whereas the rat isomers tend to be cleaved first at the A-chain junction to produce *des 64,65* proinsulin [69]. This is thought to be due to the amino acid located four residues prior to the cleavage site (P4 position [70]), where the presence of a basic lysine or arginine residue enhances substrate recognition and/or enzymatic activity [71]. Both rat isomers contain a basic arginine at P4 in the A-chain site, and both rat proinsulin I and human proinsulin contain a basic lysine at P4 in the B-chain site [72]. As a result, rat proinsulin I is more rapidly converted into insulin, and the accumulation of processing intermediates from this isomer is reduced due to the existence of basic residues at P4 in both cleavage sites [69].

Processing of human proinsulin follows a sequence favoring the prior activity of PC1/3 on the B-chain junction, followed by PC2, which has a far better affinity for *des 31,32* proinsulin than intact proinsulin [73]. Although this points to the existence of sequential cleavage through the action of both endoproteases, multiple lines of evidence indicate that PC1/3 works alone to produce mature insulin from both rat and human proinsulin isomers. While each enzyme possesses the catalytic ability to cleave at both dibasic sites [74], PC1/3 achieves this far more efficiently than PC2 [75–77] and processing intermediates of proinsulin accumulate when PC1/3 expression is low [76]. The situation is different in mice, seemingly requiring the activity of both endoproteases; while the deletion of PC1/3 from mice results in an extremely pronounced block in proinsulin conversion [78], knockout of PC2 also significantly hampers insulin maturation despite the presence of PC1/3 [79]. Finally, a recent study that re-characterized the expression of PC1/3 and PC2 in human islet β-cells found an abundance of PC1/3 and an absence of PC2, suggesting that PC1/3 is sufficient for humans to produce insulin [80]. Interestingly, humans with T2D had upregulated PC1/3 and an induction of PC2 expression. The authors of this study speculated that aberrant PC2 expression could cause a processing defect that underlies the pathological state, although, it may be the case that PC2 expression is invoked by metabolic stress as a compensatory response to assist PC1/3 in proteolytic activities. Indeed, the catalytic rate of PC2 on des 31,32 proinsulin exceeds that of PC1/3 on intact proinsulin [73]. Simple overexpression of either PC1/3 or PC2 has been shown to enhance proinsulin conversion in rat insulinoma INS1 cells [81], hence, induction of PC2 activity could support proinsulin conversion when PC1/3 is overwhelmed especially considering that the A-chain junction is not preferred by PC1/3 [4,57].

Both PC1/3 and PC2 endoprotease activities are sensitive to pH and Ca^{2+}. In vitro assays using enzymes isolated from rat islets have shown that PC1/3 requires millimolar levels of Ca^{2+} and a pH close to 5.5 for activity whereas PC2 can exert activity at a micromolar levels of Ca^{2+} and over a broader pH range, although its pH optimum is also 5.5 [4]. In cells however, PC1/3 undergoes fast maturation into an active enzyme upon entry into the SG [82]. Due to the stringent regulation of PC2 by the molecular chaperone 7B2 [83–85], the low pH requirement for its autocatalytic activation [82,86], as well as its substrate-specificity to *des 31,32* proinsulin [73], its activity is likely to be restricted to later stages of SG maturation. Therefore, it appears that early PC1/3 activity at both sites could render PC2 redundant, as has been demonstrated in animal models [75–77] but not quite yet in humans. Crucially, compensatory upregulation of the endoproteases may be futile, considering the premature ISG release that occurs in β-cell failure. It has been known for some time that *des 31,32* intermediates are the predominant species of circulating proinsulin that is elevated in human T2D [87,88], therefore fast endoprotease activity is critical for systemic metabolic homeostasis. Therapeutic compounds that alter the ionic composition of the SG to bolster endoprotease activation and activity could be effective in treating T2D.

Despite a common outcome, nuances in the generation of insulin are clear between species. Their awareness may be important for translating data from model organisms to the context of human β-cell function.

Islet Amyloid Polypeptide. IAPP is a 37 amino-acid peptide stored in the MSG that is co-secreted with insulin in a 1:100 molar ratio [89–91], and can function to suppress insulin secretion and control various aspects of energy homeostasis [22,92]. Additionally, known as amylin, IAPP and its precursors and derivatives are notorious for forming fibrils that distribute extracellularly throughout islets as amyloid deposits, a pathological feature of human T2D [93]. Early observations report the occurrence of islet amyloid deposits in >90% of diabetic patients [94,95] but later studies have shown a variable prevalence depending on duration of disease and ethnicity, especially when sample size is increased [96]. The question of how IAPP remains non-pathogenic in healthy conditions and how it transitions to a pathogenic molecule has kept researchers occupied for some time. Appropriately, most work on IAPP has been focused on its secretory dynamics, processing, and amyloidogenic properties [97–99] rather than its luminal sorting. To this end, not much is known about its behavior in the early secretory pathway or the determinants of its trafficking fate.

Akin to proinsulin, human pro-IAPP is a 67 amino-acid (aa) peptide derived from pre-proIAPP that forms an intramolecular disulfide bridge in the ER and is subject to endoproteolytic processing [100–103] (Figure 2B). It is thought that PC1/3 acts first on a C-terminal proregion in the TGN [102] which is followed by CPE action to generate a 48-residue processing intermediate. A subsequent round of PC1/3 or PC2 and CPE action on the N-terminal proregion generates a 38 aa peptide with a C-terminal Gly termed amylin free acid [104]. A fourth enzyme, peptidyl-glycine alpha-amidating monooxygenase (PAM), is probably responsible for amidation at the C-terminal 38 glycine residue which may or may not occur prior to cleavage of the N-terminal pro-region, to generate the mature C-terminally amidated form of IAPP [104]. Finally, a fifth membrane-bound enzyme that localizes to the β-cell SG, β-secretase 2, can process IAPP further into smaller fragments [105].

Early reports demonstrated that PC2 can cleave at both the N- and C-terminal proregions of pro-mouse IAPP (mIAPP) in addition to PC1/3 cleavage occurring only at the C-terminal site [100–103]. However, both rat and human islets appear not to express PC2 at detectable levels normally [80]. Indeed, pro-human IAPP (hIAPP) can be fully processed by PC1/3 in PC2 null mice [106]; however, pro-rat IAPP (rIAPP) can be processed at both sites by PC2 but only at one site by PC1/3 [103]. These differences are likely due to the modified sequence at the C-terminal site (Figure 3), where the position of Ala and Val residues may determine whether PC1/3 can cleave: i.e., KR↓VA (Val at P1′ and Ala at P2′) in rIAPP compared to KR↓AV (Ala at P1′ and Val at P2′) in hIAPP and KR↓AA (Ala at P1′ and P2′) in mIAPP. Likewise, it has also been suggested that a Val common to both hIAPP and rIAPP at the N-terminal site allows cleavage by PC1/3 [80], which is not experienced by mIAPP [100] that contains Met at this residue. Thus, PC1/3 may be sufficient for full pro-IAPP processing in humans due to the sequence variations that lie between species.

```
H: -MGILKLQVFLIVLSVALNHLKATPIES---HQVEKRKCNTATCATQRLANFLVHSSNNFGAILSSTNVGSNTYGKRNAVEVLKREPLNYLPL 89
M: MMCISKLPAVLLILSVALNHLRATPVRSGSNPQMDKRKCNTATCATQRLANFLVRSSNNLGPVLPPTNVGSNTYGKRNAAGDPNRESLDFLLV 93
R: MRCISRLPAVLLILSVALGHLRATPVGSGTNPQVDKRKCNTATCATQRLANFLVRSSNNLGPVLPPTNVGSNTYGKRNVAEDPNRESLDFLLL 93
pre-proIAPP
proIAPP
proIAPP1-48
amylin free acid
IAPP
```

Figure 3. IAPP precursor/product amino acid sequence in human (H), mouse (M), and rat (R). The green glycine residue is amidated after the C-terminal cleavage site is processed. Red residues denote dibasic sites of endoproteolytic processing. Blue residues indicate a modified sequence between species at cleavage sites that could account for their differential specificity to PC enzymes.

IAPP resides in the soluble fraction of the MSG, and hIAPP is extremely fibrillogenic whereas rIAPP and mIAPP are not fibrillogenic at all [107]. It has been shown in vitro that pro-hIAPP products become increasingly more amyloidogenic with further cleavages [108], however, it is thought that the presence of ionic and molecular factors in the MSG periphery inhibits hIAPP oligomer formation in healthy conditions [109–114]. An extensive body of literature has covered the molecular mechanisms of hIAPP pathogenicity in T2D [97–99]. Considering that hIAPP is likely not a driver of SG biogenesis or function but instead plays the role of chaotic passenger, we will focus here on how secretory pathway dysfunction could precede hIAPP-mediated β-cell damage.

A current working hypothesis to explain the initiation of islet amyloid formation is that hIAPP-related peptides undergo dysregulated fibrillogenesis at some point inside the β-cell secretory pathway, potentially due to overproduction during compensation/failure [115], but this only occurs within a small subset of islet β-cells [97,98,116–118]. If pro-hIAPP overproduction is the fuel, then disruption of organellar membranes from within the cell is the spark that results in β-cell death and the deposition of extracellular amyloid. Through regular exocytosis from neighboring β-cells, released hIAPP can then add to the size of the initial deposit. An alternative (and not mutually exclusive) hypothesis suggests that release of the 48-residue pro-hIAPP (pro-hIAPP$_{1–48}$) intermediate can initiate extracellular amyloid formation by a specific interaction with heparin sulfate proteoglycans in the extracellular matrix [119]. Interestingly, large ordered fibrils that make up the bulk of the visually identified IAPP deposition are thought to be relatively inert on a cytotoxic level [99], although interruption of islet cytoarchitecture could impair coordinated islet function. Rather, it is the presence of medium-sized disordered oligomers that are thought to exert most of the cellular damage [120]. Fitting with this, recent experimental focus has instead been placed on the mechanism of medium-sized oligomer formation and cytotoxicity [121,122].

Not surprisingly, the N-terminal prosequence of hIAPP is detectable in islet amyloid deposits [123,124]. This observation resembles what is observed with proinsulin in that incompletely processed hIAPP may be released from the β-cell during T2D, and indeed, elevated serum pro-hIAPP has been observed in glucose intolerant and T2D patients [125]. If one considers that ISGs released during β-cell failure are a source of hIAPP processing intermediates (Figure 2C), we could look to luminal factors that might explain the propensity of these molecules to become pathogenic. Indeed, insulin, Zn^{2+}, H^+, Ca^{2+}, C-peptide, and proinsulin have been assessed individually or in combination, in vitro, along with hIAPP, in various studies to reason that a delicate balance of cofactors is required to inhibit hIAPP oligomerization [109–114,126]. In healthy cells, regulated exocytosis of MSGs could maintain this balance as components are released in an appropriate molar ratio. Conversely, during β-cell compensation and failure, release of the incompletely formed ISG might not replicate this outcome, and cytotoxic hIAPP oligomers could form in the extracellular microenvironment adjacent to the plasma membrane to induce membrane damage.

Dysregulated hIAPP oligomerization exacerbates the progression of T2D, so preventing β-cell death at the hands of hIAPP could limit T2D severity. Abnormal SG composition or the premature release of ISGs may be contributing factors, highlighting the importance of correctly forming the insulin SG.

Granins. The granin family of proteins (CgA, CgB, SgII, SgIII, and VGF) are ubiquitously expressed across neurons and endocrine cells and are considered major contributors to the biogenesis of regulated SGs from within the lumen (Figure 4). Their effectiveness has been displayed by several groups with findings that expression of just a single granin in cells that do not have a regulated secretory pathway is able to produce SGs that are capable of regulated release [24,127–130]. Granins are synthesized as soluble cargo precursor proteins, which are highly acidic and hydrophilic, but are prone to aggregation under mild acidity ($pH < 6.4$) and high Ca^{2+} (>1 mM) conditions [25,131]. It has been shown that both of these ionic requirements must be met for granin aggregation [131,132], which can be achieved at the initiating site of SG biogenesis in the TGN (Figure 1) where ion pumps maintain a high luminal Ca^{2+} concentration and contribute to a substantial lowering of the pH through enhanced H^+ uptake [27,133–135]. Moreover, several granins have been shown to interact with each other, and, lacking transmembrane domains themselves, some can also interact with lipid species on the luminal aspect of the secretory pathway membrane to provide a link between soluble and membrane fractions. In this way, their physical abundance, coordinated aggregation within the TGN, and binding to specific components on the membrane has been proposed to drive the segregation and sorting of peptide hormones and other proteins into the regulated secretory pathway [23], meeting the requirements of a 'refined bulk-flow' *sorting by entry* mechanism [40].

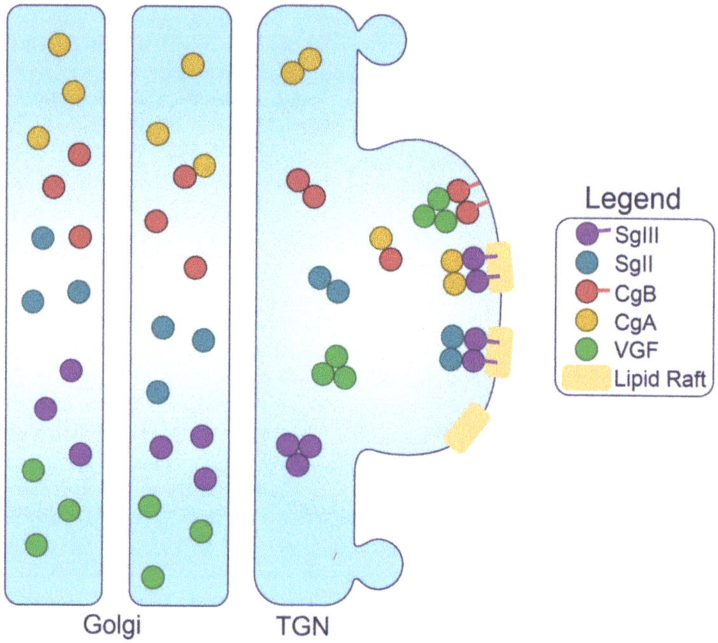

Figure 4. Trafficking of the granins. Upon exposure to mild acidity (pH < 6.4) within the Golgi apparatus, the granins will bind Ca^{2+}, which triggers their aggregation and interaction with other granin members. In the TGN, several of these members will interact with target molecules in the membrane to drive the formation of SGs at distinct sites from within the lumen.

Sphingolipid–cholesterol lipid rafts accumulate in microdomains of the TGN [136,137], and these rafts can alter the distribution of transmembrane components to create sorting stations that are essential for granule biogenesis [138,139]. These rafts are also enriched in vesicles of the regulated secretory pathway [140–143], indicating that SG membranes originate from the sorting domains where granins and other SG constituents aggregate and bind. Importantly, saturated fatty acid and cholesterol intake can change the composition of lipid species distributed among cell membranes to influence trafficking and SG morphology [144,145]. Therefore, dietary status could affect interactions between the granins and secretory pathway membranes, but this requires investigation. Finally, since granins can bind Ca^{2+} at a high capacity with low affinity [132], they are also thought to equip the SG with the ability to store and release Ca^{2+} [146,147].

It should be noted that most of the literature on granins has reported on their role in neuronal/neuroendocrine cell lines, which although share features in common with the β-cell, have secretory pathways adapted to the specific needs of neural transmission. In these settings, we can draw insight from molecular interactions that govern trafficking and behavior of the granins themselves, but specific effects of granin depletion on SG biogenesis/secretion are often subject to cell-specific variation and thus will only be discussed with respect to the β-cell.

Granins can possess multiple sorting determinants and may be targeted to several SG sub-populations (Figure 4). SgIII is membrane-associated, and contains an N-terminal lipid-binding region that is required for its sorting into SGs of AtT-20 cells [148] and for its interaction with cholesterol in INS1 and AtT-20 SGs [149]. This suggests that SgIII is sorted into the regulated secretory pathway through an interaction with TGN cholesterol [148]. N-terminal residues (48–111) of CgA can bind to SgIII to follow SgIII sorting into the regulated secretory pathway of AtT-20 cells, where it also exists in association with SG membranes [149]. Importantly, CgA also associates with INS1 granule membranes but only in the presence of SgIII [148]. These results collectively indicate that SgIII is an adaptor for CgA in β-cells, with both granins associated at least to some degree with the SG membrane, and that correct trafficking relies on the presence of an N-terminal region on SgIII that binds cholesterol. This has been further demonstrated in PC12 cells, where SgIII was specifically shown to sort large aggregates of CgA into SGs [150]. Very little is known about the trafficking determinants of SgII aside from an understanding that both the N- and C-termini contain information for sorting [151], and that it may interact with SgIII on the SG membrane [152,153]. While SgII also regulates granule biogenesis in other secretory cells [154], interactions have not been published in β-cell models, and in general, insight into SgII function in the β-cell is lacking.

Like SgIII, CgB can interact with cellular membranes via a highly conserved N-terminal 22 residue disulfide-bonded loop [155,156]. This loop is essential for CgB sorting to SGs but is not required for its aggregation within the TGN, indicating that CgB aggregates are not routed by default to SGs but are sorted through mediation of exposed N-terminal loops with the TGN membrane [155]; although, the corresponding membrane component is yet to be found. Several observations suggest that CgB trafficking is not entirely synchronous with insulin. In addition to the insulin SG, CgB also occupies distinct granules that do not contain insulin and conversely, insulin can occupy SGs devoid of CgB [157]. Additionally, CgB is present with SgII in nucleoplasmic vesicles of bovine chromaffin cells where they may have a role in regulating nuclear Ca^{2+} homeostasis [158], although this has not been studied in the β-cell. In the β-cell, CgB co-localizes and co-immunoprecipitates with VGF [159], and it has been shown in vitro that CgA and CgB can form dimers at pH 7.5 and heterotetramers at pH 5.5 [160], suggesting that CgB could traffic with either VGF or CgA. However, VGF does not immunoprecipitate with CgA [159]. Little else is known about the determinants of VGF trafficking, although a predicted alpha helix loop in its C-terminus may be required for direction into INS1 SGs [161].

A handful of studies have investigated the consequences of granin depletion in β-cells albeit with varying success, possibly due to the method of study. Transient gene silencing seems to outcompete stable knockouts for studying function, and this is probably due to the circumvention of compensatory changes that occur during development. For example, whole body CgB knockout (KO) provided an insulin secretory defect that was unable to be explained aside from a small decrease in the number of docked SGs [162], whereas adenoviral knockdown (KD) of CgB in INS1-832/3 insulinoma cells and isolated mouse islets resulted in marked insulin secretory defects that could be explained by a defect to SG biogenesis [159]. Similarly, islets from whole body CgA KOs actually have enhanced insulin secretion with no defects to SG generation [163], whereas siRNA KD of CgA in the human β-cell line, EndoC-βH1, resulted in reduced basal and glucose-stimulated insulin secretion (GSIS) as well as cellular insulin content [164]. CgA KO mice had compensatory doubling in CgB expression and tripling in SgII expression [163], which may explain the absence of a secretory phenotype in CgA KOs. Islets from whole-body SgIII KO mice have impaired GSIS but only when subject to a high-fat-high-sucrose diet. This is associated with reduced insulin and increased proinsulin content, but there were no reported ultrastructural granule abnormalities [165]. Interestingly, in this study, CgA levels failed to increase when SgIII KO mice were put on diet but did so in the islets of wild-type mice [165]. As discussed previously, SgIII is a known adaptor for CgA and therefore its absence could result in CgA mis-sorting and thus the failure of compensatory upregulation. Finally, VGF depletion has also been assessed via KD of its mRNA in INS1-832/3 cells and a tamoxifen-inducible KO from mouse islets [166]. This study noted reduced GSIS in both models, associated with a loss of total and docked SGs, and a reduction to their size in line with an increased cellular proinsulin-to-insulin ratio and delays to proinsulin conversion [166]. This study concluded that VGF depletion caused a granule replenishment defect, hampering the secretion of newly synthesized granules during the sustained second phase of GSIS [166].

In summary, the granins are critical components of ISG formation, driving the formation of regulated carriers from within the lumen through aggregating and binding to distinct sites of the TGN membrane. Their ubiquity across cells of the neuroendocrine system implies an essential role for SG function, where their combined abundance and aggregative nature may confer unique characteristics to the SG.

2.2. Luminal Enzymes and Chaperones

Several enzymes and chaperones undergo processing in the secretory pathway and are targeted to the β-cell SG to generate the diverse intragranular cocktail. PC1/3, PC2, CPE, the PC2 binding partner 7B2, PAM, furin, chaperonin 60, and β-secretase 2 are members of this functional group that have activity in the β-cell. Active furin is widespread across the trans Golgi network, cell surface and endosomes, however it traverses the ISG before being sorted out of the maturing granule [167]. β-secretase 2 is a transmembrane aspartic protease that was previously mentioned for its role in mature IAPP proteolysis [105,168]. Chaperonin 60, a heat shock related protein, has also been found to co-localize and co-immunoprecipitate with proinsulin and PC1/3 [169]; however, the functional significance of this protein has not been investigated. Since we have already discussed the activity of CPE, PC1/3 and PC2 on proinsulin and proIAPP conversion, here we will restrict their discussion to trafficking and activation.

Carboxypeptidase E. CPE, also known as CPH, exists in both soluble and membrane-associated forms in β-cells [170]. An alpha-helix in the C-terminus of CPE anchors through cholesterol rich lipid rafts of the secretory pathway membrane, leaving six residues protruding to the cytoplasm [171,172] (Figure 5). Importantly, penetration through the membrane only occurs at or below pH 6 [172], conditions reflecting the late Golgi and SG compartments but not the proximal Golgi or ER [173] (Figure 1). Several lines of evidence have also demonstrated that membrane-bound forms of CPE can aggregate at this pH with at least 1 mM Ca^{2+} [174], and co-immunoprecipitate in these conditions with both pro-opiomelanocortin and insulin in vitro [175]. This aggregation appears to occur independently from membrane binding, since treatment with Triton X-100 at pH < 6 to interfere with aggregation does not dissociate CPE from the membrane [174]. Collectively, pH-dependent lipid-raft insertion and aggregation provide a means by which CPE is concentrated along the TGN membrane for targeting to the SG. Moreover, it has been shown that CPE interacts with SgIII in both INS1 and AtT-20 cells [176], providing more control over targeting. Following SG entry, CPE is cleaved by an endoprotease at its C-terminus to generate the soluble, major enzymatic form of CPE [177] (Figure 5). Immunostaining reveals its predominant localization to the SG at steady state [178], where its enzymatic activity operates in a narrow pH optimum between 5.0 and 5.5 [179].

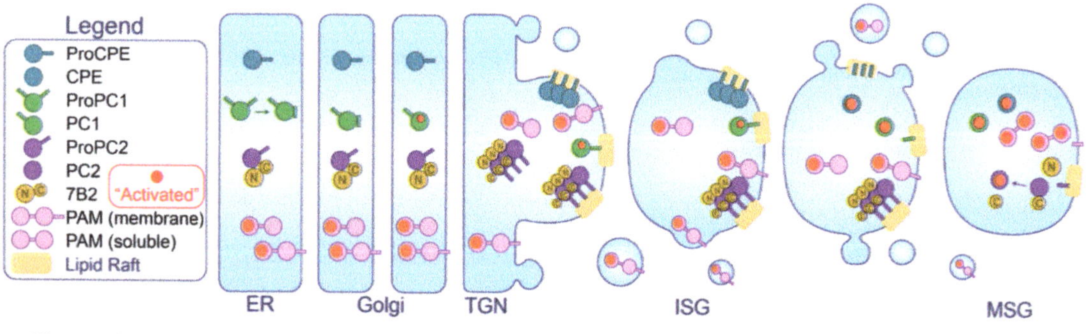

Figure 5. Enzyme suppression and activation. The major processing enzymes of the insulin SG are synthesized as inactive precursors in the ER. Through mechanisms unique to each member, their activity (indicated in red) is suppressed during transit. Activation is principally driven by ionic changes; several enzymes require certain conditions for trafficking into the ISG and conformational activation, and all members require a low pH for optimal enzymatic activity. This will naturally play out through the TGN and the maturing SG as the luminal composition is modified.

PC1/3. The trafficking and activation of PC1/3 is considerably less complicated than PC2. In the ER, pre-proPC1/3 undergoes cotranslational signal peptide removal to generate a 94 kDa pro-form of PC1/3 [180,181]. This precursor harbors enzymatic activity but only toward its own N-terminal pro-region, which is cleaved in the ER [180,181] and thought to remain associated with the catalytic site of mature 87 kDa PC1/3 to prevent activity in the early secretory pathway [182] (Figure 5). This will later dissociate after a second cleavage by PC1/3 [182]; however, the location of this event (TGN or ISG) is not settled yet. Moreover, the C-terminal region partially inhibits PC1/3 activity [183,184], and befittingly, PC1/3 can process certain substrates prior to its entry and complete activation inside the SG [185,186]. The C-terminal region also contains a predicted alpha helix required for sorting into the regulated secretory pathway [187], likely through an interaction with membrane lipid rafts [188]. Following entry into the SG, the inhibitory C-terminal region is cleaved (possibly by itself) to generate fully active 74 and 66 kDa products [183,184,189] (Figure 5), providing a relatively simple activation mechanism in the SG. 87 kDa PC1/3 exhibits a pH optimum between 5.5 and 6.5 [190], respective conditions reflecting the SG and the TGN (Figure 1). Both 74 and 66 kDa products exhibit pH optima between 5.0 and 5.5 [4,189], reflecting the MSG. All forms also have a high Ca^{2+}-dependence [189,190], so the luminal ionic composition must be optimized for PC1/3 activity.

PC2 and 7B2. In the ER, pre-proPC2 undergoes cotranslational signal peptide cleavage to generate a 76 kDa proPC2. Unlike proPC1/3, its inhibitory N-terminal pro-region is not cleaved and remains associated with the catalytic subunit until it reaches the SG [191]. ProPC2 traffics through the secretory pathway together with its chaperon 7B2 [192]; after synthesis and folding, proPC2 can bind pro7B2 in the ER where it accelerates proPC2 trafficking to the Golgi [83,193,194]. Pro7B2 is proteolytically cleaved at Arg152 (a furin cleavage site) in the TGN to generate a 21 kDa N-terminal (7B2-NT) and a 5 kDa C-terminal peptide (7B2-CT) [85,195], both of which remain associated with proPC2 [86]. While the 7B2-NT appears to maintain proPC2 folding and trafficking, 7B2-CT is a well-established PC2 inhibitor in vitro [196] (Figure 5)

Both proPC2 and 7B2-NT can aggregate under pH- and Ca^{2+}- conditions mimicking the TGN [197,198]. Residues 45–84 in proPC2 have been shown to mediate its association with TGN membranes [199]. Here, proPC2 likely interacts with sphingolipids in the TGN membrane since sphingolipid depletion causes re-routing of transfected mature PC2 to the constitutive pathway [199]. ProPC2 also requires 7B2 for direction [200], so 7B2 depletion will cause proPC2 to traffic constitutively [201]. Therefore, in the absence of 7B2-peptides, unfolded, improperly aggregated PC2 could route constitutively [192,201]. In the SG, proPC2 does not undergo full autocatalytic maturation until the luminal pH drops to 5.2 [86,191], conditions reflecting the MSG (Figure 1), and once fully mature PC2 cleaves and removes the inhibitory 7B2-CT fragment [84]. Thus, although PC2 is active in vitro over a broad range of pHs and Ca^{2+} [4], its fully active form is restricted to the MSG within cells.

PAM. PAM is a bifunctional enzyme consisting of two contiguous catalytic domains, peptidylglycine alpha-hydroxylating monooxygenase (PHM) and peptidyl-α-hydroxyglycine α-amidating lyase (PAL) [202] (Figure 5). Sequential action of PHM followed by PAL functions to amidate glycine at the carboxyl terminus of peptides [203], which removes charge to confer full biological activity to the peptide [204]. Cargoes that have already been processed by PC1/3, PC2 and CPE to yield C-terminal glycine residues are generally subject to this reaction [205]. While PHM is active over a broad acidity range [206], the stability of the intermediate formed by PHM declines at pH levels >6.0 [207] and the pH optimum for PAL is around 5.0. At least 50% of all biologically active peptides require amidation for full biological activity [208], and thus far, PAM is the only enzyme identified to be responsible for this reaction in vivo. In both human and mouse islets, PAM co-localizes with insulin, glucagon, and somatostatin [164,209], and while insulin is not a target of PAM, IAPP is a potential target [104] and CgA was recently verified as a PAM substrate in β-cells [164]. PAM depletion can affect insulin content and GSIS which may be mediated by its downstream targets including CgA [164]; however, PAM haploinsufficiency in mice does not accelerate diet- or IAPP-induced diabetes [209]. Notably, T2D-associated PAM risk alleles exist that are associated with reduced insulinogenic index [164,209], thus, PAM activity appears to have relevance to β-cell function.

Several isoforms of PAM differentially expressed between tissues [210] are produced by alternative mRNA splicing; some are soluble and others are type-I transmembrane enzymes [211]. Human β-cells only express soluble isoforms whereas mouse β-cells only express transmembrane isoforms [209]. Isoform differences are due to the presence/absence of specific regions, including a C-terminal transmembrane region, a linker region between the two contiguous enzymes PHM and PAL, as well as each of the two enzymes themselves [210]. Both forms are enzymatically active although integral membrane PAM will presumably have more restricted access to substrates [212].

Both soluble and integral forms of PAM traffic simultaneously through the early secretory pathway, but then diverge upon entry into the ISG [213]. The enzymatic domains of PAM contain information for direction toward the regulated secretory pathway since expressed soluble PHM or PAL traffic correctly into ISGs [214], however, the transmembrane/cytosolic aspect of integral membrane PAM can override luminal trafficking information for transit via an independent route [213,215]. While both forms enter the ISG efficiently and are retained to some degree within maturing SGs, they can both exit the ISG through the constitutive-like pathway in unstimulated conditions [213]. Integral membrane PAM exits the SG to a greater extent [213], and has been shown to cycle through the PM where it may be retrieved to the TGN [216] or the MSG [217,218].

Since human β-cells only express PAM3 [209], a soluble isoform, its trafficking is relatively simple. The situation is more complicated in mouse β-cells which express transmembrane isoforms PAM1 and PAM2 [209], and thus, are subject to additional trafficking fates and require endoproteolytic cleavage within the SG to generate soluble PAM to access substrates more readily [212]. In addition to a low pH [206,212], Zn^{2+} and Ca^{2+} [219], PAM requires Cu^{2+} [220] and ascorbate [203] for activity.

Collectively, enzymes control the intragranular landscape by modifying proteins and altering their properties, operating as the focal regulatory units of the SG lumen. This culminates in the main transformative event of the granule interior—the crystallization of proteolytically generated insulin—which creates an extremely dense proteinaceous core. Though deploying enzymes that require such specific conditions for activity, it appears that the β-cell strikes a balance between rapid processing and orderly aggregation to ensure the safe generation of a functional product.

2.3. *Ions, Transporters, and Channels*

Transporters and channels embedded into the secretory pathway membrane control the composition of the luminal milieu to facilitate cargo sorting and processing. They also control the release of ions from the SG store to regulate cytosolic events. In this section, we will discuss the coordinated function of transporters and channels that modulate important features of the intergranular lumen.

Activation: H^+. Foundational studies from the Hutton lab established the low pH of the β-cell SG [221] and found activity of an ATP-dependent pump responsible for translocating H^+ into the granule lumen [26]. The vacuolar H^+-ATPase (V-ATPase) localizes to the β-cell SG [133] and is the major component responsible for endoprotease activation and cargo processing in regulated secretory cells [27,134]. Due to the influx of positive charge, a complimentary influx or efflux of anions or cations, respectively, would be required to maintain a normal electrochemical gradient across the SG membrane and this is normally achieved by the counterregulatory actions of Cl- transporters [222–226]. Moreover, other transporters can utilize the proton gradient generated by V-ATPase to exchange H^+ for cytosolic materials. For example, vesicular monoamine transporter type-2 (VMAT2) can exchange intragranular H^+ for cytosolic monoamines, thus functioning to regulate the luminal pH [227].

V-ATPase is widely distributed throughout the endo-lysosomal system and the plasma membrane. It consists of two complexes, V_1 (cytosolic) and V_0 (membrane-associated) (Figure 6), which, respectively, contain eight and nine subunits [228]. ATP hydrolysis by V_1 provides energy for V_0 to rotate and pump 2–4 H^+ molecules from the cytosol into luminal or extracellular spaces [229], and this can be controlled by several factors. Dissociation of V_1 from V_0 is the primary mechanism of pump regulation [230] and is sensitive to glucose exposure [231]. Complex assembly is affected by the membrane lipid composition; the β-cell SG contains an abundance of enriched lipid species, and sphingolipids are thought to stabilize assembly and facilitate ATP hydrolysis [232]. Localization and density of the pump is obviously limiting for compartmental acidification, thus, regarding the β-cell SG, an abundance of V-ATPase is critical for luminal protein processing.

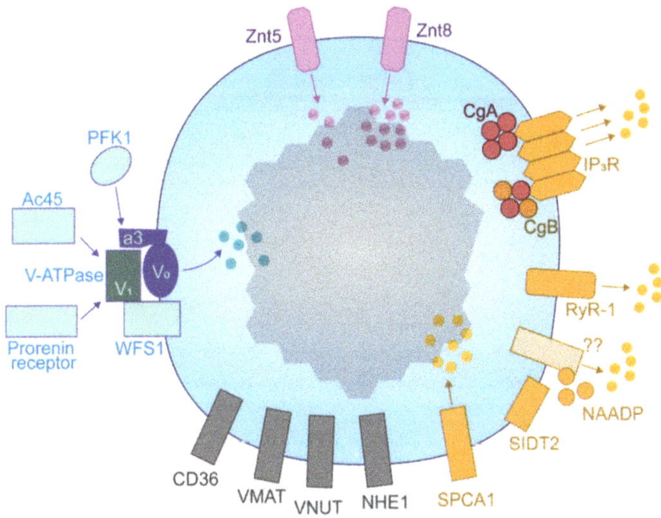

Figure 6. Channels and transporters of the insulin SG. An array of transmembrane components controls the luminal composition of the insulin SG and transform the SG into a responsive store that is utilized by the beta cell for cytosolic signaling.

In addition, H^+ pumping is sub-optimally coupled to ATP hydrolysis [233,234], providing room for further V-ATPase modulation. This may contribute to the establishment of a secretory pathway H^+ gradient or allow V-ATPase to respond to environmental stimuli. In yeast, V-ATPase efficiency is associated with the *a*-subunit of the V_0 domain, which is also likely to determine its cellular localization [235]. Subunit *a* is situated in an ideal position to modulate the pump. It is embedded into the membrane adjacent to a proteolipid ring formed by the *c* subunits of V_0 (through which H^+ passes) and extends toward the cytosol to interact with V_1 [228]. Yeast co-express two homologs to the mammalian *a*-subunit and these have been shown to affect both the localization and activity of V-ATPase [235].

Four isoforms of subunit *a* exist in mammals (*a*1–*a*4) making it the most diverse member of the V-ATPase complex, suggesting that this component can endow compartmental specificity regarding the localization and efficiency of the pump. *a*1 appears to localize to the Golgi, *a*2 to both lysosomes and the Golgi and *a*3 is mostly expressed on β-cell SGs [133]. Interestingly, the *a*4 subunit has been shown to interact with 1-phosphofructokinase (PFK1) in the human kidney [236], coupling nutrient sensing to V-ATPase activity [237]. SGs with low pH are required for GSIS [238]; therefore, glucose-mediated acidification could facilitate both SG maturation and release.

Accessory subunits Ac45 and the prorenin receptor, encoded by *ATP6ap1* and *ATP6ap2* genes, respectively, are also associated with the V-ATPase to assist SG acidification in β-cells [239–242]. Ac45 is subject to processing by furin [239,242] and the prorenin receptor interacts with and may act downstream of the GLP1 receptor [240,241]. A V-ATPase interactor, Wolfram syndrome 1 (WFS1), resides in β-cell ER and SG membranes [243] and has been shown to bind the V_1A subunit via an N-terminal region to assist SG acidification [244]. To conclude, the V-ATPase and/or its regulatory subunits are potential targets for enhancing insulin production during high demand.

Crystallization: Zn^{2+}. Insulin SGs hold high levels of Zn^{2+}, with some estimates approaching 30 mM [29,245]. Zn^{2+} can alter the structure of its bound proteins, cofactor for enzyme activity and also serve as an extracellular signaling molecule [246]. Measurement of insulin and Zn^{2+} released during GSIS reveals that the total amount of Zn^{2+} in the SG is at least double that which is expected based on the stoichiometric composition of the insulin hexamer [247], probably due to the existence of an additional Zn^{2+} that displaces water within crystallized hexamers [28]. Co-secreted Zn^{2+} has autocrine [248,249] and potentially paracrine [250,251] effects on other islet cells, and may also travel to the liver to inhibit insulin receptor endocytosis and thus hepatic insulin uptake [252]. One study using computer modelling has also suggested that Zn^{2+} could maintain insulin in an oligomeric state during secretion, and that this would limit the availability of Zn^{2+} and monomeric insulin to act as anti-fibrillogenic agents against hIAPP-related proteins [253]. Early estimates, however, report that exposure to the extracellular environment would likely dissipate the insulin-zinc hexamer seconds after complex dilution outside of the β-cell [64], allowing monomeric insulin to circulate and signal. Notably, reductions in circulating and pancreatic Zn^{2+} levels are implicated in those with excess fat and T2D [254,255], and Zn^{2+} supplementation can enhance insulin secretion to better control glycemia during the insulin resistant state [256]. Thus, intracellular changes in Zn^{2+} could drive aspects of β-cell dysfunction.

Secretory pathway Zn^{2+} uptake is under the control of the Zn^{2+} transporter (ZnT) family. ZnTs dimerize to localize and function and likely do so as H^+/Zn^{2+} antiporters [246,257], so therefore the establishment of a luminal proton gradient by V-ATPase may be permissive for Zn^{2+} uptake. ZnT5 and Znt7 localize to the β-cell Golgi apparatus whereas ZnT5 and ZnT8 are in the SG [28] (Figure 6). ZnT3 has been shown to colocalize with insulin in INS1 cells [258], however it appears to be absent in β-cells from mouse islets [259]. ZnT8 is the most abundantly expressed β-cell ZnT but has minimal expression in other tissues [28].

The current literature on ZnT8-mediated Zn^{2+} homeostasis is deep and interesting, owing to the existence of multiple T2D susceptibility loci covering the *SLC30A8* gene [260–262]. Importantly, loss of function mutations at residue 325 tend to favor a reduced risk of T2D, and opposite deleterious effects are seen with a gain of function at this residue. These may be mediated by changes to proinsulin conversion and insulin secretion, raising the question of whether Zn^{2+} is important for β-cell SG biogenesis. Surprisingly, guinea pigs express a proinsulin molecule that lacks histidine at the B10 residue and thus cannot bind Zn^{2+}, similar to the human mutation discussed in Section 1, yet they can generate SGs, process proinsulin, and secrete insulin [263]. However, despite the inability of insulin to bind Zn^{2+}, the presence of Zn^{2+} in these settings is not changed as is the case with altered ZnT8 transporter activity.

Studies of the effect of ZnT8 depletion in β-cell lines and rodent models have provided results that are difficult to synthesize thus far [261,262,264]. In general, experimental depletion of ZnT8 does not lead to major impairments in proinsulin processing or insulin content; however, SGs tend to be void of electron-dense spheres, appearing instead as electron-dense rods or as pale ISGs. This agrees with the expression of other ZnT members through the secretory pathway and on the SG, providing sufficient Zn^{2+} to sustain intragranular functions albeit with impaired insulin crystallization [28,265]. Assessment of insulin secretion in various ZnT8-deficient models has shown a mixed bag of results, with reports contrasting between mildly reduced, unchanged, or mildly enhanced effects [261,262,264]. Recognition of variable factors such as the mode of deletion, and mouse age, sex, and genetic background are hoped to assist in study design to clear up the matter [261]. Indeed, a study published in 2020 revealed both positive and negative age-dependent consequences of

ZnT8 deletion in mice [266]. Moreover, assessing the importance of Zn^{2+} for SG biogenesis is complicated by the presence of multiple ZnTs. For example, an investigation comparing the effects of a whole-body single ZnT7 KO to a double ZnT7/ZnT8 KO, revealed that the double KO could provide a significant secretory defect not seen by ZnT7 KOs alone or by previous reports in ZnT8 KOs [267].

Several lines of evidence indicate that ZnT8-mediated Zn^{2+} import affects proinsulin processing, which could explain, at least in part, the protective or deleterious impacts of *SLC30A8* mutations. It has been shown that humans with the arginine risk variant of ZnT8 at residue 325, thought to be a gain-of-function mutation [268], display increased circulating proinsulin compared to the protective tryptophan variant [269]. Moreover, mice overexpressing the arginine 325 variant have an increased β-cell proinsulin content and release [270,271]. Conversely, in INS1E cells expressing the loss-of-function tryptophan 325 variant, cyclosporin A-induced β-cell dysfunction was attenuated when compared to the gain-of-function arginine 325 variant [272]. Finally, humans heterozygous for a ZnT8 variant truncated at residue 138, which impairs transporter synthesis and thus results in ZnT8 haploinsufficiency, are equipped with increased proinsulin conversion and circulating insulin, and improved glucose tolerance [270,271]. Since ZnTs are likely to function as H^+/Zn^{2+} exchangers [246,257], enhanced flux through ZnT8 could buffer the reduction in luminal pH during SG maturation to limit endoprotease activity. Accelerated packing of the granule matrix in the presence of high Zn^{2+} could also reduce the accessibility of proinsulin to the endoproteases for conversion. In the case of reduced ZnT8 activity or haploinsufficiency, the presence of adequate Zn^{2+} supplied by the single copy of ZnT8 or other ZnTs in the secretory pathway would probably maintain hexameric proinsulin transit and cofactor for CPE activity, while the event of insulin crystallization does not appear to be necessary for regulated release. Therefore, in the setting of β-cell compensation, a ready supply of mature insulin could help to control circulating glucose. Importantly, *SLC30A8* risk-alleles are particularly noted to confer T2D susceptibility to lean individuals [262,273], so in this case, limited proinsulin conversion in newly synthesized granules that are preferentially secreted could reduce the insulin response to glucose. Nonetheless, there is still plenty to be learned about the impact that Zn^{2+} has inside the insulin SG. While the consequence of *SLC30A8* gene variants have been of interest, little is known about Znt8 structure and mechanism, and the factors that regulate it.

Modulation: Ca^{2+}. Ca^{2+} is concentrated within the secretory pathway relative to the cytosol, where it controls luminal protein sorting and processing among other activities. Travelling proximal to distal, the amount of free Ca^{2+} decreases whereas the total amount of Ca^{2+} increases (Figure 1), in line with several proteins harboring an increased affinity to Ca^{2+} when the pH is reduced [132,274]. Quantitatively, the β-cell SG can hold between 50 to 100 mM Ca^{2+} [29], although measurement of the free concentration yields values of around 50 μM (~0.05% of the total amount) [31]. Therefore, the SG compartment, endowed with transporters and channels for Ca^{2+}, should possess a high capacity to buffer cytosolic Ca^{2+}. This dynamic reservoir can be utilized during β-cell stimulation to sequester and release Ca^{2+}, modulating the cytosolic signals that underlie insulin secretion. Indeed, it has been shown that depletion of Ca^{2+} from the SG compartment impairs exocytosis [275]. β-cell SGs have been shown to take up Ca^{2+} during nutrient exposure [31], and Ca^{2+} release from the SG is required for insulin secretion [276]. Ca^{2+}-dependent effector proteins are in proximity to the SG and sites of exocytosis [37], and therefore localized Ca^{2+} released from the SG could facilitate its own trafficking and exocytosis.

Several transporters and channels act in concert to coordinate luminal and cytosolic events that are regulated by Ca^{2+} (Figure 6). The secretory pathway Ca^{2+} ATPase 1 (SPCA1) is an uptake pump that regulates a Ca^{2+}-dependent secretory protein sorting mechanism at the TGN in constitutively secreting cells [277,278]. In INS1 cells, SPCA1 may sequester cytosolic Ca^{2+} into secretory pathway compartments during glucose stimulation, accounting for around 20% of the total SG Ca^{2+} uptake [279]. Its depletion thus enhances insulin secretion [279]. In these cells, SPCA1 was shown to fractionate strongly with insulin suggesting its localization throughout the early and late secretory pathway [279].

Ryanodine receptors (RyRs) respond directly to cytosolic Ca^{2+} to release Ca^{2+} from intracellular stores but are inactive at low and high concentrations of Ca^{2+}. RyR-1 was identified as the lone RyR subtype expressed on the β-cell SG, but it is also expressed on the ER where RyR-2 also locates [276]. Pharmacological inhibition of SG RyR-1 reduces Ca^{2+} efflux and impairs GSIS [276].

NAADP is a potent signaling molecule produced during β-cell glucose metabolism [280] that binds to unidentified NAADP-sensitive sites to elicit Ca^{2+} release from intracellular stores, including the SG, during GSIS [31,281]. SIDT2 is located on insulin SGs and may mediate this mechanism [281]. Whole-body SIDT2 KO mice are glucose intolerant and have an insulin secretory defect [282]. The requirement of either NAADP or RyR-1-mediated Ca^{2+} release for GSIS suggests that the SG is an important reservoir of Ca^{2+} that is utilized during secretory functions.

After initial confusion [283,284], it was recently verified that all three subtypes of the IP_3R are expressed on β-cell SGs at a level two-fold higher than the ER [30]. IP_3Rs require tetramerization and binding of each member to inositol 1,4,5-triphosphate (IP_3) in order for the channel to open and release Ca^{2+} from stores, and these tetramers can be formed by any combination of IP_3R subtypes [285]. In the absence of IP_3, Ca^{2+} inhibits the IP_3R, although, Ca^{2+} must bind to the IP_3R with IP_3 for the channel to open [286]. IP_3Rs localized on non-β-cell SGs have been estimated to be more sensitive than those of the ER [287], although one study reported that β-cell SGs do not release Ca^{2+} in response to IP_3 [31]. Several caveats are apparent in this study; low levels of Ca^{2+} were incubated with permeabilized mouse insulinoma MIN6 cells exposed to IP_3 despite a high requirement for maximal IP_3R activation [288], and the membrane glutamine receptor (mGlu5) is unable to raise cytosolic Ca^{2+} in conjunction with its function to generate IP_3 [289] as would be seen during glucose stimulation [290]. Therefore, the role of IP_3R-mediated Ca^{2+} release from the insulin SG should be reevaluated.

On the inside, both CgA and CgB can interact with the intraluminal loop of all three IP_3R subtypes at pH 5.5 to stabilize IP_3 binding and channel rigidity [147,287,291–293]. These granins possess an extremely high capacity to bind Ca^{2+} with low affinity at the acidic pH of the MSG [132,146], and it has been suggested that IP_3R conformational changes can alter that of CgA and CgB to release bound Ca^{2+} through the IP_3R channel [293,294]. The relative abundance of IP_3R isoforms and their tetrameric composition as well as that of the interacting granin species are thought to underlie SG Ca^{2+} balance, such that equal amounts of IP_3 can stimulate varying amounts of Ca^{2+} release [295]. Therefore, the distribution of components within the SG during maturation, idling, and in primed states could modulate SG Ca^{2+} release. Indeed, CgB can undergo redistribution to the MSG periphery upon glucose stimulation [157]. Since intragranular acidification is essential for chromogranin-IP_3R interactions, it appears that the utility of the SG as an IP_3-sensitive Ca^{2+} store is reliant on its proper maturation. In conclusion, a host of components are responsible for handling Ca^{2+} for use both inside and outside of the SG.

Other transporters. SGs in the β-cell also contain a truncated form of the NHE1 Na^+/H^+ exchanger [296], fatty acid translocase (FAT/CD36) [297,298], and a vesicular nucleotide transporter (VNUT) [299]. A role for NHE1 has not yet been determined. CD36, which is predominantly expressed on the PM, has multiple roles throughout the body [300], with a general cellular function to facilitate fatty acid uptake. In β-cells, CD36 is localized to the PM and the SG and mediates the acute and chronic effects of free fatty acids on insulin secretion [297]. CD36 is upregulated in the islets of obese humans with T2D, where altered lipid handling may impair the action of exocytotic proteins [298]. VNUT is a transporter required for ATP uptake into the SG lumen [299] and its depletion results in reduced basal and glucose-stimulated ATP release and insulin secretion [299]. In addition, a Golgi-localized magnesium transporter NIPAL1 was recently shown to positively regulate insulin content and secretion in MIN6 cells [301], albeit through an unknown mechanism.

The activity of channels and transporters provide overarching control on SG maturation and serve as conduits that alter luminal responses based on events transpiring in the cytosol. A prime example is the ability of glucose to drive V-ATPase subunit association to enhance acidification. Together, their presence transforms the SG into a dynamic hub used for ionic signaling and buffering.

2.4. Sorting Receptors

Early efforts centered around identifying a '*sorting by entry*' receptor for proinsulin proved futile, possibly because soluble secretory cargo destined for the β-cell MSG are not sorted through receptor-mediated recognition. CPE was once entertained as a candidate for proinsulin sorting from the TGN [302,303], but this was contested [304]. Conversely, the mannose 6 phosphate receptor (M6PR) has been identified as a critical component of '*sorting by exit*' which serves to refine the SG during maturation. It functions to target proteins modified with a mannose 6 phosphate carbohydrate group from the TGN and the ISG to the lysosome, as well as from the plasma membrane for endosomal retrieval [32,305–307]. The M6PR specifically contributes to the sorting of the luminal lysosomal hydrolase procathepsin B in β-cells [7,32] (Figure 7).

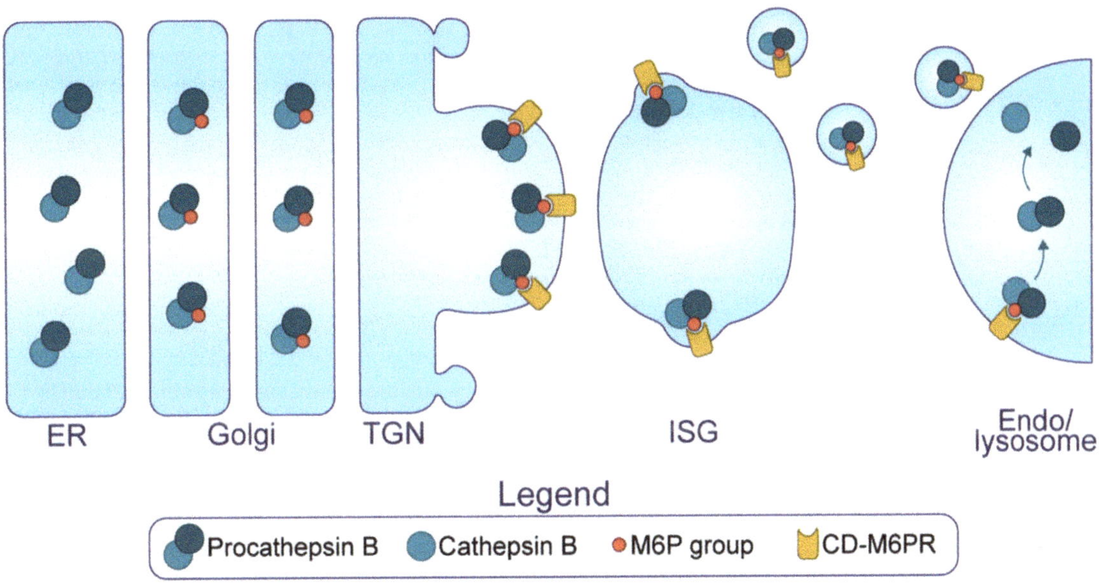

Figure 7. Granule refinement. The cation dependent mannose 6 phosphate receptor (M6PR) binds to certain proteins modified with a mannose 6 phosphate group (namely, procathepsin B) in the TGN, and this complex enters the ISG. During maturation, this complex will exit the SG via small transport vesicles and traffic to the endolysosomal system. A byproduct of this process may be the removal of soluble components in the general vicinity of the budding vesicle.

Two forms of the M6PR exist, a cation independent (CI) ~300 kDa isoform and a ~46 kDa isoform that dimerizes and requires cationic binding (CD—cation dependent) to function. The CD-M6PR locates to the TGN and the ISG where it resides in proximity to clathrin-coated patches and is found on small transport vesicles but not in the MSG [32]. Islets lacking the CD-M6PR contain four-fold more cathepsin B in the β-cell MSG [7]. While the CD-M6PR has a defined role in cathepsin B sorting, a role for the CI-M6PR in the β-cell SG has not yet been determined. To conclude here, little is known about other luminal components that are actively removed from the ISG.

3. Concluding Remarks

At its essence, insulin synthesis requires the entry of proinsulin and its processing enzymes into ISGs, followed by enzyme activation and then the complete execution of enzymatic activity. Each of these steps are facilitated by ions that are supplied by channels and transporters, which exert influence on luminal proteins by altering their behavior. First, ion-facilitated granin aggregation and binding to distinct lipid rafts of the TGN membrane may provide the luminal driving force to generate SGs, forming what is currently an ill-defined path of entry for protein to flow through. This is evidenced by observations that granin depletion can limit the biogenesis of nascent SGs. Several resident enzymes of the SG also aggregate specifically in the TGN and bind to lipid rafts to provide a common pathway for sorting, however the trafficking route that key proteins such as proinsulin and proIAPP take are yet to be fully defined. *Sorting by entry* into the ISG is efficient but may not be entirely specific, so the presence of sorting receptors and active ion uptake transporters provide the ISG with two simultaneous quality control mechanisms for refinement—*sorting by exit* and *sorting by retention*. Here, proteins specifically recognized by sorting receptors, and some unwitting bystanders, are escorted from the SG via small transport vesicles, but MSG-destined proteins will bind luminal ions to reduce their solubility and prevent their egression via these carriers. It is currently unclear how important the two post-Golgi sorting mechanisms are for generating insulin itself, but irregular SG maturation could impair overall SG composition and function and encourage the pathological formation of hIAPP oligomers. As the lumen is progressively modified by continual ion uptake, enzymes will begin to exert activity on proinsulin and other peptides to form the complete intragranular cocktail, which will develop into an extremely insoluble ion-bound crystal core surrounded by a halo of soluble components. Here, human T2D-associated Znt8 variants are the exemplar of how ionic alterations can affect parameters of SG maturation, and conceptually illustrate how seemingly small differences can precipitate the chronic disease. Once mature, the SG functions as an intracellular signaling compartment in addition to its role in holding and releasing insulin.

In T2D there is a loss of insulin SGs, GSIS is impaired, and the secretion of proinsulin and its processing intermediates is elevated. Recent subcellular evidence links these phenotypes to a diversion away from SG maturation toward premature ISG secretion, suggesting that MSG formation is a primary limiting factor for insulin secretion in T2D. It is therefore conceivable that defects within the secretory pathway could predispose individuals to the disease by creating an upstream bottleneck to MSG production, delaying the synthesis of MSGs. This might be fine in the healthy state when production is not limiting but could compromise secretion when insulin content is depleted during the chronic condition.

Currently, there is an abundance of knowledge about the distal stages of exocytosis, but a gap in our understanding of events that occur through the late Golgi and the maturing SG. Here, we have provided a comprehensive summary of what happens inside the lumen during the formative hours of the insulin SG. In doing so, it becomes clear that generating a SG that is rich in insulin is quite arduous, and therefore, prone to perturbations that may affect the capacity of the β-cell to adapt to chronic demand.

Author Contributions: Conceptualization, M.G. and M.A.K.; writing—original draft preparation, M.G.; writing—review and editing, B.Y., A.G., M.T. and M.A.K. All authors have read and agreed to the published version of the manuscript.

Funding: M.G. was supported by the Australian Government Research Training Program Scholarship. A.G. and M.T. were both supported by the Charles Perkins Centre, Summer Scholarship Program.

Institutional Review Board Statement: Not applicable.

Informed Consent Statement: Not applicable.

Data Availability Statement: Not applicable.

Conflicts of Interest: The authors declare no conflict of interest.

References

1. Orci, L.; Halban, P.; Amherdt, M.; Ravazzola, M.; Vassalli, J.D.; Perrelet, A. Nonconverted, Amino Acid Analog-Modified Proinsulin Stays in a Golgi-Derived Clathrin-Coated Membrane Compartment. *J. Cell Biol.* **1984**, *99*, 2187–2192. [CrossRef] [PubMed]
2. Rhodes, C.J.; Halban, P.A. Newly Synthesized Proinsulin/insulin and Stored Insulin Are Released from Pancreatic B Cells Predominantly via a Regulated, rather than a Constitutive, Pathway. *J. Cell Biol.* **1987**, *105*, 145–153. [CrossRef] [PubMed]
3. Michael, J.; Carroll, R.; Swift, H.H.; Steiner, D.F. Studies on the Molecular Organization of Rat Insulin Secretory Granules. *J. Biol. Chem.* **1987**, *262*, 16531–16535. [CrossRef]
4. Davidson, H.W.; Rhodes, C.J.; Hutton, J.C. Intraorganellar Calcium and pH Control Proinsulin Cleavage in the Pancreatic Beta Cell via Two Distinct Site-Specific Endopeptidases. *Nature* **1988**, *333*, 93–96. [CrossRef]
5. Smeekens, S.P.; Montag, A.G.; Thomas, G.; Albiges-Rizo, C.; Carroll, R.; Benig, M.; Phillips, L.A.; Martin, S.; Ohagi, S.; Gardner, P. Proinsulin Processing by the Subtilisin-Related Proprotein Convertases Furin, PC2, and PC3. *Proc. Natl. Acad. Sci. USA* **1992**, *89*, 8822–8826. [CrossRef]
6. Orci, L.; Ravazzola, M.; Storch, M.J.; Anderson, R.G.; Vassalli, J.D.; Perrelet, A. Proteolytic Maturation of Insulin Is a Post-Golgi Event Which Occurs in Acidifying Clathrin-Coated Secretory Vesicles. *Cell* **1987**, *49*, 865–868. [CrossRef]
7. Kuliawat, R.; Klumperman, J.; Ludwig, T.; Arvan, P. Differential Sorting of Lysosomal Enzymes out of the Regulated Secretory Pathway in Pancreatic Beta-Cells. *J. Cell Biol.* **1997**, *137*, 595–608. [CrossRef]
8. Baker, E.N.; Blundell, T.L.; Cutfield, J.F.; Dodson, E.J.; Dodson, G.G.; Hodgkin, D.M.C.; Hubbard, R.E.; Isaacs, N.W.; Reynolds, C.D.; Sakabe, K.; et al. The Structure of 2Zn Pig Insulin Crystals at 1.5 Å Resolution. *Philos. Trans. R. Soc. Lond. B Biol. Sci.* **1988**, *319*, 369–456.
9. Halban, P.A. Inhibition of Proinsulin to Insulin Conversion in Rat Islets Using Arginine and Lysine Analogs. Lack of Effect on Rate of Release of Modified Products. *J. Biol. Chem.* **1982**, *257*, 13177–13180. [CrossRef]
10. Seaquist, E.R.; Kahn, S.E.; Clark, P.M.; Hales, C.N.; Porte, D., Jr.; Robertson, R.P. Hyperproinsulinemia Is Associated with Increased Beta Cell Demand after Hemipancreatectomy in Humans. *J. Clin. Investig.* **1996**, *97*, 455–460. [CrossRef]
11. Mezza, T.; Ferraro, P.M.; Sun, V.A.; Moffa, S.; Cefalo, C.M.A.; Quero, G.; Cinti, F.; Sorice, G.P.; Pontecorvi, A.; Folli, F.; et al. Increased β-Cell Workload Modulates Proinsulin-to-Insulin Ratio in Humans. *Diabetes* **2018**, *67*, 2389–2396. [CrossRef]
12. Alarcon, C.; Boland, B.B.; Uchizono, Y.; Moore, P.C.; Peterson, B.; Rajan, S.; Rhodes, O.S.; Noske, A.B.; Haataja, L.; Arvan, P.; et al. Pancreatic β-Cell Adaptive Plasticity in Obesity Increases Insulin Production but Adversely Affects Secretory Function. *Diabetes* **2016**, *65*, 438–450. [CrossRef]
13. Ward, W.K.; LaCava, E.C.; Paquette, T.L.; Beard, J.C.; Wallum, B.J.; Porte, D., Jr. Disproportionate Elevation of Immunoreactive Proinsulin in Type 2 (non-Insulin-Dependent) Diabetes Mellitus and in Experimental Insulin Resistance. *Diabetologia* **1987**, *30*, 698–702. [CrossRef]
14. Yoshino, H.; Kawakami, K.; Yoshino, G.; Hirose, T. Age-Related Changes of Proinsulin Processing in Diabetic and Non-Diabetic Japanese Individuals. *Geriatr. Gerontol. Int.* **2018**, *18*, 1046–1050. [CrossRef]
15. Pfützner, A.; Standl, E.; Hohberg, C.; Konrad, T.; Strotmann, H.-J.; Lübben, G.; Langenfeld, M.R.; Schulze, J.; Forst, T. IRIS II Study: Intact Proinsulin Is Confirmed as a Highly Specific Indicator for Insulin Resistance in a Large Cross-Sectional Study Design. *Diabetes Technol. Ther.* **2005**, *7*, 478–486. [CrossRef]
16. Pradhan, A.D.; Manson, J.E.; Meigs, J.B.; Rifai, N.; Buring, J.E.; Liu, S.; Ridker, P.M. Insulin, Proinsulin, Proinsulin:insulin Ratio, and the Risk of Developing Type 2 Diabetes Mellitus in Women. *Am. J. Med.* **2003**, *114*, 438–444. [CrossRef]
17. Nijpels, G.; Popp-Snijders, C.; Kostense, P.J.; Bouter, L.M.; Heine, R.J. Fasting Proinsulin and 2-H Post-Load Glucose Levels Predict the Conversion to NIDDM in Subjects with Impaired Glucose Tolerance: The Hoorn Study. *Diabetologia* **1996**, *39*, 113–118.
18. Vangipurapu, J.; Stančáková, A.; Kuulasmaa, T.; Kuusisto, J.; Laakso, M. Both Fasting and Glucose-Stimulated Proinsulin Levels Predict Hyperglycemia and Incident Type 2 Diabetes: A Population-Based Study of 9,396 Finnish Men. *PLoS ONE* **2015**, *10*, e0124028. [CrossRef]
19. Yang, Y.; Wang, M.; Tong, J.; Dong, Z.; Deng, M.; Ren, X.; Li, H.; Yang, J.; Meng, Z.; Sun, J.; et al. Impaired Glucose-Stimulated Proinsulin Secretion Is an Early Marker of β-Cell Impairment Before Prediabetes Stage. *J. Clin. Endocrinol. Metab.* **2019**, *104*, 4341–4346. [CrossRef]
20. Porte, D., Jr.; Kahn, S.E. Hyperproinsulinemia and Amyloid in NIDDM. Clues to Etiology of Islet Beta-Cell Dysfunction? *Diabetes* **1989**, *38*, 1333–1336. [CrossRef]
21. Rhodes, C.J.; Alarcón, C. What Beta-Cell Defect Could Lead to Hyperproinsulinemia in NIDDM? Some Clues from Recent Advances Made in Understanding the Proinsulin-Processing Mechanism. *Diabetes* **1994**, *43*, 511–517. [CrossRef]
22. Ohsawa, H.; Kanatsuka, A.; Yamaguchi, T.; Makino, H.; Yoshida, S. Islet Amyloid Polypeptide Inhibits Glucose-Stimulated Insulin Secretion from Isolated Rat Pancreatic Islets. *Biochem. Biophys. Res. Commun.* **1989**, *160*, 961–967. [CrossRef]
23. Natori, S.; Huttner, W.B. Chromogranin B (secretogranin I) Promotes Sorting to the Regulated Secretory Pathway of Processing Intermediates Derived from a Peptide Hormone Precursor. *Proc. Natl. Acad. Sci. USA* **1996**, *93*, 4431–4436. [CrossRef] [PubMed]

24. Carmon, O.; Laguerre, F.; Riachy, L.; Delestre-Delacour, C.; Wang, Q.; Tanguy, E.; Jeandel, L.; Cartier, D.; Thahouly, T.; Haeberlé, A.-M.; et al. Chromogranin A Preferential Interaction with Golgi Phosphatidic Acid Induces Membrane Deformation and Contributes to Secretory Granule Biogenesis. *FASEB J.* **2020**. [CrossRef]
25. Bartolomucci, A.; Possenti, R.; Mahata, S.K.; Fischer-Colbrie, R.; Loh, Y.P.; Salton, S.R.J. The Extended Granin Family: Structure, Function, and Biomedical Implications. *Endocr. Rev.* **2011**, *32*, 755–797. [CrossRef]
26. Hutton, J.C.; Peshavaria, M. Proton-Translocating Mg^{2+}-Dependent ATPase Activity in Insulin-Secretory Granules. *Biochem. J.* **1982**, *204*, 161–170. [CrossRef]
27. Schoonderwoert, V.T.; Holthuis, J.C.; Tanaka, S.; Tooze, S.A.; Martens, G.J. Inhibition of the Vacuolar H^+-ATPase Perturbs the Transport, Sorting, Processing and Release of Regulated Secretory Proteins. *Eur. J. Biochem.* **2000**, *267*, 5646–5654. [CrossRef]
28. Lemaire, K.; Chimienti, F.; Schuit, F. Zinc Transporters and Their Role in the Pancreatic β-Cell. *J. Diabetes Investig.* **2012**, *3*, 202–211. [CrossRef]
29. Hutton, J.C.; Penn, E.J.; Peshavaria, M. Low-Molecular-Weight Constituents of Isolated Insulin-Secretory Granules. Bivalent Cations, Adenine Nucleotides and Inorganic Phosphate. *Biochem. J.* **1983**, *210*, 297–305. [CrossRef]
30. Hur, Y.S.; Yoo, S.H. Distribution Profile of Inositol 1,4,5-Trisphosphate Receptor/Ca^{2+} Channels in α and β Cells of Pancreas: Dominant Localization in Secretory Granules and Common Error in Identification of Secretory Granule Membranes. *Pancreas* **2015**, *44*, 158–165. [CrossRef] [PubMed]
31. Mitchell, K.J.; Pinton, P.; Varadi, A.; Tacchetti, C.; Ainscow, E.K.; Pozzan, T.; Rizzuto, R.; Rutter, G.A. Dense Core Secretory Vesicles Revealed as a Dynamic $Ca^{(2+)}$ Store in Neuroendocrine Cells with a Vesicle-Associated Membrane Protein Aequorin Chimaera. *J. Cell Biol.* **2001**, *155*, 41–51. [CrossRef] [PubMed]
32. Klumperman, J.; Kuliawat, R.; Griffith, J.M.; Geuze, H.J.; Arvan, P. Mannose 6-Phosphate Receptors Are Sorted from Immature Secretory Granules via Adaptor Protein AP-1, Clathrin, and Syntaxin 6-Positive Vesicles. *J. Cell Biol.* **1998**, *141*, 359–371. [CrossRef] [PubMed]
33. Orci, L.; Ravazzola, M.; Perrelet, A. (Pro)insulin Associates with Golgi Membranes of Pancreatic B Cells. *Proc. Natl. Acad. Sci. USA* **1984**, *81*, 6743–6746. [CrossRef]
34. Kahn, S.E.; Halban, P.A. Release of Incompletely Processed Proinsulin Is the Cause of the Disproportionate Proinsulinemia of NIDDM. *Diabetes* **1997**, *46*, 1725–1732. [CrossRef]
35. Alarcón, C.; Leahy, J.L.; Schuppin, G.T.; Rhodes, C.J. Increased Secretory Demand rather than a Defect in the Proinsulin Conversion Mechanism Causes Hyperproinsulinemia in a Glucose-Infusion Rat Model of Non-Insulin-Dependent Diabetes Mellitus. *J. Clin. Investig.* **1995**, *95*, 1032–1039. [CrossRef]
36. Omar-Hmeadi, M.; Idevall-Hagren, O. Insulin Granule Biogenesis and Exocytosis. *Cell. Mol. Life Sci.* **2020**. [CrossRef]
37. Thurmond, D.C.; Gaisano, H.Y. Recent Insights into Beta-Cell Exocytosis in Type 2 Diabetes. *J. Mol. Biol.* **2020**, *432*, 1310–1325. [CrossRef]
38. Norris, N.; Yau, B.; Kebede, M.A. Isolation and Proteomics of the Insulin Secretory Granule. *Metabolites* **2021**, *11*, 288. [CrossRef]
39. Magro, M.G.; Solimena, M. Regulation of β-Cell Function by RNA-Binding Proteins. *Mol. Metab.* **2013**, *2*, 348–355. [CrossRef]
40. Arvan, P.; Halban, P.A. Sorting Ourselves out: Seeking Consensus on Trafficking in the Beta-Cell. *Traffic* **2004**, *5*, 53–61. [CrossRef]
41. Arvan, P.; Kuliawat, R.; Prabakaran, D.; Zavacki, A.M.; Elahi, D.; Wang, S.; Pilkey, D. Protein Discharge from Immature Secretory Granules Displays Both Regulated and Constitutive Characteristics. *J. Biol. Chem.* **1991**, *266*, 14171–14174. [CrossRef]
42. Kuliawat, R.; Arvan, P. Protein Targeting via the "Constitutive-like" Secretory Pathway in Isolated Pancreatic Islets: Passive Sorting in the Immature Granule Compartment. *J. Cell Biol.* **1992**, *118*, 521–529. [CrossRef]
43. Kuliawat, R.; Arvan, P. Distinct Molecular Mechanisms for Protein Sorting within Immature Secretory Granules of Pancreatic Beta-Cells. *J. Cell Biol.* **1994**, *126*, 77–86. [CrossRef]
44. Arvan, P.; Castle, D. Sorting and Storage during Secretory Granule Biogenesis: Looking Backward and Looking Forward. *Biochem. J.* **1998**, *332 Pt 3*, 593–610. [CrossRef]
45. Molinete, M.; Irminger, J.C.; Tooze, S.A.; Halban, P.A. Trafficking/sorting and Granule Biogenesis in the Beta-Cell. *Semin. Cell Dev. Biol.* **2000**, *11*, 243–251. [CrossRef]
46. Hummer, B.H.; Maslar, D.; Soltero-Gutierrez, M.; de Leeuw, N.F.; Asensio, C.S. Differential Sorting Behavior for Soluble and Transmembrane Cargoes at the Trans-Golgi Network in Endocrine Cells. *Mol. Biol. Cell* **2020**, *31*, 157–166. [CrossRef]
47. Ericson, L.E.; Håkanson, R.; Lundquist, I. Accumulation of Dopamine in Mouse Pancreatic B-Cells Following Injection of L-DOPA. Localization to Secretory Granules and Inhibition of Insulin Secretion. *Diabetologia* **1977**, *13*, 117–124. [CrossRef]
48. Ekholm, R.; Ericson, L.E.; Lundquist, I. Monoamines in the Pancreatic Islets of the Mouse. Subcellular Localization of 5-Hydroxytryptamine by Electron Microscopic Autoradiography. *Diabetologia* **1971**, *7*, 339–348. [CrossRef]
49. Lundquist, I.; Ekholm, R.; Ericson, L.E. Monoamines in the Pancreatic Islets of the Mouse. 5-Hydroxytryptamine as an Intracellular Modifier of Insulin Secretion, and the Hypoglycaemic Action of Monoamine Oxidase Inhibitors. *Diabetologia* **1971**, *7*, 414–422. [CrossRef]
50. Sakamoto, S.; Miyaji, T.; Hiasa, M.; Ichikawa, R.; Uematsu, A.; Iwatsuki, K.; Shibata, A.; Uneyama, H.; Takayanagi, R.; Yamamoto, A.; et al. Impairment of Vesicular ATP Release Affects Glucose Metabolism and Increases Insulin Sensitivity. *Sci. Rep.* **2014**, *4*, 6689. [CrossRef]
51. Henquin, J.-C. Paracrine and Autocrine Control of Insulin Secretion in Human Islets: Evidence and Pending Questions. *Am. J. Physiol. Endocrinol. Metab.* **2021**, *320*, E78–E86. [CrossRef] [PubMed]

52. Liu, M.; Lara-Lemus, R.; Shan, S.-O.; Wright, J.; Haataja, L.; Barbetti, F.; Guo, H.; Larkin, D.; Arvan, P. Impaired Cleavage of Preproinsulin Signal Peptide Linked to Autosomal-Dominant Diabetes. *Diabetes* **2012**, *61*, 828–837. [CrossRef] [PubMed]
53. Huang, X.F.; Arvan, P. Intracellular Transport of Proinsulin in Pancreatic Beta-Cells. Structural Maturation Probed by Disulfide Accessibility. *J. Biol. Chem.* **1995**, *270*, 20417–20423. [CrossRef] [PubMed]
54. Haataja, L.; Snapp, E.; Wright, J.; Liu, M.; Hardy, A.B.; Wheeler, M.B.; Markwardt, M.L.; Rizzo, M.; Arvan, P. Proinsulin Intermolecular Interactions during Secretory Trafficking in Pancreatic β Cells. *J. Biol. Chem.* **2013**, *288*, 1896–1906. [CrossRef]
55. Frank, B.H.; Veros, A.J. Interaction of Zinc with Proinsulin. *Biochem. Biophys. Res. Commun.* **1970**, *38*, 284–289. [CrossRef]
56. Kiselar, J.G.; Datt, M.; Chance, M.R.; Weiss, M.A. Structural Analysis of Proinsulin Hexamer Assembly by Hydroxyl Radical Footprinting and Computational Modeling. *J. Biol. Chem.* **2011**, *286*, 43710–43716. [CrossRef]
57. Bailyes, E.M.; Bennett, D.L.; Hutton, J.C. Proprotein-Processing Endopeptidases of the Insulin Secretory Granule. *Enzyme* **1991**, *45*, 301–313. [CrossRef]
58. Davidson, H.W.; Hutton, J.C. The Insulin-Secretory-Granule Carboxypeptidase H. Purification and Demonstration of Involvement in Proinsulin Processing. *Biochem. J.* **1987**, *245*, 575–582. [CrossRef]
59. Weiss, M.; Steiner, D.F.; Philipson, L.H. Insulin Biosynthesis, Secretion, Structure, and Structure-Activity Relationships. In *Endotext*; Feingold, K.R., Anawalt, B., Boyce, A., Chrousos, G., Dungan, K., Grossman, A., Hershman, J.M., Kaltsas, G., Koch, C., Kopp, P., et al., Eds.; MDText.com, Inc.: South Dartmouth, MA, USA, 2014.
60. Norrman, M.; Schluckebier, G. Crystallographic Characterization of Two Novel Crystal Forms of Human Insulin Induced by Chaotropic Agents and a Shift in pH. *BMC Struct. Biol.* **2007**, *7*, 83. [CrossRef]
61. Steiner, D.F. Cocrystallization of Proinsulin and Insulin. *Nature* **1973**, *243*, 528–530. [CrossRef]
62. Landreh, M.; Alvelius, G.; Willander, H.; Stukenborg, J.-B.; Söder, O.; Johansson, J.; Jörnvall, H. Insulin Solubility Transitions by pH-Dependent Interactions with Proinsulin C-Peptide. *FEBS J.* **2012**, *279*, 4589–4597. [CrossRef]
63. Verchere, C.B.; Paoletta, M.; Neerman-Arbez, M.; Rose, K.; Irminger, J.C.; Gingerich, R.L.; Kahn, S.E.; Halban, P.A. Des-(27–31)C-Peptide. A Novel Secretory Product of the Rat Pancreatic Beta Cell Produced by Truncation of Proinsulin Connecting Peptide in Secretory Granules. *J. Biol. Chem.* **1996**, *271*, 27475–27481. [CrossRef]
64. Gold, G.; Grodsky, G.M. Kinetic Aspects of Compartmental Storage and Secretion of Insulin and Zinc. *Experientia* **1984**, *40*, 1105–1114. [CrossRef]
65. Chan, S.J.; Seino, S.; Gruppuso, P.A.; Schwartz, R.; Steiner, D.F. A Mutation in the B Chain Coding Region Is Associated with Impaired Proinsulin Conversion in a Family with Hyperproinsulinemia. *Proc. Natl. Acad. Sci. USA* **1987**, *84*, 2194–2197. [CrossRef]
66. Carroll, R.J.; Hammer, R.E.; Chan, S.J.; Swift, H.H.; Rubenstein, A.H.; Steiner, D.F. A Mutant Human Proinsulin Is Secreted from Islets of Langerhans in Increased Amounts via an Unregulated Pathway. *Proc. Natl. Acad. Sci. USA* **1988**, *85*, 8943–8947. [CrossRef]
67. Halban, P.A.; Irminger, J.-C. Mutant Proinsulin That Cannot Be Converted Is Secreted Efficiently from Primary Rat Beta-Cells via the Regulated Pathway. *Mol. Biol. Cell* **2003**, *14*, 1195–1203. [CrossRef]
68. Sizonenko, S.; Irminger, J.C.; Buhler, L.; Deng, S.; Morel, P.; Halban, P.A. Kinetics of Proinsulin Conversion in Human Islets. *Diabetes* **1993**, *42*, 933–936. [CrossRef]
69. Sizonenko, S.V.; Halban, P.A. Differential Rates of Conversion of Rat Proinsulins I and II. Evidence for Slow Cleavage at the B-chain/C-Peptide Junction of Proinsulin II. *Biochem. J.* **1991**, *278*, 621–625. [CrossRef]
70. Schechter, I.; Berger, A. On the Size of the Active Site in Proteases. I. Papain. *Biochem. Biophys. Res. Commun.* **1967**, *27*, 157–162. [CrossRef]
71. Kaufmann, J.E.; Irminger, J.C.; Halban, P.A. Sequence Requirements for Proinsulin Processing at the B-chain/C-Peptide Junction. *Biochem. J.* **1995**, *310 Pt 3*, 869–874. [CrossRef]
72. Halban, P.A. Proinsulin Processing in the Regulated and the Constitutive Secretory Pathway. *Diabetologia* **1994**, *37* (Suppl. 2), S65–S72. [CrossRef] [PubMed]
73. Rhodes, C.J.; Lincoln, B.; Shoelson, S.E. Preferential Cleavage of Des-31,32-Proinsulin over Intact Proinsulin by the Insulin Secretory Granule Type II Endopeptidase. Implication of a Favored Route for Prohormone Processing. *J. Biol. Chem.* **1992**, *267*, 22719–22727. [CrossRef]
74. Kaufmann, J.E.; Irminger, J.C.; Mungall, J.; Halban, P.A. Proinsulin Conversion in GH3 Cells after Coexpression of Human Proinsulin with the Endoproteases PC2 And/or PC3. *Diabetes* **1997**, *46*, 978–982. [CrossRef] [PubMed]
75. Irminger, J.C.; Vollenweider, F.M.; Neerman-Arbez, M.; Halban, P.A. Human Proinsulin Conversion in the Regulated and the Constitutive Pathways of Transfected AtT20 Cells. *J. Biol. Chem.* **1994**, *269*, 1756–1762. [CrossRef]
76. Neerman-Arbez, M.; Sizonenko, S.V.; Halban, P.A. Slow Cleavage at the Proinsulin B-Chain/connecting Peptide Junction Associated with Low Levels of Endoprotease PC1/3 in Transformed Beta Cells. *J. Biol. Chem.* **1993**, *268*, 16098–16100. [CrossRef]
77. Neerman-Arbez, M.; Cirulli, V.; Halban, P.A. Levels of the Conversion Endoproteases PC1 (PC3) and PC2 Distinguish between Insulin-Producing Pancreatic Islet Beta Cells and Non-Beta Cells. *Biochem. J.* **1994**, *300 Pt 1*, 57–61. [CrossRef]
78. Zhu, X.; Orci, L.; Carroll, R.; Norrbom, C.; Ravazzola, M.; Steiner, D.F. Severe Block in Processing of Proinsulin to Insulin Accompanied by Elevation of Des-64,65 Proinsulin Intermediates in Islets of Mice Lacking Prohormone Convertase 1/3. *Proc. Natl. Acad. Sci. USA* **2002**, *99*, 10299–10304. [CrossRef]

79. Furuta, M.; Carroll, R.; Martin, S.; Swift, H.H.; Ravazzola, M.; Orci, L.; Steiner, D.F. Incomplete Processing of Proinsulin to Insulin Accompanied by Elevation of Des-31,32 Proinsulin Intermediates in Islets of Mice Lacking Active PC2. *J. Biol. Chem.* **1998**, *273*, 3431–3437. [CrossRef]
80. Ramzy, A.; Asadi, A.; Kieffer, T.J. Revisiting Proinsulin Processing: Evidence That Human β-Cells Process Proinsulin with Prohormone Convertase (PC) 1/3 but Not PC2. *Diabetes* **2020**, *69*, 1451–1462. [CrossRef]
81. Irminger, J.C.; Meyer, K.; Halban, P. Proinsulin Processing in the Rat Insulinoma Cell Line INS after Overexpression of the Endoproteases PC2 or PC3 by Recombinant Adenovirus. *Biochem. J.* **1996**, *320 Pt 1*, 11–15. [CrossRef]
82. Shennan, K.I.; Taylor, N.A.; Jermany, J.L.; Matthews, G.; Docherty, K. Differences in pH Optima and Calcium Requirements for Maturation of the Prohormone Convertases PC2 and PC3 Indicates Different Intracellular Locations for These Events. *J. Biol. Chem.* **1995**, *270*, 1402–1407. [CrossRef]
83. Muller, L.; Zhu, X.; Lindberg, I. Mechanism of the Facilitation of PC2 Maturation by 7B2: Involvement in ProPC2 Transport and Activation but Not Folding. *J. Cell Biol.* **1997**, *139*, 625–638. [CrossRef]
84. Zhu, X.; Rouille, Y.; Lamango, N.S.; Steiner, D.F.; Lindberg, I. Internal Cleavage of the Inhibitory 7B2 Carboxyl-Terminal Peptide by PC2: A Potential Mechanism for Its Inactivation. *Proc. Natl. Acad. Sci. USA* **1996**, *93*, 4919–4924. [CrossRef]
85. Hwang, J.R.; Lindberg, I. Inactivation of the 7B2 Inhibitory CT Peptide Depends on a Functional Furin Cleavage Site. *J. Neurochem.* **2001**, *79*, 437–444. [CrossRef]
86. Lamango, N.S.; Apletalina, E.; Liu, J.; Lindberg, I. The Proteolytic Maturation of Prohormone Convertase 2 (PC2) Is a pH-Driven Process. *Arch. Biochem. Biophys.* **1999**, *362*, 275–282. [CrossRef]
87. Ostrega, D.; Polonsky, K.; Nagi, D.; Yudkin, J.; Cox, L.J.; Clark, P.M.; Hales, C.N. Measurement of Proinsulin and Intermediates. Validation of Immunoassay Methods by High-Performance Liquid Chromatography. *Diabetes* **1995**, *44*, 437–440. [CrossRef]
88. Clark, P.M.; Levy, J.C.; Cox, L.; Burnett, M.; Turner, R.C.; Hales, C.N. Immunoradiometric Assay of Insulin, Intact Proinsulin and 32–33 Split Proinsulin and Radioimmunoassay of Insulin in Diet-Treated Type 2 (non-Insulin-Dependent) Diabetic Subjects. *Diabetologia* **1992**, *35*, 469–474. [CrossRef]
89. Lukinius, A.; Wilander, E.; Westermark, G.T.; Engström, U.; Westermark, P. Co-Localization of Islet Amyloid Polypeptide and Insulin in the B Cell Secretory Granules of the Human Pancreatic Islets. *Diabetologia* **1989**, *32*, 240–244. [CrossRef]
90. Nakazato, M.; Miyazato, M.; Asai, J.; Mitsukawa, T.; Kangawa, K.; Matsuo, H.; Matsukura, S. Islet Amyloid Polypeptide, a Novel Pancreatic Peptide, Is a Circulating Hormone Secreted under Glucose Stimulation. *Biochem. Biophys. Res. Commun.* **1990**, *169*, 713–718. [CrossRef]
91. Stridsberg, M.; Sandler, S.; Wilander, E. Cosecretion of Islet Amyloid Polypeptide (IAPP) and Insulin from Isolated Rat Pancreatic Islets Following Stimulation or Inhibition of Beta-Cell Function. *Regul. Pept.* **1993**, *45*, 363–370. [CrossRef]
92. Lutz, T.A. The Role of Amylin in the Control of Energy Homeostasis. *Am. J. Physiol. Regul. Integr. Comp. Physiol.* **2010**, *298*, R1475–R1484. [CrossRef] [PubMed]
93. Hull, R.L.; Westermark, G.T.; Westermark, P.; Kahn, S.E. Islet Amyloid: A Critical Entity in the Pathogenesis of Type 2 Diabetes. *J. Clin. Endocrinol. Metab.* **2004**, *89*, 3629–3643. [CrossRef] [PubMed]
94. Westermark, P.; Grimelius, L. The Pancreatic Islet Cells in Insular Amyloidosis in Human Diabetic and Non-Diabetic Adults. *Acta Pathol. Microbiol. Scand. A* **1973**, *81*, 291–300. [CrossRef] [PubMed]
95. Westermark, P.; Wilander, E. The Influence of Amyloid Deposits on the Islet Volume in Maturity Onset Diabetes Mellitus. *Diabetologia* **1978**, *15*, 417–421. [CrossRef]
96. Zhao, H.-L.; Lai, F.M.M.; Tong, P.C.Y.; Zhong, D.-R.; Yang, D.; Tomlinson, B.; Chan, J.C.N. Prevalence and Clinicopathological Characteristics of Islet Amyloid in Chinese Patients with Type 2 Diabetes. *Diabetes* **2003**, *52*, 2759–2766. [CrossRef]
97. Westermark, P.; Andersson, A.; Westermark, G.T. Islet Amyloid Polypeptide, Islet Amyloid, and Diabetes Mellitus. *Physiol. Rev.* **2011**, *91*, 795–826. [CrossRef]
98. Haataja, L.; Gurlo, T.; Huang, C.J.; Butler, P.C. Islet Amyloid in Type 2 Diabetes, and the Toxic Oligomer Hypothesis. *Endocr. Rev.* **2008**, *29*, 303–316. [CrossRef]
99. Raleigh, D.; Zhang, X.; Hastoy, B.; Clark, A. The β-Cell Assassin: IAPP Cytotoxicity. *J. Mol. Endocrinol.* **2017**, *59*, R121–R140. [CrossRef]
100. Wang, J.; Xu, J.; Finnerty, J.; Furuta, M.; Steiner, D.F.; Verchere, C.B. The Prohormone Convertase Enzyme 2 (PC2) Is Essential for Processing pro-Islet Amyloid Polypeptide at the NH2-Terminal Cleavage Site. *Diabetes* **2001**, *50*, 534–539. [CrossRef]
101. Marzban, L.; Soukhatcheva, G.; Verchere, C.B. Role of Carboxypeptidase E in Processing of Pro-Islet Amyloid Polypeptide in β-Cells. *Endocrinology* **2005**, *146*, 1808–1817. [CrossRef]
102. Marzban, L.; Trigo-Gonzalez, G.; Verchere, C.B. Processing of pro-Islet Amyloid Polypeptide in the Constitutive and Regulated Secretory Pathways of Beta Cells. *Mol. Endocrinol.* **2005**, *19*, 2154–2163. [CrossRef]
103. Marzban, L.; Trigo-Gonzalez, G.; Zhu, X.; Rhodes, C.J.; Halban, P.A.; Steiner, D.F.; Verchere, C.B. Role of Beta-Cell Prohormone Convertase (PC)1/3 in Processing of pro-Islet Amyloid Polypeptide. *Diabetes* **2004**, *53*, 141–148. [CrossRef]
104. Chen, Y.-C.; Taylor, A.J.; Verchere, C.B. Islet Prohormone Processing in Health and Disease. *Diabetes Obes. Metab.* **2018**, *20* (Suppl. 2), 64–76. [CrossRef]
105. Rulifson, I.C.; Cao, P.; Miao, L.; Kopecky, D.; Huang, L.; White, R.D.; Samayoa, K.; Gardner, J.; Wu, X.; Chen, K.; et al. Identification of Human Islet Amyloid Polypeptide as a BACE2 Substrate. *PLoS ONE* **2016**, *11*, e0147254. [CrossRef]

106. Courtade, J.A.; Wang, E.Y.; Yen, P.; Dai, D.L.; Soukhatcheva, G.; Orban, P.C.; Verchere, C.B. Loss of Prohormone Convertase 2 Promotes Beta Cell Dysfunction in a Rodent Transplant Model Expressing Human pro-Islet Amyloid Polypeptide. *Diabetologia* **2017**, *60*, 453–463. [CrossRef]
107. Westermark, P.; Engström, U.; Johnson, K.H.; Westermark, G.T.; Betsholtz, C. Islet Amyloid Polypeptide: Pinpointing Amino Acid Residues Linked to Amyloid Fibril Formation. *Proc. Natl. Acad. Sci. USA* **1990**, *87*, 5036–5040. [CrossRef]
108. Yonemoto, I.T.; Kroon, G.J.A.; Dyson, H.J.; Balch, W.E.; Kelly, J.W. Amylin Proprotein Processing Generates Progressively More Amyloidogenic Peptides That Initially Sample the Helical State. *Biochemistry* **2008**, *47*, 9900–9910. [CrossRef]
109. Jaikaran, E.T.A.S.; Nilsson, M.R.; Clark, A. Pancreatic Beta-Cell Granule Peptides Form Heteromolecular Complexes Which Inhibit Islet Amyloid Polypeptide Fibril Formation. *Biochem. J.* **2004**, *377*, 709–716. [CrossRef]
110. Westermark, P.; Li, Z.C.; Westermark, G.T.; Leckström, A.; Steiner, D.F. Effects of Beta Cell Granule Components on Human Islet Amyloid Polypeptide Fibril Formation. *FEBS Lett.* **1996**, *379*, 203–206. [CrossRef]
111. Brender, J.R.; Hartman, K.; Nanga, R.P.R.; Popovych, N.; de la Salud Bea, R.; Vivekanandan, S.; Marsh, E.N.G.; Ramamoorthy, A. Role of Zinc in Human Islet Amyloid Polypeptide Aggregation. *J. Am. Chem. Soc.* **2010**, *132*, 8973–8983. [CrossRef]
112. Khemtémourian, L.; Doménech, E.; Doux, J.P.F.; Koorengevel, M.C.; Killian, J.A. Low pH Acts as Inhibitor of Membrane Damage Induced by Human Islet Amyloid Polypeptide. *J. Am. Chem. Soc.* **2011**, *133*, 15598–15604. [CrossRef] [PubMed]
113. Janciauskiene, S.; Eriksson, S.; Carlemalm, E.; Ahrén, B. B Cell Granule Peptides Affect Human Islet Amyloid Polypeptide (IAPP) Fibril Formation in Vitro. *Biochem. Biophys. Res. Commun.* **1997**, *236*, 580–585. [CrossRef] [PubMed]
114. Kudva, Y.C.; Mueske, C.; Butler, P.C.; Eberhardt, N.L. A Novel Assay in Vitro of Human Islet Amyloid Polypeptide Amyloidogenesis and Effects of Insulin Secretory Vesicle Peptides on Amyloid Formation. *Biochem. J.* **1998**, *331 Pt 3*, 809–813. [CrossRef] [PubMed]
115. Johnson, K.H.; O'Brien, T.D.; Jordan, K.; Westermark, P. Impaired Glucose Tolerance Is Associated with Increased Islet Amyloid Polypeptide (IAPP) Immunoreactivity in Pancreatic Beta Cells. *Am. J. Pathol.* **1989**, *135*, 245–250.
116. Gurlo, T.; Ryazantsev, S.; Huang, C.-J.; Yeh, M.W.; Reber, H.A.; Hines, O.J.; O'Brien, T.D.; Glabe, C.G.; Butler, P.C. Evidence for Proteotoxicity in Beta Cells in Type 2 Diabetes: Toxic Islet Amyloid Polypeptide Oligomers Form Intracellularly in the Secretory Pathway. *Am. J. Pathol.* **2010**, *176*, 861–869. [CrossRef]
117. Lin, C.-Y.; Gurlo, T.; Kayed, R.; Butler, A.E.; Haataja, L.; Glabe, C.G.; Butler, P.C. Toxic Human Islet Amyloid Polypeptide (h-IAPP) Oligomers Are Intracellular, and Vaccination to Induce Anti-Toxic Oligomer Antibodies Does Not Prevent H-IAPP-Induced Beta-Cell Apoptosis in H-IAPP Transgenic Mice. *Diabetes* **2007**, *56*, 1324–1332. [CrossRef]
118. Yagui, K.; Yamaguchi, T.; Kanatsuka, A.; Shimada, F.; Huang, C.I.; Tokuyama, Y.; Ohsawa, H.; Yamamura, K.; Miyazaki, J.; Mikata, A. Formation of Islet Amyloid Fibrils in Beta-Secretory Granules of Transgenic Mice Expressing Human Islet Amyloid Polypeptide/amylin. *Eur. J. Endocrinol.* **1995**, *132*, 487–496. [CrossRef]
119. Wang, H.; Raleigh, D.P. The Ability of Insulin to Inhibit the Formation of Amyloid by pro-Islet Amyloid Polypeptide Processing Intermediates Is Significantly Reduced in the Presence of Sulfated Glycosaminoglycans. *Biochemistry* **2014**, *53*, 2605–2614. [CrossRef]
120. Janson, J.; Ashley, R.H.; Harrison, D.; McIntyre, S.; Butler, P.C. The Mechanism of Islet Amyloid Polypeptide Toxicity Is Membrane Disruption by Intermediate-Sized Toxic Amyloid Particles. *Diabetes* **1999**, *48*, 491–498. [CrossRef]
121. Brender, J.R.; Lee, E.L.; Cavitt, M.A.; Gafni, A.; Steel, D.G.; Ramamoorthy, A. Amyloid Fiber Formation and Membrane Disruption Are Separate Processes Localized in Two Distinct Regions of IAPP, the Type-2-Diabetes-Related Peptide. *J. Am. Chem. Soc.* **2008**, *130*, 6424–6429. [CrossRef]
122. Brender, J.R.; Salamekh, S.; Ramamoorthy, A. Membrane Disruption and Early Events in the Aggregation of the Diabetes Related Peptide IAPP from a Molecular Perspective. *Acc. Chem. Res.* **2012**, *45*, 454–462. [CrossRef]
123. Westermark, P.; Engström, U.; Westermark, G.T.; Johnson, K.H.; Permerth, J.; Betsholtz, C. Islet Amyloid Polypeptide (IAPP) and pro-IAPP Immunoreactivity in Human Islets of Langerhans. *Diabetes Res. Clin. Pract.* **1989**, *7*, 219–226. [CrossRef]
124. Westermark, G.T.; Steiner, D.F.; Gebre-Medhin, S.; Engström, U.; Westermark, P. Pro Islet Amyloid Polypeptide (ProIAPP) Immunoreactivity in the Islets of Langerhans. *Ups. J. Med. Sci.* **2000**, *105*, 97–106. [CrossRef]
125. Zheng, X.; Ren, W.; Zhang, S.; Liu, J.; Li, S.; Li, J.; Yang, P.; He, J.; Su, S.; Li, P. Serum Levels of Proamylin and Amylin in Normal Subjects and Patients with Impaired Glucose Regulation and Type 2 Diabetes Mellitus. *Acta Diabetol.* **2010**, *47*, 265–270. [CrossRef]
126. Xu, J.; Wijesekara, N.; Regeenes, R.; Rijjal, D.A.; Piro, A.L.; Song, Y.; Wu, A.; Bhattacharjee, A.; Liu, Y.; Marzban, L.; et al. Pancreatic β Cell-Selective Zinc Transporter 8 Insufficiency Accelerates Diabetes Associated with Islet Amyloidosis. *JCI Insight* **2021**, *6*. [CrossRef]
127. Elias, S.; Delestre, C.; Ory, S.; Marais, S.; Courel, M.; Vazquez-Martinez, R.; Bernard, S.; Coquet, L.; Malagon, M.M.; Driouich, A.; et al. Chromogranin A Induces the Biogenesis of Granules with Calcium- and Actin-Dependent Dynamics and Exocytosis in Constitutively Secreting Cells. *Endocrinology* **2012**, *153*, 4444–4456. [CrossRef]
128. Montero-Hadjadje, M.; Elias, S.; Chevalier, L.; Benard, M.; Tanguy, Y.; Turquier, V.; Galas, L.; Yon, L.; Malagon, M.M.; Driouich, A.; et al. Chromogranin A Promotes Peptide Hormone Sorting to Mobile Granules in Constitutively and Regulated Secreting Cells: Role of Conserved N- and C-Terminal Peptides. *J. Biol. Chem.* **2009**, *284*, 12420–12431. [CrossRef]
129. Inomoto, C.; Umemura, S.; Egashira, N.; Minematsu, T.; Takekoshi, S.; Itoh, Y.; Itoh, J.; Taupenot, L.; O'Connor, D.T.; Osamura, R.Y. Granulogenesis in Non-Neuroendocrine COS-7 Cells Induced by EGFP-Tagged Chromogranin A Gene Transfection: Identical and Distinct Distribution of CgA and EGFP. *J. Histochem. Cytochem.* **2007**, *55*, 487–493. [CrossRef]

130. Huh, Y.H.; Jeon, S.H.; Yoo, S.H. Chromogranin B-Induced Secretory Granule Biogenesis: Comparison with the Similar Role of Chromogranin A. *J. Biol. Chem.* **2003**, *278*, 40581–40589. [CrossRef]
131. Chanat, E.; Huttner, W.B. Milieu-Induced, Selective Aggregation of Regulated Secretory Proteins in the Trans-Golgi Network. *J. Cell Biol.* **1991**, *115*, 1505–1519. [CrossRef]
132. Yoo, S.H.; Albanesi, J.P. High Capacity, Low Affinity Ca^{2+} Binding of Chromogranin A. Relationship between the pH-Induced Conformational Change and Ca^{2+} Binding Property. *J. Biol. Chem.* **1991**, *266*, 7740–7745. [CrossRef]
133. Sun-Wada, G.-H.; Toyomura, T.; Murata, Y.; Yamamoto, A.; Futai, M.; Wada, Y. The a3 Isoform of V-ATPase Regulates Insulin Secretion from Pancreatic Beta-Cells. *J. Cell Sci.* **2006**, *119 Pt 21*, 4531–4540. [CrossRef]
134. Taupenot, L.; Harper, K.L.; O'Connor, D.T. Role of H^+-ATPase-Mediated Acidification in Sorting and Release of the Regulated Secretory Protein Chromogranin A: Evidence for a Vesiculogenic Function. *J. Biol. Chem.* **2005**, *280*, 3885–3897. [CrossRef]
135. Lissandron, V.; Podini, P.; Pizzo, P.; Pozzan, T. Unique Characteristics of Ca^{2+} Homeostasis of the Trans-Golgi Compartment. *Proc. Natl. Acad. Sci. USA* **2010**, *107*, 9198–9203. [CrossRef]
136. Klemm, R.W.; Ejsing, C.S.; Surma, M.A.; Kaiser, H.-J.; Gerl, M.J.; Sampaio, J.L.; de Robillard, Q.; Ferguson, C.; Proszynski, T.J.; Shevchenko, A.; et al. Segregation of Sphingolipids and Sterols during Formation of Secretory Vesicles at the Trans-Golgi Network. *J. Cell Biol.* **2009**, *185*, 601–612. [CrossRef]
137. Orci, L.; Montesano, R.; Meda, P.; Malaisse-Lagae, F.; Brown, D.; Perrelet, A.; Vassalli, P. Heterogeneous Distribution of Filipin–Cholesterol Complexes across the Cisternae of the Golgi Apparatus. *Proc. Natl. Acad. Sci. USA* **1981**, *78*, 293–297. [CrossRef]
138. von Blume, J.; Hausser, A. Lipid-Dependent Coupling of Secretory Cargo Sorting and Trafficking at the Trans-Golgi Network. *FEBS Lett.* **2019**, *593*, 2412–2427. [CrossRef]
139. Wang, Y.; Thiele, C.; Huttner, W.B. Cholesterol Is Required for the Formation of Regulated and Constitutive Secretory Vesicles from the Trans-Golgi Network. *Traffic* **2000**, *1*, 952–962. [CrossRef]
140. Wang, T.Y.; Silvius, J.R. Different Sphingolipids Show Differential Partitioning into Sphingolipid/cholesterol-Rich Domains in Lipid Bilayers. *Biophys. J.* **2000**, *79*, 1478–1489. [CrossRef]
141. Kreutzberger, A.J.B.; Kiessling, V.; Doyle, C.A.; Schenk, N.; Upchurch, C.M.; Elmer-Dixon, M.; Ward, A.E.; Preobraschenski, J.; Hussein, S.S.; Tomaka, W.; et al. Distinct Insulin Granule Subpopulations Implicated in the Secretory Pathology of Diabetes Types 1 and 2. *eLife* **2020**, *9*. [CrossRef] [PubMed]
142. Hussain, S.S.; Harris, M.T.; Kreutzberger, A.J.B.; Inouye, C.M.; Doyle, C.A.; Castle, A.M.; Arvan, P.; Castle, J.D. Control of Insulin Granule Formation and Function by the ABC Transporters ABCG1 and ABCA1 and by Oxysterol Binding Protein OSBP. *Mol. Biol. Cell* **2018**, *29*, 1238–1257. [CrossRef] [PubMed]
143. Tsuchiya, M.; Hosaka, M.; Moriguchi, T.; Zhang, S.; Suda, M.; Yokota-Hashimoto, H.; Shinozuka, K.; Takeuchi, T. Cholesterol Biosynthesis Pathway Intermediates and Inhibitors Regulate Glucose-Stimulated Insulin Secretion and Secretory Granule Formation in Pancreatic Beta-Cells. *Endocrinology* **2010**, *151*, 4705–4716. [CrossRef]
144. Bogan, J.S.; Xu, Y.; Hao, M. Cholesterol Accumulation Increases Insulin Granule Size and Impairs Membrane Trafficking. *Traffic* **2012**, *13*, 1466–1480. [CrossRef] [PubMed]
145. Payet, L.-A.; Pineau, L.; Snyder, E.C.R.; Colas, J.; Moussa, A.; Vannier, B.; Bigay, J.; Clarhaut, J.; Becq, F.; Berjeaud, J.-M.; et al. Saturated Fatty Acids Alter the Late Secretory Pathway by Modulating Membrane Properties. *Traffic* **2013**, *14*, 1228–1241. [CrossRef] [PubMed]
146. Yoo, S.H.; Oh, Y.S.; Kang, M.K.; Huh, Y.H.; So, S.H.; Park, H.S.; Park, H.Y. Localization of Three Types of the Inositol 1,4,5-Trisphosphate Receptor/Ca^{2+} Channel in the Secretory Granules and Coupling with the Ca^{2+} Storage Proteins Chromogranins A and B*. *J. Biol. Chem.* **2001**, *276*, 45806–45812. [CrossRef]
147. Yoo, S.H.; Lewis, M.S. Thermodynamic Study of the pH-Dependent Interaction of Chromogranin A with an Intraluminal Loop Peptide of the Inositol 1,4,5-Trisphosphate Receptor. *Biochemistry* **1995**, *34*, 632–638. [CrossRef]
148. Hosaka, M.; Suda, M.; Sakai, Y.; Izumi, T.; Watanabe, T.; Takeuchi, T. Secretogranin III Binds to Cholesterol in the Secretory Granule Membrane as an Adapter for Chromogranin A. *J. Biol. Chem.* **2004**, *279*, 3627–3634. [CrossRef]
149. Hosaka, M.; Watanabe, T.; Sakai, Y.; Uchiyama, Y.; Takeuchi, T. Identification of a Chromogranin A Domain That Mediates Binding to Secretogranin III and Targeting to Secretory Granules in Pituitary Cells and Pancreatic Beta-Cells. *Mol. Biol. Cell* **2002**, *13*, 3388–3399. [CrossRef]
150. Han, L.; Suda, M.; Tsuzuki, K.; Wang, R.; Ohe, Y.; Hirai, H.; Watanabe, T.; Takeuchi, T.; Hosaka, M. A Large Form of Secretogranin III Functions as a Sorting Receptor for Chromogranin a Aggregates in PC12 Cells. *Mol. Endocrinol.* **2008**, *22*, 1935–1949. [CrossRef]
151. Courel, M.; Vasquez, M.S.; Hook, V.Y.; Mahata, S.K.; Taupenot, L. Sorting of the Neuroendocrine Secretory Protein Secretogranin II into the Regulated Secretory Pathway: Role of N- and C-Terminal Alpha-Helical Domains. *J. Biol. Chem.* **2008**, *283*, 11807–11822. [CrossRef]
152. Sun, M.; Watanabe, T.; Bochimoto, H.; Sakai, Y.; Torii, S.; Takeuchi, T.; Hosaka, M. Multiple Sorting Systems for Secretory Granules Ensure the Regulated Secretion of Peptide Hormones. *Traffic* **2013**, *14*, 205–218. [CrossRef]
153. Hotta, K.; Hosaka, M.; Tanabe, A.; Takeuchi, T. Secretogranin II Binds to Secretogranin III and Forms Secretory Granules with Orexin, Neuropeptide Y, and POMC. *J. Endocrinol.* **2009**, *202*, 111–121.
154. Courel, M.; Soler-Jover, A.; Rodriguez-Flores, J.L.; Mahata, S.K.; Elias, S.; Montero-Hadjadje, M.; Anouar, Y.; Giuly, R.J.; O'Connor, D.T.; Taupenot, L. Pro-Hormone Secretogranin II Regulates Dense Core Secretory Granule Biogenesis in Catecholaminergic Cells. *J. Biol. Chem.* **2010**, *285*, 10030–10043. [CrossRef]

155. Glombik, M.M.; Krömer, A.; Salm, T.; Huttner, W.B.; Gerdes, H.H. The Disulfide-Bonded Loop of Chromogranin B Mediates Membrane Binding and Directs Sorting from the Trans-Golgi Network to Secretory Granules. *EMBO J.* **1999**, *18*, 1059–1070. [CrossRef]
156. Pimplikar, S.W.; Huttner, W.B. Chromogranin B (secretogranin I), a Secretory Protein of the Regulated Pathway, Is Also Present in a Tightly Membrane-Associated Form in PC12 Cells. *J. Biol. Chem.* **1992**, *267*, 4110–4118. [CrossRef]
157. Giordano, T.; Brigatti, C.; Podini, P.; Bonifacio, E.; Meldolesi, J.; Malosio, M.L. Beta Cell Chromogranin B Is Partially Segregated in Distinct Granules and Can Be Released Separately from Insulin in Response to Stimulation. *Diabetologia* **2008**, *51*, 997–1007. [CrossRef]
158. Yoo, S.H.; Chu, S.Y.; Kim, K.D.; Huh, Y.H. Presence of Secretogranin II and High-Capacity, Low-Affinity Ca^{2+} Storage Role in Nucleoplasmic Ca^{2+} Store Vesicles. *Biochemistry* **2007**, *46*, 14663–14671. [CrossRef]
159. Bearrows, S.C.; Bauchle, C.J.; Becker, M.; Haldeman, J.M.; Swaminathan, S.; Stephens, S.B. Chromogranin B Regulates Early-Stage Insulin Granule Trafficking from the Golgi in Pancreatic Islet β-Cells. *J. Cell Sci.* **2019**, *132*. [CrossRef]
160. Yoo, S.H.; Lewis, M.S. Effects of pH and Ca^{2+} on Heterodimer and Heterotetramer Formation by Chromogranin A and Chromogranin B. *J. Biol. Chem.* **1996**, *271*, 17041–17046. [CrossRef]
161. Garcia, A.L.; Han, S.-K.; Janssen, W.G.; Khaing, Z.Z.; Ito, T.; Glucksman, M.J.; Benson, D.L.; Salton, S.R.J. A Prohormone Convertase Cleavage Site within a Predicted α-Helix Mediates Sorting of the Neuronal and Endocrine Polypeptide VGF into the Regulated Secretory Pathway *. *J. Biol. Chem.* **2005**, *280*, 41595–41608. [CrossRef]
162. Obermüller, S.; Calegari, F.; King, A.; Lindqvist, A.; Lundquist, I.; Salehi, A.; Francolini, M.; Rosa, P.; Rorsman, P.; Huttner, W.B.; et al. Defective Secretion of Islet Hormones in Chromogranin-B Deficient Mice. *PLoS ONE* **2010**, *5*, e8936. [CrossRef] [PubMed]
163. Wollam, J.; Mahata, S.; Riopel, M.; Hernandez-Carretero, A.; Biswas, A.; Bandyopadhyay, G.K.; Chi, N.-W.; Eiden, L.E.; Mahapatra, N.R.; Corti, A.; et al. Chromogranin A Regulates Vesicle Storage and Mitochondrial Dynamics to Influence Insulin Secretion. *Cell Tissue Res.* **2017**, *368*, 487–501. [CrossRef] [PubMed]
164. Thomsen, S.K.; Raimondo, A.; Hastoy, B.; Sengupta, S.; Dai, X.-Q.; Bautista, A.; Censin, J.; Payne, A.J.; Umapathysivam, M.M.; Spigelman, A.F.; et al. Type 2 Diabetes Risk Alleles in PAM Impact Insulin Release from Human Pancreatic β-Cells. *Nat. Genet.* **2018**, *50*, 1122–1131. [CrossRef] [PubMed]
165. Maeda, Y.; Kudo, S.; Tsushima, K.; Sato, E.; Kubota, C.; Kayamori, A.; Bochimoto, H.; Koga, D.; Torii, S.; Gomi, H.; et al. Impaired Processing of Prohormones in Secretogranin III-Null Mice Causes Maladaptation to an Inadequate Diet and Stress. *Endocrinology* **2018**, *159*, 1213–1227. [CrossRef] [PubMed]
166. Stephens, S.B.; Edwards, R.J.; Sadahiro, M.; Lin, W.-J.; Jiang, C.; Salton, S.R.; Newgard, C.B. The Prohormone VGF Regulates β Cell Function via Insulin Secretory Granule Biogenesis. *Cell Rep.* **2017**, *20*, 2480–2489. [CrossRef]
167. Dittié, A.S.; Thomas, L.; Thomas, G.; Tooze, S.A. Interaction of Furin in Immature Secretory Granules from Neuroendocrine Cells with the AP-1 Adaptor Complex Is Modulated by Casein Kinase II Phosphorylation. *EMBO J.* **1997**, *16*, 4859–4870. [CrossRef]
168. Finzi, G.; Franzi, F.; Placidi, C.; Acquati, F.; Palumbo, E.; Russo, A.; Taramelli, R.; Sessa, F.; La Rosa, S. BACE2 Is Stored in Secretory Granules of Mouse and Rat Pancreatic Beta Cells. *Ultrastruct. Pathol.* **2008**, *32*, 246–251. [CrossRef]
169. Arias, A.E.; Vélez-Granell, C.S.; Mayer, G.; Bendayan, M. Colocalization of Chaperone Cpn60, Proinsulin and Convertase PC1 within Immature Secretory Granules of Insulin-Secreting Cells Suggests a Role for Cpn60 in Insulin Processing. *J. Cell Sci.* **2000**, *113 Pt 11*, 2075–2083. [CrossRef]
170. Docherty, K.; Hutton, J.C. Carboxypeptidase Activity in the Insulin Secretory Granule. *FEBS Lett.* **1983**, *162*, 137–141. [CrossRef]
171. Dhanvantari, S.; Arnaoutova, I.; Snell, C.R.; Steinbach, P.J.; Hammond, K.; Caputo, G.A.; London, E.; Loh, Y.P. Carboxypeptidase E, a Prohormone Sorting Receptor, Is Anchored to Secretory Granules via a C-Terminal Transmembrane Insertion. *Biochemistry* **2002**, *41*, 52–60. [CrossRef]
172. Fricker, L.D.; Das, B.; Angeletti, R.H. Identification of the pH-Dependent Membrane Anchor of Carboxypeptidase E (EC 3.4.17.10). *J. Biol. Chem.* **1990**, *265*, 2476–2482. [CrossRef]
173. Demaurex, N.; Furuya, W.; D'Souza, S.; Bonifacino, J.S.; Grinstein, S. Mechanism of Acidification of the Trans-Golgi Network (TGN). In Situ Measurements of pH Using Retrieval of TGN38 and Furin from the Cell Surface. *J. Biol. Chem.* **1998**, *273*, 2044–2051. [CrossRef]
174. Song, L.; Fricker, L.D. Calcium- and pH-Dependent Aggregation of Carboxypeptidase E. *J. Biol. Chem.* **1995**, *270*, 7963–7967. [CrossRef]
175. Rindler, M.J. Carboxypeptidase E, a Peripheral Membrane Protein Implicated in the Targeting of Hormones to Secretory Granules, Co-Aggregates with Granule Content Proteins at Acidic pH. *J. Biol. Chem.* **1998**, *273*, 31180–31185. [CrossRef]
176. Hosaka, M.; Watanabe, T.; Sakai, Y.; Kato, T.; Takeuchi, T. Interaction between Secretogranin III and Carboxypeptidase E Facilitates Prohormone Sorting within Secretory Granules. *J. Cell Sci.* **2005**, *118 Pt 20*, 4785–4795. [CrossRef]
177. Guest, P.C.; Arden, S.D.; Rutherford, N.G.; Hutton, J.C. The Post-Translational Processing and Intracellular Sorting of Carboxypeptidase H in the Islets of Langerhans. *Mol. Cell. Endocrinol.* **1995**, *113*, 99–108. [CrossRef]
178. Chu, K.Y.; Briggs, M.J.L.; Albrecht, T.; Drain, P.F.; Johnson, J.D. Differential Regulation and Localization of Carboxypeptidase D and Carboxypeptidase E in Human and Mouse β-Cells. *Islets* **2011**, *3*, 155–165. [CrossRef]
179. Greene, D.; Das, B.; Fricker, L.D. Regulation of Carboxypeptidase E. Effect of pH, Temperature and Co2+ on Kinetic Parameters of Substrate Hydrolysis. *Biochem. J.* **1992**, *2*, 613–618. [CrossRef]

180. Benjannet, S.; Rondeau, N.; Paquet, L.; Boudreault, A.; Lazure, C.; Chrétien, M.; Seidah, N.G. Comparative Biosynthesis, Covalent Post-Translational Modifications and Efficiency of Prosegment Cleavage of the Prohormone Convertases PC1 and PC2: Glycosylation, Sulphation and Identification of the Intracellular Site of Prosegment Cleavage of PC1 and PC2. *Biochem. J.* **1993**, *294 Pt 3*, 735–743.
181. Zhou, A.; Mains, R.E. Endoproteolytic Processing of Proopiomelanocortin and Prohormone Convertases 1 and 2 in Neuroendocrine Cells Overexpressing Prohormone Convertases 1 or 2. *J. Biol. Chem.* **1994**, *269*, 17440–17447. [CrossRef]
182. Boudreault, A.; Gauthier, D.; Lazure, C. Proprotein Convertase PC1/3-Related Peptides Are Potent Slow Tight-Binding Inhibitors of Murine PC1/3 and Hfurin*. *J. Biol. Chem.* **1998**, *273*, 31574–31580. [CrossRef] [PubMed]
183. Jutras, I.; Seidah, N.G.; Reudelhuber, T.L.; Brechler, V. Two Activation States of the Prohormone Convertase PC1 in the Secretory Pathway. *J. Biol. Chem.* **1997**, *272*, 15184–15188. [CrossRef] [PubMed]
184. Rabah, N.; Gauthier, D.; Dikeakos, J.D.; Reudelhuber, T.L.; Lazure, C. The C-Terminal Region of the Proprotein Convertase 1/3 (PC1/3) Exerts a Bimodal Regulation of the Enzyme Activity in Vitro. *FEBS J.* **2007**, *274*, 3482–3491. [CrossRef] [PubMed]
185. Benjannet, S.; Rondeau, N.; Day, R.; Chrétien, M.; Seidah, N.G. PC1 and PC2 Are Proprotein Convertases Capable of Cleaving Proopiomelanocortin at Distinct Pairs of Basic Residues. *Proc. Natl. Acad. Sci. USA* **1991**, *88*, 3564–3568. [CrossRef]
186. Thomas, L.; Leduc, R.; Thorne, B.A.; Smeekens, S.P.; Steiner, D.F.; Thomas, G. Kex2-like Endoproteases PC2 and PC3 Accurately Cleave a Model Prohormone in Mammalian Cells: Evidence for a Common Core of Neuroendocrine Processing Enzymes. *Proc. Natl. Acad. Sci. USA* **1991**, *88*, 5297–5301. [CrossRef]
187. Jutras, I.; Seidah, N.G.; Reudelhuber, T.L. A Predicted Alpha -Helix Mediates Targeting of the Proprotein Convertase PC1 to the Regulated Secretory Pathway. *J. Biol. Chem.* **2000**, *275*, 40337–40343. [CrossRef]
188. Blázquez, M.; Docherty, K.; Shennan, K.I. Association of Prohormone Convertase 3 with Membrane Lipid Rafts. *J. Mol. Endocrinol.* **2001**, *27*, 107–116. [CrossRef]
189. Zhou, Y.; Lindberg, I. Enzymatic Properties of Carboxyl-Terminally Truncated Prohormone Convertase 1 (PC1/SPC3) and Evidence for Autocatalytic Conversion. *J. Biol. Chem.* **1994**, *269*, 18408–18413. [CrossRef]
190. Zhou, Y.; Lindberg, I. Purification and Characterization of the Prohormone Convertase PC1(PC3). *J. Biol. Chem.* **1993**, *268*, 5615–5623. [CrossRef]
191. Muller, L.; Cameron, A.; Fortenberry, Y.; Apletalina, E.V.; Lindberg, I. Processing and Sorting of the Prohormone Convertase 2 Propeptide *. *J. Biol. Chem.* **2000**, *275*, 39213–39222. [CrossRef]
192. Lee, S.-N.; Lindberg, I. 7B2 Prevents Unfolding and Aggregation of Prohormone Convertase 2. *Endocrinology* **2008**, *149*, 4116–4127. [CrossRef]
193. Benjannet, S.; Savaria, D.; Chrétien, M.; Seidah, N.G. 7B2 Is a Specific Intracellular Binding Protein of the Prohormone Convertase PC2. *J. Neurochem.* **1995**, *64*, 2303–2311. [CrossRef]
194. Benjannet, S.; Mamarbachi, A.M.; Hamelin, J.; Savaria, D.; Munzer, J.S.; Chrétien, M.; Seidah, N.G. Residues Unique to the pro-Hormone Convertase PC2 Modulate Its Autoactivation, Binding to 7B2 and Enzymatic Activity. *FEBS Lett.* **1998**, *428*, 37–42. [CrossRef]
195. Paquet, L.; Bergeron, F.; Boudreault, A.; Seidah, N.G.; Chrétien, M.; Mbikay, M.; Lazure, C. The Neuroendocrine Precursor 7B2 Is a Sulfated Protein Proteolytically Processed by a Ubiquitous Furin-like Convertase. *J. Biol. Chem.* **1994**, *269*, 19279–19285. [CrossRef]
196. Braks, J.A.; Van Horssen, A.M.; Martens, G.J. Dissociation of the Complex between the Neuroendocrine Chaperone 7B2 and Prohormone Convertase PC2 Is Not Associated with proPC2 Maturation. *Eur. J. Biochem.* **1996**, *238*, 505–510. [CrossRef]
197. Shennan, K.I.; Taylor, N.A.; Docherty, K. Calcium- and pH-Dependent Aggregation and Membrane Association of the Precursor of the Prohormone Convertase PC2. *J. Biol. Chem.* **1994**, *269*, 18646–18650. [CrossRef]
198. Linard, C.G.; Tadros, H.; Sirois, F.; Mbikay, M. Calcium-Induced Aggregation of Neuroendocrine Protein 7B2 in Vitro and Its Modulation by ATP. *Mol. Cell. Biochem.* **1995**, *151*, 39–47. [CrossRef]
199. Blázquez, M.; Thiele, C.; Huttner, W.B.; Docherty, K.; Shennan, K.I. Involvement of the Membrane Lipid Bilayer in Sorting Prohormone Convertase 2 into the Regulated Secretory Pathway. *Biochem. J.* **2000**, *349 Pt 3*, 843–852. [CrossRef]
200. Rovère, C.; Mort, J.S.; Chrétien, M.; Seidah, N.G. Cathepsin-B Fusion Proteins Misroute Secretory Protein Partners such as the Proprotein Convertase PC2-7B2 Complex toward the Lysosomal Degradation Pathways. *Biochem. Biophys. Res. Commun.* **2000**, *276*, 594–599. [CrossRef]
201. Barbero, P.; Kitabgi, P. Protein 7B2 Is Essential for the Targeting and Activation of PC2 into the Regulated Secretory Pathway of rMTC 6-23 Cells. *Biochem. Biophys. Res. Commun.* **1999**, *257*, 473–479. [CrossRef]
202. Perkins, S.N.; Husten, E.J.; Eipper, B.A. The 108-kDA Peptidylglycine Alpha-Amidating Monooxygenase Precursor Contains Two Separable Enzymatic Activities Involved in Peptide Amidation. *Biochem. Biophys. Res. Commun.* **1990**, *171*, 926–932. [CrossRef]
203. Bradbury, A.F.; Finnie, M.D.; Smyth, D.G. Mechanism of C-Terminal Amide Formation by Pituitary Enzymes. *Nature* **1982**, *298*, 686–688. [CrossRef] [PubMed]
204. Merkler, D.J. C-Terminal Amidated Peptides: Production by the in Vitro Enzymatic Amidation of Glycine-Extended Peptides and the Importance of the Amide to Bioactivity. *Enzym. Microb. Technol.* **1994**, *16*, 450–456. [CrossRef]
205. Bousquet-Moore, D.; Mains, R.E.; Eipper, B.A. Peptidylgycine α-Amidating Monooxygenase and Copper: A Gene-Nutrient Interaction Critical to Nervous System Function. *J. Neurosci. Res.* **2010**, *88*, 2535–2545. [CrossRef]

206. Eipper, B.A.; Perkins, S.N.; Husten, E.J.; Johnson, R.C.; Keutmann, H.T.; Mains, R.E. Peptidyl-Alpha-Hydroxyglycine Alpha-Amidating Lyase. Purification, Characterization, and Expression. *J. Biol. Chem.* **1991**, *266*, 7827–7833. [CrossRef]
207. Bundgaard, H.; Kahns, A.H. Chemical Stability and Plasma-Catalyzed Dealkylation of Peptidyl-α-Hydroxyglycine derivatives—Intermediates in Peptide α-Amidation. *Peptides* **1991**, *12*, 745–748. [CrossRef]
208. Kyun-Hwan, K.; Baik, L.S. Peptide Amidation: Production of Peptide Hormones in Vivo And in Vitro. *Biotechnol. Bioproce. Eng.* **2001**, *6*, 244–251.
209. Chen, Y.-C.; Mains, R.E.; Eipper, B.A.; Hoffman, B.G.; Czyzyk, T.A.; Pintar, J.E.; Verchere, C.B. PAM Haploinsufficiency Does Not Accelerate the Development of Diet- and Human IAPP-Induced Diabetes in Mice. *Diabetologia* **2020**, *63*, 561–576. [CrossRef]
210. Kumar, D.; Mains, R.E.; Eipper, B.A. 60 YEARS OF POMC: From POMC and α-MSH to PAM, Molecular Oxygen, Copper, and Vitamin C. *J. Mol. Endocrinol.* **2016**, *56*, T63–T76. [CrossRef]
211. Oyarce, A.M.; Eipper, B.A. Neurosecretory Vesicles Contain Soluble and Membrane-Associated Monofunctional and Bifunctional Peptidylglycine Alpha-Amidating Monooxygenase Proteins. *J. Neurochem.* **1993**, *60*, 1105–1114. [CrossRef]
212. Husten, E.J.; Eipper, B.A. Purification and Characterization of PAM-1, an Integral Membrane Protein Involved in Peptide Processing. *Arch. Biochem. Biophys.* **1994**, *312*, 487–492. [CrossRef]
213. Milgram, S.L.; Eipper, B.A.; Mains, R.E. Differential Trafficking of Soluble and Integral Membrane Secretory Granule-Associated Proteins. *J. Cell Biol.* **1994**, *124*, 33–41. [CrossRef]
214. Milgram, S.L.; Johnson, R.C.; Mains, R.E. Expression of Individual Forms of Peptidylglycine Alpha-Amidating Monooxygenase in AtT-20 Cells: Endoproteolytic Processing and Routing to Secretory Granules. *J. Cell Biol.* **1992**, *117*, 717–728. [CrossRef]
215. Milgram, S.L.; Mains, R.E.; Eipper, B.A. COOH-Terminal Signals Mediate the Trafficking of a Peptide Processing Enzyme in Endocrine Cells. *J. Cell Biol.* **1993**, *121*, 23–36. [CrossRef]
216. Milgram, S.L.; Mains, R.E.; Eipper, B.A. Identification of Routing Determinants in the Cytosolic Domain of a Secretory Granule-Associated Integral Membrane Protein. *J. Biol. Chem.* **1996**, *271*, 17526–17535. [CrossRef]
217. El Meskini, R.; Galano, G.J.; Marx, R.; Mains, R.E.; Eipper, B.A. Targeting of Membrane Proteins to the Regulated Secretory Pathway in Anterior Pituitary Endocrine Cells. *J. Biol. Chem.* **2001**, *276*, 3384–3393. [CrossRef]
218. Marx, R.; Mains, R.E. Routing of Membrane Proteins to Large Dense Core Vesicles in PC12 Cells. *J. Mol. Neurosci.* **2002**, *18*, 113–127. [CrossRef]
219. Chufán, E.E.; De, M.; Eipper, B.A.; Mains, R.E.; Amzel, L.M. Amidation of Bioactive Peptides: The Structure of the Lyase Domain of the Amidating Enzyme. *Structure* **2009**, *17*, 965–973. [CrossRef]
220. Kulathila, R.; Consalvo, A.P.; Fitzpatrick, P.F.; Freeman, J.C.; Snyder, L.M.; Villafranca, J.J.; Merkler, D.J. Bifunctional Peptidylglcine Alpha-Amidating Enzyme Requires Two Copper Atoms for Maximum Activity. *Arch. Biochem. Biophys.* **1994**, *311*, 191–195. [CrossRef]
221. Hutton, J.C. The Internal pH and Membrane Potential of the Insulin-Secretory Granule. *Biochem. J.* **1982**, *204*, 171–178. [CrossRef]
222. Barg, S.; Huang, P.; Eliasson, L.; Nelson, D.J.; Obermüller, S.; Rorsman, P.; Thévenod, F.; Renström, E. Priming of Insulin Granules for Exocytosis by Granular Cl(-) Uptake and Acidification. *J. Cell Sci.* **2001**, *114*, 2145–2154. [CrossRef]
223. Deriy, L.V.; Gomez, E.A.; Jacobson, D.A.; Wang, X.; Hopson, J.A.; Liu, X.Y.; Zhang, G.; Bindokas, V.P.; Philipson, L.H.; Nelson, D.J. The Granular Chloride Channel ClC-3 Is Permissive for Insulin Secretion. *Cell Metab.* **2009**, *10*, 316–323. [CrossRef]
224. Li, D.-Q.; Jing, X.; Salehi, A.; Collins, S.C.; Hoppa, M.B.; Rosengren, A.H.; Zhang, E.; Lundquist, I.; Olofsson, C.S.; Mörgelin, M.; et al. Suppression of Sulfonylurea- and Glucose-Induced Insulin Secretion in Vitro and In Vivo in Mice Lacking the Chloride Transport Protein ClC-3. *Cell Metab.* **2009**, *10*, 309–315. [CrossRef]
225. Maritzen, T.; Keating, D.J.; Neagoe, I.; Zdebik, A.A.; Jentsch, T.J. Role of the Vesicular Chloride Transporter ClC-3 in Neuroendocrine Tissue. *J. Neurosci.* **2008**, *28*, 10587–10598. [CrossRef]
226. Jentsch, T.J.; Maritzen, T.; Keating, D.J.; Zdebik, A.A.; Thévenod, F. ClC-3–a Granular Anion Transporter Involved in Insulin Secretion? *Cell Metab.* **2010**, *12*, 307–308. [CrossRef]
227. Blackmore, C.G.; Varro, A.; Dimaline, R.; Bishop, L.; Gallacher, D.V.; Dockray, G.J. Measurement of Secretory Vesicle pH Reveals Intravesicular Alkalinization by Vesicular Monoamine Transporter Type 2 Resulting in Inhibition of Prohormone Cleavage. *J. Physiol.* **2001**, *531 Pt 3*, 605–617. [CrossRef]
228. Holliday, L.S. Vacuolar H^+-ATPase: An Essential Multitasking Enzyme in Physiology and Pathophysiology. *New J. Sci.* **2014**, *2014*. [CrossRef]
229. Maxson, M.E.; Grinstein, S. The Vacuolar-Type H^+-ATPase at a Glance—More than a Proton Pump. *J. Cell Sci.* **2014**, *127 Pt 23*, 4987–4993. [CrossRef]
230. Lafourcade, C.; Sobo, K.; Kieffer-Jaquinod, S.; Garin, J.; van der Goot, F.G. Regulation of the V-ATPase along the Endocytic Pathway Occurs through Reversible Subunit Association and Membrane Localization. *PLoS ONE* **2008**, *3*, e2758. [CrossRef]
231. Kane, P.M. Disassembly and Reassembly of the Yeast Vacuolar H(+)-ATPase in Vivo. *J. Biol. Chem.* **1995**, *270*, 17025–17032. [CrossRef]
232. Chung, J.-H.; Lester, R.L.; Dickson, R.C. Sphingolipid Requirement for Generation of a Functional V1 Component of the Vacuolar ATPase*. *J. Biol. Chem.* **2003**, *278*, 28872–28881. [CrossRef] [PubMed]
233. Kettner, C.; Bertl, A.; Obermeyer, G.; Slayman, C.; Bihler, H. Electrophysiological Analysis of the Yeast V-Type Proton Pump: Variable Coupling Ratio and Proton Shunt. *Biophys. J.* **2003**, *85*, 3730–3738. [CrossRef]

234. Cross, R.L.; Müller, V. The Evolution of A-, F-, and V-Type ATP Synthases and ATPases: Reversals in Function and Changes in the H+/ATP Coupling Ratio. *FEBS Lett.* **2004**, *576*, 1–4. [CrossRef] [PubMed]
235. Kawasaki-Nishi, S.; Bowers, K.; Nishi, T.; Forgac, M.; Stevens, T.H. The Amino-Terminal Domain of the Vacuolar Proton-Translocating ATPase a Subunit Controls Targeting and in Vivo Dissociation, and the Carboxyl-Terminal Domain Affects Coupling of Proton Transport and ATP Hydrolysis*. *J. Biol. Chem.* **2001**, *276*, 47411–47420. [CrossRef]
236. Su, Y.; Zhou, A.; Al-Lamki, R.S.; Karet, F.E. The a-Subunit of the V-Type H+-ATPase Interacts with Phosphofructokinase-1 in Humans. *J. Biol. Chem.* **2003**, *278*, 20013–20018. [CrossRef]
237. Banerjee, S.; Kane, P.M. Regulation of V-ATPase Activity and Organelle pH by Phosphatidylinositol Phosphate Lipids. *Front. Cell Dev. Biol.* **2020**, *8*, 510. [CrossRef]
238. Stiernet, P.; Guiot, Y.; Gilon, P.; Henquin, J.-C. Glucose Acutely Decreases pH of Secretory Granules in Mouse Pancreatic Islets. Mechanisms and Influence on Insulin Secretion. *J. Biol. Chem.* **2006**, *281*, 22142–22151. [CrossRef]
239. Louagie, E.; Taylor, N.A.; Flamez, D.; Roebroek, A.J.M.; Bright, N.A.; Meulemans, S.; Quintens, R.; Herrera, P.L.; Schuit, F.; Van de Ven, W.J.M.; et al. Role of Furin in Granular Acidification in the Endocrine Pancreas: Identification of the V-ATPase Subunit Ac45 as a Candidate Substrate. *Proc. Natl. Acad. Sci. USA* **2008**, *105*, 12319–12324. [CrossRef]
240. Dai, F.F.; Bhattacharjee, A.; Liu, Y.; Batchuluun, B.; Zhang, M.; Wang, X.S.; Huang, X.; Luu, L.; Zhu, D.; Gaisano, H.; et al. A Novel GLP1 Receptor Interacting Protein ATP6ap2 Regulates Insulin Secretion in Pancreatic Beta Cells. *J. Biol. Chem.* **2015**, *290*, 25045–25061. [CrossRef]
241. Binger, K.J.; Neukam, M.; Tattikota, S.G.; Qadri, F.; Puchkov, D.; Willmes, D.M.; Wurmsee, S.; Geisberger, S.; Dechend, R.; Raile, K.; et al. Atp6ap2 Deletion Causes Extensive Vacuolation That Consumes the Insulin Content of Pancreatic β Cells. *Proc. Natl. Acad. Sci. USA* **2019**, *116*, 19983–19988. [CrossRef]
242. Brouwers, B.; Coppola, I.; Vints, K.; Dislich, B.; Jouvet, N.; Van Lommel, L.; Segers, C.; Gounko, N.V.; Thorrez, L.; Schuit, F.; et al. Loss of Furin in β-Cells Induces an mTORC1-ATF4 Anabolic Pathway That Leads to β-Cell Dysfunction. *Diabetes* **2021**, *70*, 492–503. [CrossRef]
243. Hatanaka, M.; Tanabe, K.; Yanai, A.; Ohta, Y.; Kondo, M.; Akiyama, M.; Shinoda, K.; Oka, Y.; Tanizawa, Y. Wolfram Syndrome 1 Gene (WFS1) Product Localizes to Secretory Granules and Determines Granule Acidification in Pancreatic Beta-Cells. *Hum. Mol. Genet.* **2011**, *20*, 1274–1284. [CrossRef]
244. Gharanei, S.; Zatyka, M.; Astuti, D.; Fenton, J.; Sik, A.; Nagy, Z.; Barrett, T.G. Vacuolar-Type H+-ATPase V1A Subunit Is a Molecular Partner of Wolfram Syndrome 1 (WFS1) Protein, Which Regulates Its Expression and Stability. *Hum. Mol. Genet.* **2013**, *22*, 203–217. [CrossRef]
245. Vinkenborg, J.L.; Nicolson, T.J.; Bellomo, E.A.; Koay, M.S.; Rutter, G.A.; Merkx, M. Genetically Encoded FRET Sensors to Monitor Intracellular Zn^{2+} Homeostasis. *Nat. Methods* **2009**, *6*, 737–740. [CrossRef]
246. Bafaro, E.; Liu, Y.; Xu, Y.; Dempski, R.E. The Emerging Role of Zinc Transporters in Cellular Homeostasis and Cancer. *Signal Transd. Target. Ther.* **2017**, 2. [CrossRef]
247. Lubag, A.J.M.; De Leon-Rodriguez, L.M.; Burgess, S.C.; Sherry, A.D. Noninvasive MRI of β-Cell Function Using a Zn^{2+}-Responsive Contrast Agent. *Proc. Natl. Acad. Sci. USA* **2011**, *108*, 18400–18405. [CrossRef]
248. Slepchenko, K.G.; Daniels, N.A.; Guo, A.; Li, Y.V. Autocrine Effect of Zn^{2+} on the Glucose-Stimulated Insulin Secretion. *Endocrine* **2015**, *50*, 110–122. [CrossRef]
249. Slepchenko, K.G.; James, C.B.L.; Li, Y.V. Inhibitory Effect of Zinc on Glucose-Stimulated Zinc/insulin Secretion in an Insulin-Secreting β-Cell Line. *Exp. Physiol.* **2013**, *98*, 1301–1311. [CrossRef]
250. Zhou, H.; Zhang, T.; Harmon, J.S.; Bryan, J.; Robertson, R.P. Zinc, Not Insulin, Regulates the Rat Alpha-Cell Response to Hypoglycemia in Vivo. *Diabetes* **2007**, *56*, 1107–1112. [CrossRef]
251. Hardy, A.B.; Serino, A.S.; Wijesekara, N.; Chimienti, F.; Wheeler, M.B. Regulation of Glucagon Secretion by Zinc: Lessons from the β Cell-Specific Znt8 Knockout Mouse Model. *Diabetes Obes. Metab.* **2011**, *13*, 112–117. [CrossRef]
252. Tamaki, M.; Fujitani, Y.; Hara, A.; Uchida, T.; Tamura, Y.; Takeno, K.; Kawaguchi, M.; Watanabe, T.; Ogihara, T.; Fukunaka, A.; et al. The Diabetes-Susceptible Gene SLC30A8/ZnT8 Regulates Hepatic Insulin Clearance. *J. Clin. Investig.* **2013**, *123*, 4513–4524. [CrossRef]
253. Nedumpully-Govindan, P.; Ding, F. Inhibition of IAPP Aggregation by Insulin Depends on the Insulin Oligomeric State Regulated by Zinc Ion Concentration. *Sci. Rep.* **2015**, *5*, 8240. [CrossRef]
254. Banaszak, M.; Górna, I.; Przysławski, J. Zinc and the Innovative Zinc-α2-Glycoprotein Adipokine Play an Important Role in Lipid Metabolism: A Critical Review. *Nutrients* **2021**, *13*, 2023. [CrossRef]
255. Rutter, G.A.; Chabosseau, P.; Bellomo, E.A.; Maret, W.; Mitchell, R.K.; Hodson, D.J.; Solomou, A.; Hu, M. Intracellular Zinc in Insulin Secretion and Action: A Determinant of Diabetes Risk? *Proc. Nutr. Soc.* **2016**, *75*, 61–72. [CrossRef]
256. Cooper-Capetini, V.; de Vasconcelos, D.A.A.; Martins, A.R.; Hirabara, S.M.; Donato, J., Jr.; Carpinelli, A.R.; Abdulkader, F. Zinc Supplementation Improves Glucose Homeostasis in High Fat-Fed Mice by Enhancing Pancreatic β-Cell Function. *Nutrients* **2017**, *9*, 1150. [CrossRef]
257. Daniels, M.J.; Jagielnicki, M.; Yeager, M. Structure/Function Analysis of Human ZnT8 (SLC30A8): A Diabetes Risk Factor and Zinc Transporter. *Curr. Res. Struct. Biol.* **2020**, *2*, 144–155. [CrossRef]

258. Smidt, K.; Larsen, A.; Brønden, A.; Sørensen, K.S.; Nielsen, J.V.; Praetorius, J.; Martensen, P.M.; Rungby, J. The Zinc Transporter ZNT3 Co-Localizes with Insulin in INS-1E Pancreatic Beta Cells and Influences Cell Survival, Insulin Secretion Capacity, and ZNT8 Expression. *Biometals* **2016**, *29*, 287–298. [CrossRef]
259. Bellomo, E.A.; Meur, G.; Rutter, G.A. Glucose Regulates Free Cytosolic Zn^{2+} Concentration, Slc39 (ZiP), and Metallothionein Gene Expression in Primary Pancreatic Islet β-Cells. *J. Biol. Chem.* **2011**, *286*, 25778–25789. [CrossRef]
260. Barragán-Álvarez, C.P.; Padilla-Camberos, E.; Díaz, N.F.; Cota-Coronado, A.; Hernández-Jiménez, C.; Bravo-Reyna, C.C.; Díaz-Martínez, N.E. Loss of Znt8 Function in Diabetes Mellitus: Risk or Benefit? *Mol. Cell. Biochem.* **2021**. [CrossRef]
261. Rutter, G.; Chimienti, F. SLC30A8 Mutations in Type 2 Diabetes. *Diabetologia* **2014**, *58*, 31–36. [CrossRef]
262. Davidson, H.W.; Wenzlau, J.M.; O'Brien, R.M. Zinc Transporter 8 (ZnT8) and β Cell Function. *Trends Endocrinol. Metab.* **2014**, *25*, 415–424. [CrossRef] [PubMed]
263. Zimmerman, A.E.; Kells, D.I.; Yip, C.C. Physical and Biological Properties of Guinea Pig Insulin. *Biochem. Biophys. Res. Commun.* **1972**, *46*, 2127–2133. [CrossRef]
264. O'Halloran, T.V.; Kebede, M.; Philips, S.J.; Attie, A.D. Zinc, Insulin, and the Liver: A Ménage à Trois. *J. Clin. Investig.* **2013**, *123*, 4136–4139. [CrossRef] [PubMed]
265. Lemaire, K.; Ravier, M.A.; Schraenen, A.; Creemers, J.W.M.; Van de Plas, R.; Granvik, M.; Van Lommel, L.; Waelkens, E.; Chimienti, F.; Rutter, G.A.; et al. Insulin Crystallization Depends on Zinc Transporter ZnT8 Expression, but Is Not Required for Normal Glucose Homeostasis in Mice. *Proc. Natl. Acad. Sci. USA* **2009**, *106*, 14872–14877. [CrossRef]
266. Syring, K.E.; Bosma, K.J.; Goleva, S.B.; Singh, K.; Oeser, J.K.; Lopez, C.A.; Skaar, E.P.; McGuinness, O.P.; Davis, L.K.; Powell, D.R.; et al. Potential Positive and Negative Consequences of ZnT8 Inhibition. *J. Endocrinol.* **2020**, *246*, 189–205. [CrossRef]
267. Syring, K.E.; Boortz, K.A.; Oeser, J.K.; Ustione, A.; Platt, K.A.; Shadoan, M.K.; McGuinness, O.P.; Piston, D.W.; Powell, D.R.; O'Brien, R.M. Combined Deletion of Slc30a7 and Slc30a8 Unmasks a Critical Role for ZnT8 in Glucose-Stimulated Insulin Secretion. *Endocrinology* **2016**, *157*, 4534–4541. [CrossRef]
268. Merriman, C.; Huang, Q.; Rutter, G.A.; Fu, D. Lipid-Tuned Zinc Transport Activity of Human ZnT8 Protein Correlates with Risk for Type-2 Diabetes. *J. Biol. Chem.* **2016**, *291*, 26950–26957. [CrossRef]
269. Kirchhoff, K.; Machicao, F.; Haupt, A.; Schäfer, S.A.; Tschritter, O.; Staiger, H.; Stefan, N.; Häring, H.-U.; Fritsche, A. Polymorphisms in the TCF7L2, CDKAL1 and SLC30A8 Genes Are Associated with Impaired Proinsulin Conversion. *Diabetologia* **2008**, *51*, 597–601. [CrossRef]
270. Tumarada, N.; Li, L.; Bai, S.; Sheline, C.T. hZnT8 (Slc30a8) Transgenic Mice That Overexpress the R325W Polymorph Have Reduced Islet Zn2+ and Proinsulin Levels, Increased Glucose Tolerance After a High-Fat Diet, and Altered Levels of Pancreatic Zinc Binding Proteins. *Diabetes* **2017**, *66*, 551–559. [CrossRef]
271. Dwivedi, O.P.; Lehtovirta, M.; Hastoy, B.; Chandra, V.; Krentz, N.A.J.; Kleiner, S.; Jain, D.; Richard, A. M.; Abaitua, F.; Beer, N.L.; et al. Loss of ZnT8 Function Protects against Diabetes by Enhanced Insulin Secretion. *Nat. Genet.* **2019**, *51*, 1596–1606. [CrossRef]
272. Kim, I.; Kang, E.S.; Yim, Y.S.; Ko, S.J.; Jeong, S.-H.; Rim, J.H.; Kim, Y.S.; Ahn, C.W.; Cha, B.S.; Lee, H.C.; et al. A Low-Risk ZnT-8 Allele (W325) for Post-Transplantation Diabetes Mellitus Is Protective against Cyclosporin A-Induced Impairment of Insulin Secretion. *Pharm. J.* **2011**, *11*, 191–198. [CrossRef]
273. Cauchi, S.; Nead, K.T.; Choquet, H.; Horber, F.; Potoczna, N.; Balkau, B.; Marre, M.; Charpentier, G.; Froguel, P.; Meyre, D. The Genetic Susceptibility to Type 2 Diabetes May Be Modulated by Obesity Status: Implications for Association Studies. *BMC Med. Genet.* **2008**, *9*, 45. [CrossRef]
274. Gerdes, H.H.; Rosa, P.; Phillips, E.; Baeuerle, P.A.; Frank, R.; Argos, P.; Huttner, W.B. The Primary Structure of Human Secretogranin II, a Widespread Tyrosine-Sulfated Secretory Granule Protein That Exhibits Low pH- and Calcium-Induced Aggregation. *J. Biol. Chem.* **1989**, *264*, 12009–12015. [CrossRef]
275. Scheenen, W.J.; Wollheim, C.B.; Pozzan, T.; Fasolato, C. Ca^{2+} Depletion from Granules Inhibits Exocytosis. A Study with Insulin-Secreting Cells. *J. Biol. Chem.* **1998**, *273*, 19002–19008. [CrossRef]
276. Mitchell, K.J.; Lai, F.A.; Rutter, G.A. Ryanodine Receptor Type I and Nicotinic Acid Adenine Dinucleotide Phosphate Receptors Mediate Ca^{2+} Release from Insulin-Containing Vesicles in Living Pancreatic Beta-Cells (MIN6). *J. Biol. Chem.* **2003**, *278*, 11057–11064. [CrossRef]
277. von Blume, J.; Alleaume, A.-M.; Kienzle, C.; Carreras-Sureda, A.; Valverde, M.; Malhotra, V. Cab45 Is Required for $Ca^{(2+)}$-Dependent Secretory Cargo Sorting at the Trans-Golgi Network. *J. Cell Biol.* **2012**, *199*, 1057–1066. [CrossRef]
278. von Blume, J.; Alleaume, A.-M.; Cantero-Recasens, G.; Curwin, A.; Carreras-Sureda, A.; Zimmermann, T.; van Galen, J.; Wakana, Y.; Valverde, M.A.; Malhotra, V. ADF/cofilin Regulates Secretory Cargo Sorting at the TGN via the Ca^{2+} ATPase SPCA1. *Dev. Cell* **2011**, *20*, 652–662. [CrossRef]
279. Mitchell, K.J.; Tsuboi, T.; Rutter, G.A. Role for Plasma Membrane-Related Ca^{2+}-ATPase-1 (ATP2C1) in Pancreatic Beta-Cell Ca^{2+} Homeostasis Revealed by RNA Silencing. *Diabetes* **2004**, *53*, 393–400. [CrossRef]
280. Masgrau, R.; Churchill, G.C.; Morgan, A.J.; Ashcroft, S.J.H.; Galione, A. NAADP: A New Second Messenger for Glucose-Induced Ca^{2+} Responses in Clonal Pancreatic Beta Cells. *Curr. Biol.* **2003**, *13*, 247–251. [CrossRef]
281. Chang, G.; Yang, R.; Cao, Y.; Nie, A.; Gu, X.; Zhang, H. SIDT2 Is Involved in the NAADP-Mediated Release of Calcium from Insulin Secretory Granules. *J. Mol. Endocrinol.* **2016**, *56*, 249–259. [CrossRef]
282. Gao, J.; Gu, X.; Mahuran, D.J.; Wang, Z.; Zhang, H. Impaired Glucose Tolerance in a Mouse Model of sidt2 Deficiency. *PLoS ONE* **2013**, *8*, e66139. [CrossRef]

283. Blondel, O.; Moody, M.M.; Depaoli, A.M.; Sharp, A.H.; Ross, C.A.; Swift, H.; Bell, G.I. Localization of Inositol Trisphosphate Receptor Subtype 3 to Insulin and Somatostatin Secretory Granules and Regulation of Expression in Islets and Insulinoma Cells. *Proc. Natl. Acad. Sci. USA* **1994**, *91*, 7777–7781. [CrossRef]
284. Ravazzola, M.; Halban, P.A.; Orci, L. Inositol 1,4,5-Trisphosphate Receptor Subtype 3 in Pancreatic Islet Cell Secretory Granules Revisited. *Proc. Natl. Acad. Sci. USA* **1996**, *93*, 2745–2748. [CrossRef]
285. Taylor, C.W.; Konieczny, V. IP3 Receptors: Take Four IP3 to Open. *Sci. Signal.* **2016**, *9*, e1. [CrossRef]
286. Taylor, C.W.; Tovey, S.C. IP(3) Receptors: Toward Understanding Their Activation. *Cold Spring Harb. Perspect. Biol.* **2010**, *2*, a004010. [CrossRef]
287. Huh, Y.H.; Kim, K.D.; Yoo, S.H. Comparison of and Chromogranin Effect on Inositol 1,4,5-Trisphosphate Sensitivity of Cytoplasmic and Nucleoplasmic Inositol 1,4,5-Trisphosphate receptor/Ca^{2+} Channels. *Biochemistry* **2007**, *46*, 14032–14043. [CrossRef]
288. Marshall, I.C.; Taylor, C.W. Biphasic Effects of Cytosolic Ca^{2+} on Ins(1,4,5)P3-Stimulated Ca^{2+} Mobilization in Hepatocytes. *J. Biol. Chem.* **1993**, *268*, 13214–13220. [CrossRef]
289. Storto, M.; Capobianco, L.; Battaglia, G.; Molinaro, G.; Gradini, R.; Riozzi, B.; Di Mambro, A.; Mitchell, K.J.; Bruno, V.; Vairetti, M.P.; et al. Insulin Secretion Is Controlled by mGlu5 Metabotropic Glutamate Receptors. *Mol. Pharmacol.* **2006**, *69*, 1234–1241. [CrossRef] [PubMed]
290. Takahashi, H.; Yokoi, N.; Seino, S. Glutamate as Intracellular and Extracellular Signals in Pancreatic Islet Functions. *Proc. Jpn. Acad. Ser. B Phys. Biol. Sci.* **2019**, *95*, 246–260. [CrossRef] [PubMed]
291. Yoo, S.H.; Lewis, M.S. Interaction of Chromogranin B and the near N-Terminal Region of Chromogranin B with an Intraluminal Loop Peptide of the Inositol 1,4, 5-Trisphosphate Receptor. *J. Biol. Chem.* **2000**, *275*, 30293–30300. [CrossRef]
292. Yoo, S.H.; Jeon, C.J. Inositol 1,4,5-Trisphosphate receptor/Ca^{2+} Channel Modulatory Role of Chromogranin A, a Ca^{2+} Storage Protein of Secretory Granules. *J. Biol. Chem.* **2000**, *275*, 15067–15073. [CrossRef]
293. Yoo, S.H.; So, S.H.; Kweon, H.S.; Lee, J.S.; Kang, M.K.; Jeon, C.J. Coupling of the Inositol 1,4,5-Trisphosphate Receptor and Chromogranins A and B in Secretory Granules. *J. Biol. Chem.* **2000**, *275*, 12553–12559. [CrossRef]
294. Yoo, S.H. Secretory Granules in Inositol 1,4,5-Trisphosphate-Dependent Ca^{2+} Signaling in the Cytoplasm of Neuroendocrine Cells. *FASEB J.* **2010**, *24*, 653–664. [CrossRef]
295. Yoo, S.H. Coupling of the IP3 receptor/Ca^{2+} Channel with Ca^{2+} Storage Proteins Chromogranins A and B in Secretory Granules. *Trends Neurosci.* **2000**, *23*, 424–428. [CrossRef]
296. Moulin, P.; Guiot, Y.; Jonas, J.-C.; Rahier, J.; Devuyst, O.; Henquin, J.-C. Identification and Subcellular Localization of the Na^+/H^+ Exchanger and a Novel Related Protein in the Endocrine Pancreas and Adrenal Medulla. *J. Mol. Endocrinol.* **2007**, *38*, 409–422. [CrossRef]
297. Noushmehr, H.; D'Amico, E.; Farilla, L.; Hui, H.; Wawrowsky, K.A.; Mlynarski, W.; Doria, A.; Abumrad, N.A.; Perfetti, R. Fatty Acid Translocase (FAT/CD36) Is Localized on Insulin-Containing Granules in Human Pancreatic Beta-Cells and Mediates Fatty Acid Effects on Insulin Secretion. *Diabetes* **2005**, *54*, 472–481. [CrossRef]
298. Nagao, M.; Esguerra, J.L.S.; Asai, A.; Ofori, J.K.; Edlund, A.; Wendt, A.; Sugihara, H.; Wollheim, C.B.; Oikawa, S.; Eliasson, L. Potential Protection Against Type 2 Diabetes in Obesity Through Lower CD36 Expression and Improved Exocytosis in β-Cells. *Diabetes* **2020**, *69*, 1193–1205. [CrossRef]
299. Geisler, J.C.; Corbin, K.L.; Li, Q.; Feranchak, A.P.; Nunemaker, C.S.; Li, C. Vesicular Nucleotide Transporter-Mediated ATP Release Regulates Insulin Secretion. *Endocrinology* **2013**, *154*, 675–684. [CrossRef]
300. Su, X.; Abumrad, N.A. Cellular Fatty Acid Uptake: A Pathway under Construction. *Trends Endocrinol. Metab.* **2009**, *20*, 72–77. [CrossRef]
301. Manialawy, Y.; Khan, S.R.; Bhattacharjee, A.; Wheeler, M.B. The Magnesium Transporter NIPAL1 Is a Pancreatic Islet-Expressed Protein That Conditionally Impacts Insulin Secretion. *J. Biol. Chem.* **2020**. [CrossRef]
302. Cool, D.R.; Normant, E.; Shen, F.; Chen, H.C.; Pannell, L.; Zhang, Y.; Loh, Y.P. Carboxypeptidase E Is a Regulated Secretory Pathway Sorting Receptor: Genetic Obliteration Leads to Endocrine Disorders in Cpe(fat) Mice. *Cell* **1997**, *88*, 73–83. [CrossRef]
303. Normant, E.; Loh, Y.P. Depletion of Carboxypeptidase E, a Regulated Secretory Pathway Sorting Receptor, Causes Misrouting and Constitutive Secretion of Proinsulin and Proenkephalin, but Not Chromogranin A. *Endocrinology* **1998**, *139*, 2137–2145. [CrossRef]
304. Irminger, J.C.; Verchere, C.B.; Meyer, K.; Halban, P.A. Proinsulin Targeting to the Regulated Pathway Is Not Impaired in Carboxypeptidase E-Deficient Cpefat/Cpefat Mice. *J. Biol. Chem.* **1997**, *272*, 27532–27534. [CrossRef]
305. Klumperman, J.; Hille, A.; Veenendaal, T.; Oorschot, V.; Stoorvogel, W.; von Figura, K.; Geuze, H.J. Differences in the Endosomal Distributions of the Two Mannose 6-Phosphate Receptors. *J. Cell Biol.* **1993**, *121*, 997–1010. [CrossRef]
306. Dittié, A.S.; Klumperman, J.; Tooze, S.A. Differential Distribution of Mannose-6-Phosphate Receptors and Furin in Immature Secretory Granules. *J. Cell Sci.* **1999**, *112 Pt 22*, 3955–3966. [CrossRef]
307. Puertollano, R.; Aguilar, R.C.; Gorshkova, I.; Crouch, R.J.; Bonifacino, J.S. Sorting of Mannose 6-Phosphate Receptors Mediated by the GGAs. *Science* **2001**, *292*, 1712–1716. [CrossRef]

metabolites MDPI

Review

Isolation and Proteomics of the Insulin Secretory Granule

Nicholas Norris, **Belinda Yau *** and **Melkam Alamerew Kebede**

Charles Perkins Centre, School of Medical Sciences, University of Sydney, Camperdown, NSW 2006, Australia; nicholas.norris@sydney.edu.au (N.N.); melkam.kebede@sydney.edu.au (M.A.K.)
* Correspondence: belinda.yau@sydney.edu.au

Abstract: Insulin, a vital hormone for glucose homeostasis is produced by pancreatic beta-cells and when secreted, stimulates the uptake and storage of glucose from the blood. In the pancreas, insulin is stored in vesicles termed insulin secretory granules (ISGs). In Type 2 diabetes (T2D), defects in insulin action results in peripheral insulin resistance and beta-cell compensation, ultimately leading to dysfunctional ISG production and secretion. ISGs are functionally dynamic and many proteins present either on the membrane or in the lumen of the ISG may modulate and affect different stages of ISG trafficking and secretion. Previously, studies have identified few ISG proteins and more recently, proteomics analyses of purified ISGs have uncovered potential novel ISG proteins. This review summarizes the proteins identified in the current ISG proteomes from rat insulinoma INS-1 and INS-1E cell lines. Here, we also discuss techniques of ISG isolation and purification, its challenges and potential future directions.

Keywords: insulin secretory granule; beta-cells; granule protein purification

check for updates

Citation: Norris, N.; Yau, B.; Kebede, M.A. Isolation and Proteomics of the Insulin Secretory Granule. *Metabolites* **2021**, *11*, 288. https://doi.org/10.3390/metabo11050288

Academic Editor: Amedeo Lonardo

Received: 12 April 2021
Accepted: 28 April 2021
Published: 30 April 2021

1. Insulin Granule Biogenesis and Function

The insulin secretory granule (ISG) is the storage vesicle for insulin in pancreatic beta-cells. It was long treated as an inert carrier for insulin but is now appreciated as a regulatory structure all on its own. There is a continuous turnover of insulin granules in the beta-cell, which is highly specialised in its capacity for ISG biogenesis, and insulin represents the most abundant protein within the beta-cell at 5–10% of total cell protein mass [1]. Production of insulin first begins in the rough endoplasmic reticulum with the synthesis of preproinsulin [2]. The signal peptide of preproinsulin is cleaved to form proinsulin, which is folded and trafficked to the Golgi complex [3]. Here, proinsulin is packaged with other proteins destined for secretion into a budding immature ISG at the *trans*-Golgi network via a mechanism termed *'sorting by entry'* [4]. Following the release of these granules from the *trans*-Golgi network, maturation of the immature ISG includes acidification of the granule lumen by ATP-dependent proton pumps and promotion of endoprotease convertases (PC1/3 and PC2) activity that cleave proinsulin to form free C-peptide and mature insulin, comprised of the A and B chains bound together by two inter-chain disulfide bonds [5–7]. Through a secondary mechanism called *'sorting by retention'*, proinsulin and other proteins are retained in the immature ISG (Davidson et al., 1988), while in parallel, proteins such as clathrin are removed from the immature ISG via *'sorting by exit'* [8,9]. Finally, insulin crystallises with zinc cations (Zn^{2+}), assembling an ~300 nm dense-core mature ISG [10]. From its point of synthesis, proinsulin enters an ISG within 4 hours [5] and is processed into insulin in a mature ISG within 40 minutes [9].

The ISG has a half-life of 3–5 days within the beta-cell cytoplasm [11] (Figure 1), and are ultimately destined for secretion or degradation. Upon glucose stimulation, ISG are motivated to undergo exocytosis, which requires the coordination of cellular machinery present both on ISGs and at the plasma membrane. It is therefore likely that ISG composition contributes to exocytosis, though the variables that determine whether an ISG eventually undergoes secretion are still unclear. Only 1–2% of total ISG content is released upon a single glucose stimulation [12]. Plasma membrane proximity [12,13] and docking [14,15] have long been suggested to contribute to an ISG's secretory capacity. More recently, ISG motility [16] and age [16–18] have also been shown to significantly contribute to an ISG's propensity for translocation to the plasma membrane and its necessity for docking [16,18]. Finally, an ISG's fusion capacity–whether the ISG collapses or is recycled–may also be intrinsically regulated [19].

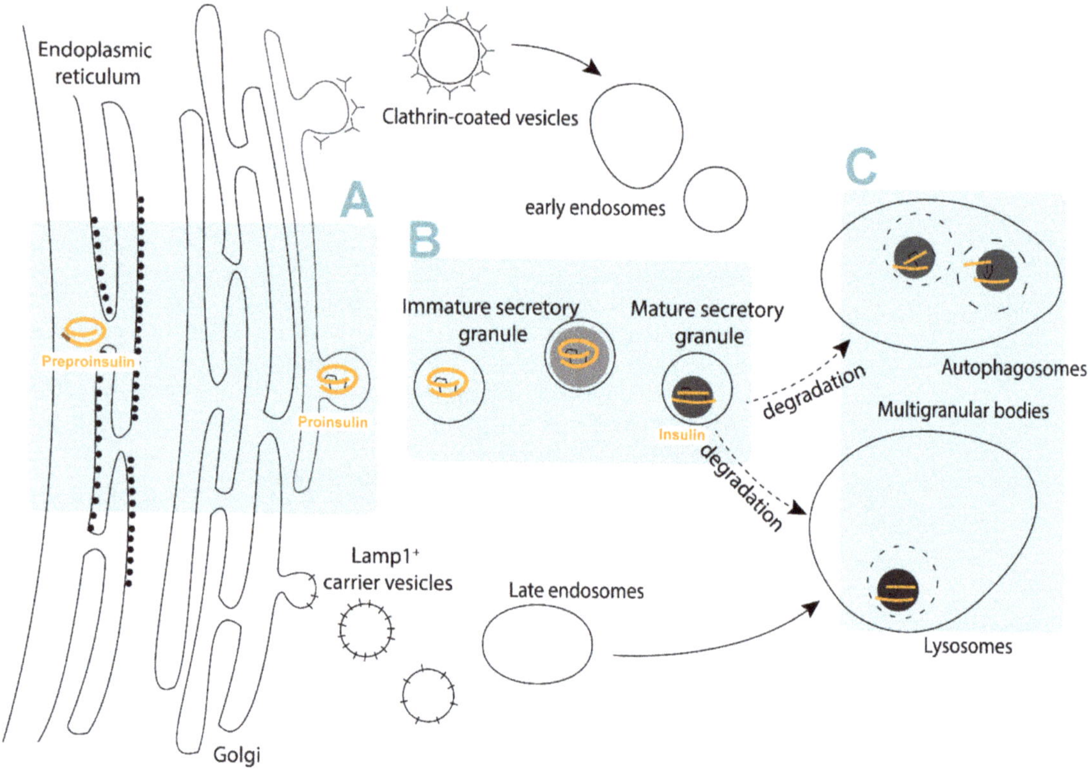

Figure 1. The insulin secretory pathway. (**A**) Proinsulin is synthesised and folded at the endoplasmic reticulum, trafficked to the trans-Golgi network and sorted into budding immature insulin secretory granules. (**B**) Immature insulin secretory granules undergo maturation where proinsulin is cleaved into mature insulin and condenses with zinc to form the dense core within the mature insulin secretory granule. (**C**) Aged mature insulin secretory granules that do not undergo secretion are trafficked for degradation within lysosomes or autophagosomes.

ISGs that are not secreted are targeted to the lysosome for degradation, either through autophagosome-dependent or independent pathways [20]. As insulin accounts for a large proportion of protein synthesis in pancreatic beta-cells [21], ISG homeostasis is essential to maintaining beta-cell function [22]. In autophagosome-dependent degradation, ISGs are engulfed by autophagosomes and subsequently fuse with the lysosomes, degrading ISG contents [22,23]. Autophagosome-independent degradation involves the fusion of ISGs with the lysosomes directly (crinophagy) [24]. Apart from whole ISG degradation, many proteases involved may also directly influence insulin turnover. For example, insulin has been shown to be degraded by insulin-degrading enzyme (IDE) in beta-cells and deletion or inhibition of this enzyme perturbs insulin secretion in beta-cells [25,26].

It is now appreciated that all these processes are not only externally regulated by the ISG environment, and proteins both in and on ISG can modulate both the processing and trafficking of ISGs, ultimately controlling granule mobility, secretion capacity, and degradation. Our current review focuses on the continuing pursuit to characterise ISG-localised proteins from pancreatic beta-cells.

The ISG is key to beta-cell identity. Pathological dysfunction related to insulin occurs at all stages, from synthesis to secretion and primarily results in diabetes. Loss of ISG in beta-cells, termed degranulation, is particularly characteristic of Type 2 diabetes (T2D) and recognised as a marker of beta-cell failure. It is most commonly visualised as a loss of insulin content [27] and seen as a precursor to beta-cell dedifferentiation [28]. Degranulation may occur at the point of SG biogenesis, such as in instances of chromogranin B (CgB) deficiency [29] or the loss of vacuolar sorting protein 41 [30], which regulate ISG budding and ISG coat formation respectively. Alternatively, degranulation may also be the result of chronic overnutrition leading to beta-cell exhaustion, where persistent hyperglycaemia driving increased insulin secretion is unable to be matched by proinsulin biosynthesis in the beta-cell [31,32]. Degranulation can also be a result of increased ISGs degradation as in the case of Sorcs1 deficiency [33]. Many genes relevant to the ISG secretory pathway have recently been reviewed extensively by Liu and colleagues in the context of pathology [34]. These include the hydrolases that function both as an endopeptidase for prohormone maturation and as lysosomal proteases [35,36], vacuolar-type H^+-transporting ATPases which regulate granule pH [37], and ZnT8 (SLC30A8), the key membrane transporter for zinc translocation into the maturing ISG [38]. Additionally, SNARE proteins and Rabs such as Vamp8 and Rab37 mediate ISG fusion at the site of insulin secretion [39]. Particularly interesting are the roles of cargo proteins within the ISG, which appear to have interdependent relationships. These include the well described soluble proteins carboxypeptidase E [40], VGF [41], the prohormone convertases PC1/3 and PC2, and the granin proteins chromogranin A [42], CgB and secretogranin II [43,44]. Figure 2 collates these ISG proteins and their localizations to immature and mature ISG. Mutations in these proteins can affect ISG formation, proinsulin processing, and glucose-stimulated insulin secretion, ultimately resulting in reduced ISG numbers and impaired secretion. However, loss of a single ISG cargo protein can drive compensatory behaviours in other ISG cargo proteins [42], suggesting ISG contents is a dynamic system.

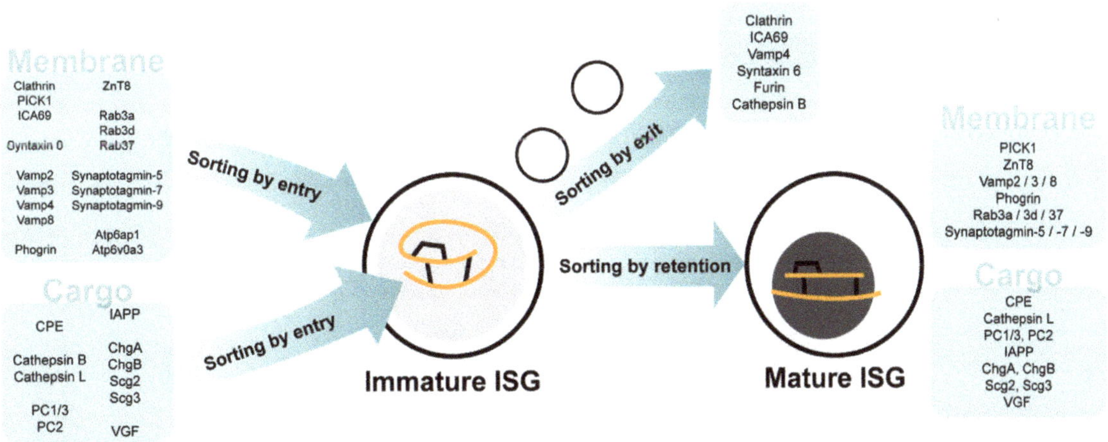

Figure 2. ISG membrane and cargo proteins of the immature and mature ISG. ISG proteins associated with the *'sorting by entry'*, *'sorting by exit'* and *'sorting by retention'* steps of ISG biogenesis and maturation.

2. Isolating the Insulin Granule

ISG isolation has long been used in the beta-cell physiology research, though the prioritisation of purity in the context of proteomic analyses is relatively new. Techniques used for the isolation of ISGs can essentially be separated into two categories, (a) differential subcellular fractionation using density gradients [45–47] and (b) immuno-based isolation [18,48–50]. In subcellular fractionation protocols, commercially available high-viscosity mediums such as Ficoll, Percoll, Optiprep and Nycodenz, or laboratory-prepared glycerol, sucrose or mannitol solutions are used to separate intact ISGs from cell lysates. These centrifugation techniques exploit physical properties including the size, density and/or shape of each subcellular compartment to separate ISGs from other organelles in collected fractions of various volumes. As these fractions are crude and undoubtedly contain contaminants, most studies employ the use of two or more subcellular fractionation steps to improve the purification of ISGs [45]. The main advantage of centrifugation techniques is that they are inexpensive and efficient [51], allowing researchers to obtain reasonably enriched ISG fractions within a few hours.

The second most common approach are immuno-based methods to enrich for ISGs [49,52], which exploit the tagging of proteins expressed in or on ISGs for isolation using immunoprecipitation. Often, this technique is used in conjunction with differential density gradients. For example, Hickey and colleagues employed the use of an Optiprep density gradient followed by Vamp2 immunoprecipitation to isolate ISGs from rat insulinoma INS-1 cells [48]. Immunoprecipitation offers the advantage of increased specificity to ISG compared to centrifugation techniques, however these methods are often more expensive and laborious, and rely on prior ISG protein knowledge. Interestingly, some proteins may be differentially expressed on ISGs. For example, CgB has heterogenous localization with insulin-positive granules in the INS-1 cell line [53]. Most importantly, immunoprecipitation of specific granule proteins that may be heterogeneously expressed would lead to the selective isolation of a specific ISG pool and the unknowing loss of information about the total ISG population.

On the other hand, immunoprecipitation could also selectively enrich for a non-ISG pool. In the same example, while Vamp2 immunoprecipitation may enrich for ISG, Vamp2 can also be expressed on Golgi recycling vesicles and endosomal membranes [54], and contamination of an ISG immunoprecipitation by these organelles cannot be disregarded. SG cargo proteins may also be present in pre-ISG compartments during the sorting process. Finally, immunoprecipitation methods also can be extended to protein pull-down studies which do not enrich ISG themselves, but instead immunoprecipitate interacting partners of known ISG proteins. Though these studies cannot offer a complete picture of the ISG proteome, they can offer an additional layer of insight into ISG protein functions and relationships [55–57].

It is likely that some combination of both immuno-based and centrifugation methods will be necessary to obtain the purest ISG fractions. Techniques used by the studies that have attempted proteomic analysis of ISG isolations are summarised in Figure 3. There is currently no consensus on the optimal strategy for intact ISG isolation from whole beta-cells. Insulin SGs are intrinsically dynamic and distribute in many compartments of the beta-cell as they traffic through their maturation, secretion and degradation pathways [52]. Isolation of a pure ISG fraction is most challenging due to the association of ISG with proteins in multiple subcellular compartments [58,59], and previous proteomic analyses of ISGs notably include contaminating proteins from pre-granule compartments such as the ER and *trans*-Golgi network (TGN), as well as cytoskeletal and lysosomal proteins associated with the trafficking and degradation of ISGs respectively [58,59]. Mitochondrial contamination present in ISG purification methods is a major problem [60] and attempts to isolate ISGs often identify different mitochondrial proteins in insulin enriched fractions [48,50,59,61–65].

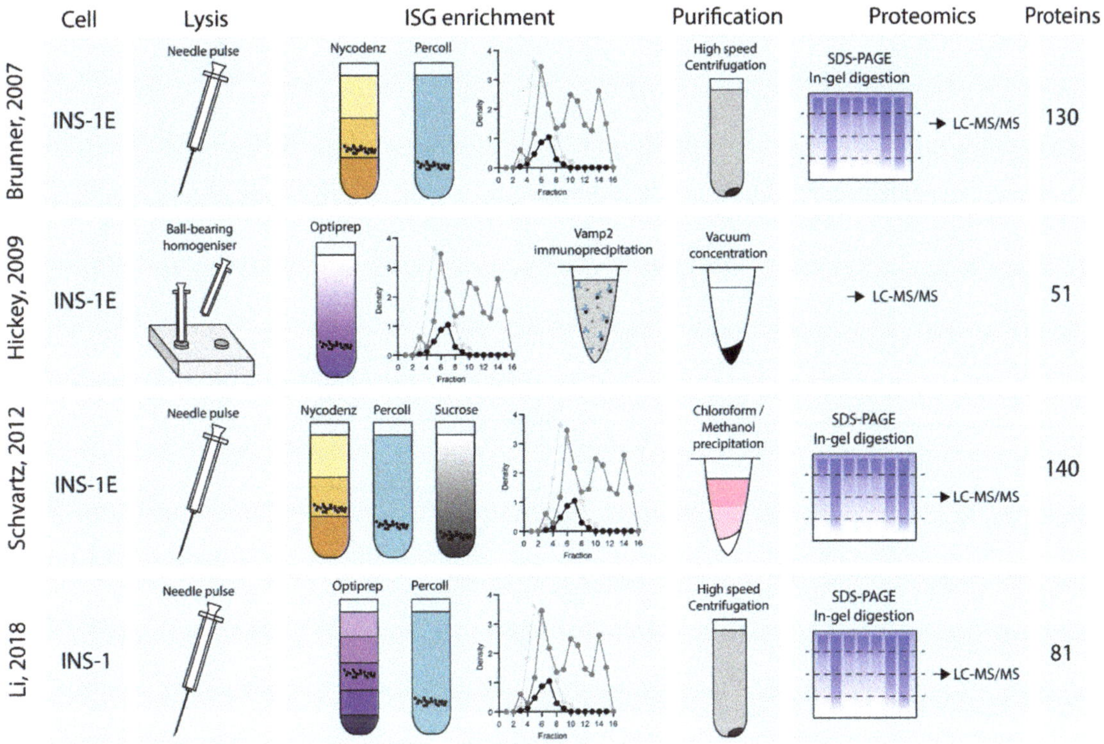

Figure 3. Schematic of insulin granule isolation and purification techniques across the 4 ISG proteomes. Rat insulinoma cells in all ISG proteomic studies are lysed before ISG enrichment through various density gradients or immunoprecipitations. ISG are then purified prior to proteomics analysis by LC-MS/MS to obtain list of ISG proteins. Data collated from Brunner et al., 2007, Hickey et al., 2009, Schvartz et al., 2012 and Li et al., 2018.

3. Identifying Insulin Granule Proteins

Only four studies have attempted to investigate ISG proteins by proteomic analysis to date [48,59,62,65]. These studies employ various combinations of density gradient centrifugations, *in silico* analyses, and immunoprecipitation techniques (Figure 3). As a result, Li and colleagues identified 81 total ISG proteins from the INS-1 rat beta-cell line, while Schvartz et al. identified 140 ISG proteins, Hickey et al. identified 51 ISG proteins, and Brunner et al. identified 130 ISG proteins from the INS-1E rat beta-cell line (Figure 4). A complete list of overlap proteins can be found in Supplementary Table S1. Proteomic data obtained from these four studies on ISG proteins from INS-1 or INS1-E cells produced a total of 5 proteins that were consistently identified. These were: Insulin-1 (Ins1), Insulin-2 (Ins2), Carboxypeptidase E (CPE), Chromogranin-A (CgA) and Prohormone convertase 2 (PC2). Rat beta-cells synthesize two different forms of insulin encoded by the *Ins*1 and *Ins*2 gene that share 90% homology [66,67], hence two insulin forms found in these proteomes. Though different isolation techniques would influence the proteins identified, one would expect that using similar cell lines would result in more than a handful of proteins consistently identified across all four studies.

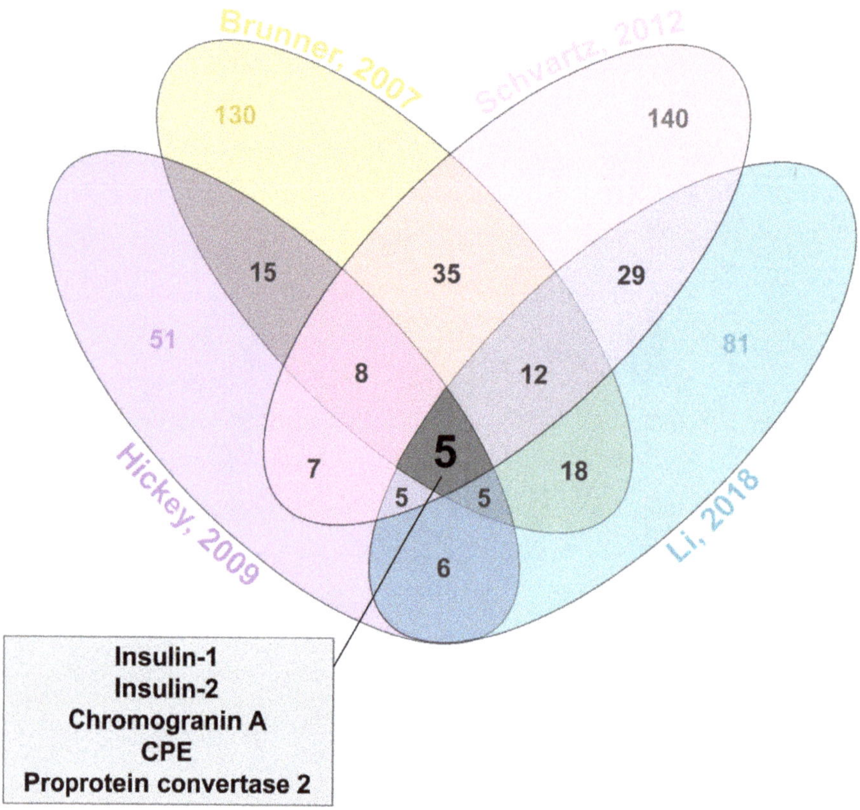

Figure 4. Venn diagram of overlapping proteins identified across the 4 ISG proteomes. Total of 5 proteins identified (box) in all proteomic analyses. Data collated from Brunner et al., 2007, Hickey et al., 2009, Schvartz et al., 2012 and Li et al., 2018.

Prior to ISG proteomics, Hutton et al. suggested ISGs may contain ~150 proteins using two-dimensional gel analysis of ISGs isolated from a rat islet tumour [46]. Approximately 30 specific proteins were described as ISG associated proteins before the first ISG proteome [62]. These proteins were individually classified primarily through cDNA screening and confocal microscopy. For example, the discovery of a well described ISG protein, the ZnT8 transporter was described as a pancreas-specific zinc transporter using RT-PCR on cDNA libraries with human tissue extracts [38]. Furthermore, ZnT8 was found to be localized specifically on ISGs through confocal microscopy of a fluorescent ZnT8 fusion protein expressed in INS-1 cells [38]. Similarly, phogrin was discovered as a membrane localized ISG protein through cDNA expression analysis and western blotting of phogrin with ISG enriched fractions [68]. These studies, among others were pivotal in uncovering different proteins that may modulate and affect insulin granule processes. Proteomics analysis of ISGs however provides an unbiased, comprehensive approach to the identification of multiple proteins simultaneously. Considering this however, all four studies lack the identification of these well-described ISG proteins such as ZnT8, any other zinc transporter and phogrin.

Here, we have classified the proteins identified in the ISG proteomes [48,59,62,65] into three groups: (i) intravesicular proteins, (ii) membrane proteins and (iii) other proteins:

3.1. *Intravesicular Proteins*

The most consistently identified intravesicular proteins in the proteomic studies were the previously well-characterised ISG proteins insulin (Ins1 and Ins2), CPE, PC2 and CgA [9,40,42,69,70]. Discovery of proinsulin processing of labelled insulin [71] and CgA [72] have allowed subsequent studies to identify localization of PC1/3 [73], PC2 [74] and CPE [75] as ISG localized enzymes. While all proteomes identified PC2 and CPE, PC1/3 was discovered only in two studies [59,65]. Other intravesicular proteins identified were from the chromogranin-secretogranin protein family. CgA in particular was identified in all four studies, with full-length CgA believed to be important for the biogenesis of granules in beta-cells [76]. Interestingly, CgA knockout mice display a reduced islet number, beta-cell to alpha-cell ratio and plasma insulin levels [77]; however, they exhibit normal blood glucose levels, as a result of compensation from other granin proteins [42]. CgB has been suggested to not be specifically involved in granule formation but instead is essential in the secretion of insulin and other islet hormones such as somatostatin and glucagon [29]. However, through pulse-chase labelling of CgB, Bearrows et al. show that in the absence of CgB, there is a delay in proinsulin trafficking from the TGN followed by a reduction in nascent ISGs at the plasma membrane [44]. CgB was identified in three of the four ISG proteomes (all but Li et al.). Significantly, aside from the full-length granins, PC1/3 and PC2 also cleave granins to form active peptides [69,78]. Beta-granin is an example of a CgA derived peptide identified by Li et al. and is proposed to inhibit insulin secretion through unknown mechanisms [79]. This emphasises technical challenges in peptide identification in proteomics analysis, to differentiate the presence and eventual function of both granins and their derived peptides in future studies.

Hydrolases were found in two of the proteomics analyses [48,62]. Cathepsins B and L were identified by Brunner et al. and are most intriguing as these proteins have been previously shown by electron microscopy to localise in immature ISGs, while cathepsin L alone remains in mature ISGs [36]. While some hydrolases have previously been described within ISGs [80,81], other hydrolases present in proteomic analysis may be appearing due to crinophagy processes of ISGs with lysosomes [62,82]. As such, further validation of hydrolase proteins will be essential to help elucidate their role in ISG biogenesis and processing. Particularly, the validation of cathepsins present in immature and mature ISGs demonstrates that these enzymes may follow sorting mechanisms out of immature ISGs via the mannose 6-phosphate receptor [36,83]. This adds weight to the *'sorting by retention'* and *'sorting by exit'* hypotheses in ISGs, in which immature ISGs may target proteins either for retention in maturing granules or exit towards the lysosome [36,62,84].

3.2. Membrane Proteins

A substantial proportion of ISG proteins identified by the proteomic analyses were membrane-bound or membrane-associated proteins. Of this group, the most commonly identified were synaptobrevin proteins (VAMPs), including Vamp3 [59,62,65], Vamp7 and Vamp8 [62]. VAMPs interact with their cognate t-SNAREs and other proteins that mediate the fusion of vesicles to the target membrane [85,86], which in turn interact with a variety of presynaptic proteins and q-SNAREs to form the complete SNARE complex [87–89]. Vamp2 was first described as an ISG localised v-SNARE protein [90] by cDNA cloning and confocal microscopy. Brunner et al. then identified Vamp2 in their proteomics analysis and following this, Hickey et al. used Vamp2 antibodies to immuno-purify ISGs. Surprisingly, Hickey et al. and Li et al. do not identify Vamp2 in their proteomes, with Hickey et al. suggesting that it and many other docking proteins potentially remained on the immunoaffinity beads [48]. If these membranal proteins were left unidentified, this may explain why fewer proteins (51) were identified in comparison to other proteomes.

Rab proteins were also found to be enriched with ISG fractions. Rab proteins are a family of GTPases from the Ras superfamily [91] that modulate several stages of vesicle trafficking and fusion of ISGs with the plasma membrane [92,93]. Through proteomic analysis and colocalisation imaging, Brunner's study illustrated that both VAMP8 and Rab37 are novel ISG associated proteins that colocalise with ISGs of INS1-E cells [62]. Previous to this, only 30 proteins were described as ISG associated proteins in beta-cells [62] and information surrounding the trafficking of ISGs was limited. Their proteomic analyses and validation of novel proteins suggested a more complex trafficking process than previously established in beta-cells. Other SNARE complex proteins present in the proteomes include syntaxin5 and 12, (Stx5, Stx12) [59] and granuphilin [62]. However, these proteins are believed to be localised to the plasma membrane [94] and not on ISG membranes, suggesting that they were present in contaminant co-purification with ISG fractions.

Many ATPase subunits were commonly identified in the four proteomic analyses, most notably the vacuolar-H^+ ATPases (V-type). These V-type ATPases have been previously shown to be localized to ISGs in beta-cells [95], and are important in producing and maintaining a proton gradient by acidifying the granule [95–97]. This facilitates the maturation of ISGs [98] as well as maintaining a suitable pH for intravesicular enzymes [8,82]. Many other subunits of ATPases identified are lysosomal isoforms and should be validated as to whether they are genuine ISG proteins or proteins co-purified with ISGs.

3.3. Other Proteins

The remaining proteins identified with non-specific or unknown localization in ISGs are often grouped in these studies. These include cytoskeletal, cytoplasmic and organelle localized proteins. The cytoplasmic proteins identified range from mis-folding chaperones [48] and isomerases (PDIA3) [62] to N-ethylmaleimide sensitive fusion protein [59,65]. Whether these proteins are genuinely ISG-associated, or technical contaminants, requires further validation. Different cytoskeleton-associated proteins are found across all four proteomes. Alpha-centractin [65], alpha and beta-actin [48] and kinesin subunits [65] are some examples of cytoskeletal associated proteins identified. ISGs are transported along microtubules by kinesins [99] and cytoskeleton remodelling is critical for ISG trafficking during glucose-stimulated insulin secretion [100]. The presence of these proteins is therefore unsurprising, though are likely present due to co-purification of these proteins through the isolation of ISGs. Indeed, the presence of proteins localized to the ER, Golgi, mitochondria and lysosomes are also commonly observed across all four studies. Examples include Erp44 (ER), Glg1 (Golgi), SHMT (mitochondria) and Lamp1 (lysosomes) [59,65]. It is difficult to prevent the copurification of these proteins using present isolation techniques and their co-localisations with ISGs need further validation.

The presence of isomerases and proteins involved in protein folding is quite surprising. Hickey et al. in particular find a striking number of chaperone proteins (~20% of proteins identified) [48]. Recent studies have shown that ER chaperone proteins are vital in proinsulin handling and insulin-like growth factor folding [101]; however, none of these ER-resident proteins have been shown to be localized in ISGs. Interestingly, Stanniocalcin-1 (STC1) or its precursors were found in three of the four proteomes (Li, Schvartz, Brunner). STC1 is found in many tissue types such as muscle, kidney, adrenal and lung [102]. Human STC1 protein is described as an uncoupler of oxidative phosphorylation in mitochondria [103], and has been implicated in apoptotic mechanisms and carcinogenesis [104]. Its function in beta-cells is not well understood, however; immunocytochemistry, and *in situ* ligand binding and hybridization [105] show that STC1 colocalizes with insulin in mouse pancreatic beta-cells. The abundance of these chaperones, alongside identification of proteins such as STC1, illustrates the importance of ISG proteomics as a rich source of data to potentially identify novel ISG proteins that may modulate different processes of ISG biogenesis, trafficking, and secretion. Altogether, these studies highlight the importance of developing improved purification techniques that restrict isolation of ISGs to granules post-sorting and packaging from the TGN, and before degradation.

4. Understanding ISG Function through the Proteome

Many aspects of ISG biosynthesis, processing, trafficking and secretion have been well reported [106], with the majority of studies focusing on individual protein effects on beta-cell function. Fewer studies use a broad view approach of ISG proteins, and their interactions and localisations. Efforts to identify exclusive ISG proteins in beta-cells remains scarce, and it is obvious that experimental methodology is the primary challenge.

Proteomic analysis is appealing because it provides an unbiased approach to uncovering new ISGs proteins, and validation of targets will help understand mechanisms underlying beta-cell function. Indeed, the proteomics-based discovery of VAMP8 and Rab37 as ISG proteins by Brunner and colleagues resulted in the detection of a novel set of proteins that regulate fusion of ISGs to the plasma membrane, and thus established the paradigm for ISG exocytosis [62]. In a similar fashion, the identification of hydrolases [36] within the ISG lumen suggests there are still many facets of ISG recycling and degradation that remain unappreciated.

Intrinsic ISG behaviour is an intriguing concept, and the evidence for functionally distinct populations of mature ISG is growing. For a long time, ISG have been believed to exist in either a 'readily releasable pool (RRP)' or 'reserve pool (RP)' of granules within the beta-cell cytoplasm [107–109]. The presence of Rab37a effector protein granuphilin on ISG appears to regulate granule docking at the plasma membrane, interacting with Syntaxin-1A-Munc18-1 complexes [110], and contributing to the RRP. However, ISG docking has been found to be a limiting step in ISG exocytosis and is not a requirement for granule fusion, as it restricts ISG motility and is dysregulated in T2D [111]. In contrast, newcomer granules from the RP have been identified to exhibit high mobility [112] and fusion competence irrelative of docking [113]. Newcomer granules abundantly express Syntaxin-3, which interacts with Munc-13-1 and Vamp8 to mediate their priming and fusion states [114]. Newcomer ISG also appear to have high calcium sensitivity, fusing away from Syntaxin-1A and L-type Ca^{2+} channels [115]. Whether these distinct subpopulations can be distinguished by their proteome will be critical to understanding the physiological relevance for granule pools in ISG function. Indeed, there is some evidence for the existence of distinct mature granule subpopulations differentiated by the expression of surface markers synaptotagmins-7 and -9 [116].These ISG populations exhibit unique lipid compositions, calcium sensitivities, and even proprotein convertase protein distribution. Most significantly, relative proportions of these subpopulations are changed in diabetes, with the specific depletion of synaptotagmin-9 ISG observed in a model of T2D [116].

Recent studies have demonstrated that ISG age plays an important role in dictating secretion and degradation [16–18,117,118], with younger ISG preferentially secreted in first-phase glucose-stimulated insulin secretion. It is possible that changes in protein composition occur in aging ISG, controlling functional differences in these younger and older populations. Two unique strategies have since been used to identify age-distinct ISGs from beta-cells. The first is a fluorescent protein timer construct, syncollin-dsREDE5TIMER, that localises to the lumen of ISGs [18,119] and changes its emission spectra over time. Integrating this construct into beta-cells and then applying a technique termed fluorescence-assisted organelle sorting (FAOS), submicron vesicles are thus fluorescently-labelled for sorting [120]. In the second, Neukam et al. and Ivanova et al. employ the use of pulse-chase labelling of ISGs using either a SNAP or CLIP tag fused to insulin or phogrin respectively, followed by immuno-purification using fluorescent dye TMR [17,49]. The advantage of techniques that track syncollin or phogrin, as opposed to insulin, lie in the resulting exclusion of pre-sorting compartments within the beta-cell. Syncollin-dsREDE5TIMER is only red fluorescent in ISG from approximately 18 hours onwards, when ISG are distinctly mature, while phogrin-fused CLIP is immuno-precipitated by TMR only after its sorting in ISG. Neither technique has yet produced proteomic samples, potentially due to the challenge of separating old ISG from degradation pathways. Mature syncollin-dsREDE5TIMER is detectable within Lamp1-positive vesicles (data not shown), while Hoboth and colleagues also visualise SNAP-tagged ISG within multigranular autophagic bodies [16]. Neukam et al. attempt to mitigate this issue with the addition of a second immunoprecipitation step with Lamp2 and Syp1 to deplete apparent lysosomal contamination [49]. The optimisation of these methodologies will help expand our current understanding of the underpinnings regulating insulin secretion and beta-cell function.

5. Moving Forward

The four proteomic analyses examined in this review used combinations of density gradients to isolate ISGs. The use of additional ISG markers by Hickey et al. to further purify ISG is desirable in theory, but practically results in additional challenges. Of note, their use of Vamp2 immuno-isolation of ISG did not result in the identification of Vamp2 (or any other Vamp proteins) within their proteome. Indeed, with only five proteins identified across all four ISG proteomes, many established ISG-exclusive proteins such as PC1/3 [74], phogrin [68] and the ZnT8 transporter (SLC30A8, [38]) were not identified consistently or at all, confirming major technical limitations. There have also been enormous leaps in mass spectrometry technology since Brunner and colleagues first established an ISG proteome in 2007, and the exceedingly increased sensitivities from mass spectrometers and improved peptide search databases currently available will allow deeper proteome depth and accuracy [121,122]. Recent proteomics studies utilising library-based analyses techniques in mouse primary islets identified over 11,000 unique proteins using minimal starting material (unpublished data), suggesting similar database searches could generated and applied to the ISG proteomes to improve protein recognition. Li et al. additionally demonstrate the potential for novel protein discovery by utilising protein correlation profiling to match candidate proteins to known ISG markers based on Euclidean distance [59]. Recent development of different protein sequencing, such as nanopore technology [123] and fluorescent "protein fingerprinting" [124] may also facilitate new ISG protein identification.

The majority of ISG isolation studies focus solely on the mature ISG. Immature ISG isolation is considerably more difficult since immature ISGs lack the dense zinc core, and more closely associate with pre-sorting compartments. Though Chen et al. demonstrate the use of fixed Percoll percentages to enrich immature ISG [45] using density, both ER and TGN membrane proteins were found to be contaminating in those fractions. There is potential that the use of immature ISG-specific proteins could be further exploited to isolate immature ISGs. For example, PICK1 and ICA69 form a protein complex on immature granules, but only PICK1 persists in mature granules [125]. Similarly, clathrin is *'sorted-by-exit'* from immature ISG, though is also present on non-ISG vesicles. Proteomic analyses of immature ISGs will improve our understanding of both sorting mechanisms at the TGN, and processing of the ISG itself during insulin maturation. Many beta-cell pathologies are intimately linked to ISG formation, despite the most common diabetes therapies targeting defective ISG secretion. Dysregulated ISG biogenesis leads to glucose intolerance *in vivo* [30,33,41,126], while increased proinsulin / insulin ratios are archetypical of diabetic patients and indicative of impaired processing within immature ISG [126,127].

Current ISG proteomes studies have only investigated rat insulinoma INS1 or INS1-E cell lines. This is most likely due to ease of culture and scaling to large starting material quantities, but it is important to consider how the ISG proteomes in mouse or human beta-cells may differ, potentially with the application of these methods to the MIN6 or EndoC-βH1 cell lines. Of the proteins with consensus across the ISG proteomes, all have human orthologs (INS, CHG, PCSK2, CPE), though humans only have a single insulin gene. Human beta-cell proteomics studies are also rare as they rely on precious and scarce material and are often subject to contamination by other endocrine cell types. Up to 707 potential beta-cell proteins have been identified [128–130], though it is yet unknown how many of those are ISG-specific.

Moreover, it will be critical to translate those techniques established within cell lines to primary cells to provide a more accurate snapshot of ISG proteins *in vivo*. There is potential to incorporate flow cytometry sorting techniques to isolate primary beta-cells, separated by an insulin-tagged fluorophore [131], zinc dyes or probes [132,133], or even NADPH autofluorescence [134,135] prior to ISG enrichment. Following this, the application of immunoprecipitation of select ISG markers [48], or dynamic fluorophores such as dsRedE5TIMER [18], will further allow ISG population separation. Once optimised, these methodologies would provide a standard for ISG proteomics that could be applied to multiple models of insulin-associated pathologies, including T2D.

A clean proteomic analysis of ISGs will provide a resource for more complete understanding of ISG sorting, processing, and trafficking. Currently, ISG proteomes are scarce, limited to rat insulinoma cell lines, and contain significant contamination. With the continued development of improved ISG isolation techniques, purification strategies and advancements in proteomics, ISG proteomes should be revisited, applied to different cell lines and ISG subpopulations to investigate and uncover novel players in the ISG secretory pathway.

Supplementary Materials: The following are available online at https://www.mdpi.com/article/10.3390/metabo11050288/s1, Table S1: Overlap of proteins identified across the 4 ISG proteomes.

Author Contributions: Conceptualization, N.N., B.Y., M.A.K.; writing—original draft preparation, N.N.; writing—review and editing, B.Y., M.A.K. All authors have read and agreed to the published version of the manuscript.

Funding: This research received no external funding.

Institutional Review Board Statement: Not applicable.

Informed Consent Statement: Not applicable.

Data Availability Statement: The figures and tables are original and not reproduced anywhere else.

Acknowledgments: We would like to thank Peter Thorn for his advice in the preparation of this review.

Conflicts of Interest: The authors declare no conflict of interest.

References

1. Rorsman, P.; Ashcroft, F.M. Pancreatic β-cell electrical activity and insulin secretion: Of mice and men. *Physiol. Rev.* **2018**, *98*, 117–214. [CrossRef] [PubMed]
2. Weiss, M.; Steiner, D.F.; Philipson, L.H. Insulin Biosynthesis, Secretion, Structure, and Structure-Activity Relationships. In *Endotext*; Feingold, K.R., Anawalt, B., Boyce, A., Chrousos, G., de Herder, W.W., Dungan, K., Grossman, A., Hershman, J.M., Hofland, H.J., Kaltsas, G., et al., Eds.; MDText.com, Inc.: South Dartmouth, MA, USA, 2000.
3. Gehart, H.; Ricci, R. Saving the neck from scission. *Commun. Integr. Biol.* **2013**, *6*, e23098. [CrossRef]
4. Arvan, P.; Halban, P.A. Sorting ourselves out: Seeking consensus on trafficking in the beta-cell. *Traffic* **2004**, *5*, 53–61. [CrossRef]
5. Davidson, H.W.; Rhodes, C.J.; Hutton, J.C. Intraorganellar calcium and pH control proinsulin cleavage in the pancreatic beta cell via two distinct site-specific endopeptidases. *Nature* **1988**, *333*, 93–96. [CrossRef]
6. Sun, J.; Cui, J.; He, Q.; Chen, Z.; Arvan, P.; Liu, M. Proinsulin misfolding and endoplasmic reticulum stress during the development and progression of diabetes. *Mol. Aspects Med.* **2015**, *42*, 105–118. [CrossRef] [PubMed]
7. Chang, S.-G.; Choi, K.-D.; Jang, S.-H.; Shin, H.-C. Role of disulfide bonds in the structure and activity of human insulin. *Mol. Cells* **2003**, *16*, 323–330. [PubMed]
8. Orci, L.; Ravazzola, M.; Amherdt, M.; Madsen, O.; Perrelet, A.; Vassalli, J.D.; Anderson, R.G. Conversion of proinsulin to insulin occurs coordinately with acidification of maturing secretory vesicles. *J. Cell Biol.* **1986**, *103*, 2273–2281. [CrossRef] [PubMed]
9. Orci, L.; Ravazzola, M.; Amherdt, M.; Madsen, O.; Vassalli, J.-D.; Perrelet, A. Direct identification of prohormone conversion site in insulin-secreting cells. *Cell* **1985**, *42*, 671–681. [CrossRef]
10. Dunn, M.F. Zinc-ligand interactions modulate assembly and stability of the insulin hexamer—A review. *Biometals* **2005**, *18*, 295–303. [CrossRef] [PubMed]
11. Marsh, B.J.; Soden, C.; Alarcón, C.; Wicksteed, B.L.; Yaekura, K.; Costin, A.J.; Morgan, G.P.; Rhodes, C.J. Regulated autophagy controls hormone content in secretory-deficient pancreatic endocrine β-cells. *Mol. Endocrinol.* **2007**, *21*, 2255–2269. [CrossRef] [PubMed]
12. Rorsman, P.; Renstrom, E. Insulin granule dynamics in pancreatic beta cells. *Diabetologia* **2003**, *46*, 1029–1045. [CrossRef] [PubMed]
13. Kasai, H.; Hatakeyama, H.; Ohno, M.; Takahashi, N. Exocytosis in islet beta-cells. *Adv. Exp. Med. Biol.* **2010**, *654*, 305–338.
14. Daniel, S.; Noda, M.; Straub, S.G.; Sharp, G. Identification of the docked granule pool responsible for the first phase of glucose-stimulated insulin secretion. *Diabetes* **1999**, *48*, 1686–1690. [CrossRef] [PubMed]
15. Yin, P.; Gandasi, N.R.; Arora, S.; Omar-Hmeadi, M.; Saras, J.; Barg, S. Syntaxin clusters at secretory granules in a munc18-bound conformation. *Mol. Biol. Cell* **2018**, *29*, 2700–2708. [CrossRef] [PubMed]
16. Hoboth, P.; Muller, A.; Ivanova, A.; Mziaut, H.; Dehghany, J.; Sonmez, A.; Lachnit, M.; Meyer-Hermann, M.; Kalaidzidis, Y.; Solimena, M. Aged insulin granules display reduced microtubule-dependent mobility and are disposed within actin-positive multigranular bodies. *Proc. Natl. Acad. Sci. USA* **2015**, *112*, E667–E676. [CrossRef] [PubMed]
17. Ivanova, A.; Kalaidzidis, Y.; Dirkx, R.; Sarov, M.; Gerlach, M.; Schroth-Diez, B.; Muller, A.; Liu, Y.; Andree, C.; Mulligan, B.; et al. Age-dependent labeling and imaging of insulin secretory granules. *Diabetes* **2013**, *62*, 3687–3696. [CrossRef]

18. Yau, B.; Hays, L.; Liang, C.; Laybutt, D.R.; Thomas, H.E.; Gunton, J.E.; Williams, L.; Hawthorne, W.J.; Thorn, P.; Rhodes, C.J.; et al. A fluorescent timer reporter enables sorting of insulin secretory granules by age. *J. Biol. Chem.* **2020**, *295*, 8901–8911. [CrossRef]
19. Hanna, S.T.; Pigeau, G.M.; Galvanovskis, J.; Clark, A.; Rorsman, P.; Macdonald, P.E. Kiss-and-run exocytosis and fusion pores of secretory vesicles in human beta-cells. *Pflugers Arch.* **2009**, *457*, 1343–1350. [CrossRef]
20. Pasquier, A.; Vivot, K.; Erbs, E.; Spiegelhalter, C.; Zhang, Z.; Aubert, V.; Liu, Z.; Senkara, M.; Maillard, E.; Pinget, M. Lysosomal degradation of newly formed insulin granules contributes to β cell failure in diabetes. *Nat. Commun.* **2019**, *10*, 1–14. [CrossRef]
21. Schuit, F.; In't Veld, P.; Pipeleers, D. Glucose stimulates proinsulin biosynthesis by a dose-dependent recruitment of pancreatic beta cells. *Proc. Natl. Acad. Sci. USA* **1988**, *85*, 3865–3869. [CrossRef] [PubMed]
22. Riahi, Y.; Wikstrom, J.D.; Bachar-Wikstrom, E.; Polin, N.; Zucker, H.; Lee, M.-S.; Quan, W.; Haataja, L.; Liu, M.; Arvan, P. Autophagy is a major regulator of beta cell insulin homeostasis. *Diabetologia* **2016**, *59*, 1480–1491. [CrossRef] [PubMed]
23. Müller, A.; Mziaut, H.; Neukam, M.; Knoch, K.P.; Solimena, M. A 4D view on insulin secretory granule turnover in the β-cell. *Diabetes Obes. Metab.* **2017**, *19*, 107–114. [CrossRef] [PubMed]
24. Orci, L.; Ravazzola, M.; Amherdt, M.; Yanaihara, C.; Yanaihara, N.; Halban, P.; Renold, A.; Perrelet, A. Insulin, not C-peptide (proinsulin), is present in crinophagic bodies of the pancreatic B-cell. *J. Cell Biol.* **1984**, *98*, 222–228. [CrossRef] [PubMed]
25. Fernández-Díaz, C.M.; Merino, B.; López-Acosta, J.F.; Cidad, P.; de la Fuente, M.A.; Lobatón, C.D.; Moreno, A.; Leissring, M.A.; Perdomo, G.; Cózar-Castellano, I. Pancreatic β-cell-specific deletion of insulin-degrading enzyme leads to dysregulated insulin secretion and β-cell functional immaturity. *Am. J. Physiol. Endocrinol. Metab.* **2019**, *317*, E805–E819. [CrossRef] [PubMed]
26. Steneberg, P.; Bernardo, L.; Edfalk, S.; Lundberg, L.; Backlund, F.; Östenson, C.-G.; Edlund, H. The type 2 diabetes–associated gene Ide is required for insulin secretion and suppression of α-synuclein levels in β-cells. *Diabetes* **2013**, *62*, 2004–2014. [CrossRef] [PubMed]
27. Marselli, L.; Suleiman, M.; Masini, M.; Campani, D.; Bugliani, M.; Syed, F.; Martino, L.; Focosi, D.; Scatena, F.; Olimpico, F.; et al. Are we overestimating the loss of beta cells in type 2 diabetes? *Diabetologia* **2014**, *57*, 362–365. [CrossRef] [PubMed]
28. Cinti, F.; Bouchi, R.; Kim-Muller, J.Y.; Ohmura, Y.; Sandoval, P.R.; Masini, M.; Marselli, L.; Suleiman, M.; Ratner, L.E.; Marchetti, P. Evidence of β-cell dedifferentiation in human type 2 diabetes. *J. Clin. Endocrinol. Metab.* **2016**, *101*, 1044–1054. [CrossRef]
29. Obermüller, S.; Calegari, F.; King, A.; Lindqvist, A.; Lundquist, I.; Salehi, A.; Francolini, M.; Rosa, P.; Rorsman, P.; Huttner, W.B. Defective secretion of islet hormones in chromogranin-B deficient mice. *PLoS ONE* **2010**, *5*, e8936. [CrossRef]
30. Burns, C.H.; Yau, B.; Rodriguez, A.; Triplett, J.; Maslar, D.; An, Y.S.; van der Welle, R.E.; Kossina, R.G.; Fisher, M.R.; Strout, G.W. Pancreatic β-Cell–Specific Deletion of VPS41 Causes Diabetes Due to Defects in Insulin Secretion. *Diabetes* **2021**, *70*, 436–448. [CrossRef]
31. White, M.G.; Shaw, J.A.; Taylor, R. Type 2 Diabetes: The Pathologic Basis of Reversible beta-Cell Dysfunction. *Diabetes Care* **2016**, *39*, 2080–2088. [CrossRef]
32. Nichols, C.G.; Remedi, M.S. The diabetic beta-cell: Hyperstimulated vs. hyperexcited. *Diabetes Obes. Metab.* **2012**, *14* (Suppl. 3), 129–135. [CrossRef] [PubMed]
33. Kebede, M.A.; Oler, A.T.; Gregg, T.; Balloon, A.J.; Johnson, A.; Mitok, K.; Rabaglia, M.; Schueler, K.; Stapleton, D.; Thorstenson, C.; et al. SORCS1 is necessary for normal insulin secretory granule biogenesis in metabolically stressed beta cells. *J. Clin. Investig.* **2014**, *124*, 4240–4256. [CrossRef] [PubMed]
34. Liu, M.; Huang, Y.; Xu, X.; Li, X.; Alam, M.; Arunagiri, A.; Haataja, L.; Ding, L.; Wang, S.; Itkin-Ansari, P. Normal and defective pathways in biogenesis and maintenance of the insulin storage pool. *J. Clin. Investig.* **2021**, *131*, e142240. [CrossRef]
35. Docherty, K.; Hutton, J.; Steiner, D. Cathepsin B-related proteases in the insulin secretory granule. *J. Biol. Chem.* **1984**, *259*, 6041–6044. [CrossRef]
36. Kuliawat, R.; Klumperman, J.; Ludwig, T.; Arvan, P. Differential sorting of lysosomal enzymes out of the regulated secretory pathway in pancreatic beta-cells. *J. Cell Biol.* **1997**, *137*, 595–608. [CrossRef]
37. Naik, A.R.; Formosa, B.J.; Pulvender, R.G.; Liyanaarachchi, A.G.; Jena, B.P. vH^+-ATPase-induced intracellular acidification is critical to glucose-stimulated insulin secretion in beta cells. *Histochem. Cell Biol.* **2020**, *153*, 279–285. [CrossRef]
38. Chimienti, F.; Devergnas, S.; Favier, A.; Seve, M. Identification and cloning of a β-cell–specific zinc transporter, ZnT-8, localized into insulin secretory granules. *Diabetes* **2004**, *53*, 2330–2337. [CrossRef]
39. Wang, Y.-S.; Tzeng, H.-T.; Tsai, C.-H.; Cheng, H.-C.; Lai, W.-W.; Liu, H.-S.; Wang, Y.-C. VAMP8, a vesicle-SNARE required for RAB37-mediated exocytosis, possesses a tumor metastasis suppressor function. *Cancer Lett.* **2018**, *437*, 79–88. [CrossRef]
40. Rindler, M.J. Carboxypeptidase E, a peripheral membrane protein implicated in the targeting of hormones to secretory granules, co-aggregates with granule content proteins at acidic pH. *J. Biol. Chem.* **1998**, *273*, 31180–31185. [CrossRef]
41. Stephens, S.B.; Edwards, R.J.; Sadahiro, M.; Lin, W.-J.; Jiang, C.; Salton, S.R.; Newgard, C.B. The prohormone VGF regulates β cell function via insulin secretory granule biogenesis. *Cell Rep.* **2017**, *20*, 2480–2489. [CrossRef]
42. Hendy, G.N.; Li, T.; Girard, M.; Feldstein, R.C.; Mulay, S.; Desjardins, R.; Day, R.; Karaplis, A.C.; Tremblay, M.L.; Canaff, L. Targeted ablation of the chromogranin a (Chga) gene: Normal neuroendocrine dense-core secretory granules and increased expression of other granins. *Mol. Endocrinol.* **2006**, *20*, 1935–1947. [CrossRef]
43. Yoshie, S.; Hagn, C.; Ehrhart, M.; Fischer-Colbrie, R.; Grube, D.; Winkler, H.; Gratzl, M. Immunological characterization of chromogranins A and B and secretogranin II in the bovine pancreatic islet. *Histochemistry* **1987**, *87*, 99–106. [CrossRef]
44. Bearrows, S.C.; Bauchle, C.J.; Becker, M.; Haldeman, J.M.; Swaminathan, S.; Stephens, S.B. Chromogranin B regulates early-stage insulin granule trafficking from the Golgi in pancreatic islet beta-cells. *J. Cell Sci.* **2019**, *132*, jcs231373. [CrossRef]

45. Chen, Y.; Xia, Z.; Wang, L.; Yu, Y.; Liu, P.; Song, E.; Xu, T. An efficient two-step subcellular fractionation method for the enrichment of insulin granules from INS-1 cells. *Biophys. Rep.* **2015**, *1*, 34–40. [CrossRef] [PubMed]
46. Hutton, J.; Penn, E.; Peshavaria, M. Isolation and characterisation of insulin secretory granules from a rat islet cell tumour. *Diabetologia* **1982**, *23*, 365–373. [CrossRef] [PubMed]
47. Jones, P.M.; Saermark, T.; Howell, S.L. A small-scale method for the isolation of insulin-containing secretory granules from islets of Langerhans. *Anal. Biochem.* **1987**, *166*, 142–149. [CrossRef]
48. Hickey, A.J.; Bradley, J.W.; Skea, G.L.; Middleditch, M.J.; Buchanan, C.M.; Phillips, A.R.; Cooper, G.J. Proteins associated with immunopurified granules from a model pancreatic islet beta-cell system: Proteomic snapshot of an endocrine secretory granule. *J. Proteome Res.* **2009**, *8*, 178–186. [CrossRef] [PubMed]
49. Neukam, M.; Gans, K.; Vasiljević, J.; Broichhagen, J.; Johnsson, K.; Kurth, T.; Solimena, M. Purification of age-distinct insulin secretory granules through antigen restriction using the CLIP-tag. *bioRxiv* **2020**. [CrossRef]
50. Guest, P.C. Multiplex Sequential Immunoprecipitation of Insulin Secretory Granule Proteins from Radiolabeled Pancreatic Islets. *Methods Mol. Biol.* **2017**, *1546*, 177–185. [PubMed]
51. Lee, Y.H.; Tan, H.T.; Chung, M.C. Subcellular fractionation methods and strategies for proteomics. *Proteomics* **2010**, *10*, 3935–3956. [CrossRef] [PubMed]
52. Ferri, G.; Digiacomo, L.; Lavagnino, Z.; Occhipinti, M.; Bugliani, M.; Cappello, V.; Caracciolo, G.; Marchetti, P.; Piston, D.W.; Cardarelli, F. Insulin secretory granules labelled with phogrin-fluorescent proteins show alterations in size, mobility and responsiveness to glucose stimulation in living beta-cells. *Sci. Rep.* **2019**, *9*, 2890. [CrossRef]
53. Giordano, T.; Brigatti, C.; Podini, P.; Bonifacio, E.; Meldolesi, J.; Malosio, M. Beta cell chromogranin B is partially segregated in distinct granules and can be released separately from insulin in response to stimulation. *Diabetologia* **2008**, *51*, 997–1007. [CrossRef] [PubMed]
54. Ramm, G.; Slot, J.W.; James, D.E.; Stoorvogel, W. Insulin recruits GLUT4 from specialized VAMP2-carrying vesicles as well as from the dynamic endosomal/trans-Golgi network in rat adipocytes. *Mol. Biol. Cell* **2000**, *11*, 4079–4091. [CrossRef] [PubMed]
55. Yasuda, T.; Shibasaki, T.; Minami, K.; Takahashi, H.; Mizoguchi, A.; Uriu, Y.; Numata, T.; Mori, Y.; Miyazaki, J.-I.; Miki, T. Rim2α determines docking and priming states in insulin granule exocytosis. *Cell Metab.* **2010**, *12*, 117–129. [CrossRef] [PubMed]
56. Fan, F.; Matsunaga, K.; Wang, H.; Ishizaki, R.; Kobayashi, E.; Kiyonari, H.; Mukumoto, Y.; Okunishi, K.; Izumi, T. Exophilin-8 assembles secretory granules for exocytosis in the actin cortex via interaction with RIM-BP2 and myosin-VIIa. *eLife* **2017**, *6*, e26174. [CrossRef]
57. Zhang, K.; Wang, T.; Liu, X.; Yuan, Q.; Xiao, T.; Yuan, X.; Zhang, Y.; Yuan, L.; Wang, Y. CASK, APBA1, and STXBP1 collaborate during insulin secretion. *Mol. Cell. Endocrinol.* **2021**, *520*, 111076. [CrossRef]
58. Tooze, S.A.; Flatmark, T.; Tooze, J.; Huttner, W.B. Characterization of the immature secretory granule, an intermediate in granule biogenesis. *J. Cell Biol.* **1991**, *115*, 1491–1503. [CrossRef]
59. Li, M.; Du, W.; Zhou, M.; Zheng, L.; Song, E.; Hou, J. Proteomic analysis of insulin secretory granules in INS-1 cells by protein correlation profiling. *Biophys. Rep.* **2018**, *4*, 329–338. [CrossRef]
60. MacDonald, M.J.; Ade, L.; Ntambi, J.M.; Ansari, I.-U.H.; Stoker, S.W. Characterization of phospholipids in insulin secretory granules and mitochondria in pancreatic beta cells and their changes with glucose stimulation. *J. Biol. Chem.* **2015**, *290*, 11075–11092. [CrossRef]
61. Albrethsen, J.; Goetze, J.P.; Johnsen, A.H. Mining the granule proteome: A potential source of endocrine biomarkers. *Biomark. Med.* **2015**, *9*, 259–265. [CrossRef]
62. Brunner, Y.; Coute, Y.; Iezzi, M.; Foti, M.; Fukuda, M.; Hochstrasser, D.F.; Wollheim, C.B.; Sanchez, J.C. Proteomics analysis of insulin secretory granules. *Mol. Cell. Proteom.* **2007**, *6*, 1007–1017. [CrossRef]
63. Chandramouli, K.; Qian, P.Y. Proteomics: Challenges, techniques and possibilities to overcome biological sample complexity. *Hum. Genom. Proteom.* **2009**, *2009*, 239204. [CrossRef]
64. Rindler, M.J.; Xu, C.F.; Gumper, I.; Smith, N.N.; Neubert, T.A. Proteomic analysis of pancreatic zymogen granules: Identification of new granule proteins. *J. Proteome Res.* **2007**, *6*, 2978–2992. [CrossRef] [PubMed]
65. Schvartz, D.; Brunner, Y.; Coute, Y.; Foti, M.; Wollheim, C.B.; Sanchez, J.C. Improved characterization of the insulin secretory granule proteomes. *J. Proteom.* **2012**, *75*, 4620–4631. [CrossRef] [PubMed]
66. Lomedico, P.; Rosenthal, N.; Efstratiadis, A.; Gilbert, W.; Kolodner, R.; Tizard, R. The structure and evolution of the two nonallelic rat preproinsulin genes. *Cell* **1979**, *18*, 545–558. [CrossRef]
67. Ullrich, A.; Shine, J.; Chirgwin, J.; Pictet, R.; Tischer, E.; Rutter, W.J.; Goodman, H.M. Rat insulin genes: Construction of plasmids containing the coding sequences. *Science* **1977**, *196*, 1313–1319. [CrossRef]
68. Wasmeier, C.; Hutton, J.C. Molecular cloning of phogrin, a protein-tyrosine phosphatase homologue localized to insulin secretory granule membranes. *J. Biol. Chem.* **1996**, *271*, 18161–18170. [CrossRef]
69. Hoflehner, J.; Eder, U.; Laslop, A.; Seidah, N.G.; Fischer-Colbrie, R.; Winkler, H. Processing of secretogranin II by prohormone convertases: Importance ofPC1 in generation of secretoneurin. *FEBS Lett.* **1995**, *360*, 294–298. [PubMed]
70. Laslop, A.; Weiss, C.; Savaria, D.; Eiter, C.; Tooze, S.A.; Seidah, N.G.; Winkler, H. Proteolytic processing of chromogranin B and secretogranin II by prohormone convertases. *J. Neurochem.* **1998**, *70*, 374–383. [CrossRef]
71. Davidson, H.W.; Peshavaria, M.; Hutton, J.C. Proteolytic conversion of proinsulin into insulin. Identification of a Ca^{2+}-dependent acidic endopeptidase in isolated insulin-secretory granules. *Biochem. J.* **1987**, *246*, 279–286. [CrossRef]

72. Hutton, J.C.; Davidson, H.W.; Peshavaria, M. Proteolytic processing of chromogranin A in purified insulin granules. Formation of a 20 kDa N-terminal fragment (betagranin) by the concerted action of a Ca^{2+}-dependent endopeptidase and carboxypeptidase H (EC 3.4. 17.10). *Biochem. J.* **1987**, *244*, 457–464. [CrossRef]
73. Muller, L.; Lindberg, I. The cell biology of the prohormone convertases PCI and PC2. *Prog. Nucleic Acid Res. Mol. Biol.* **1999**, *63*, 69–108. [PubMed]
74. Bennett, D.L.; Bailyes, E.; Nielsen, E.; Guest, P.; Rutherford, N.; Arden, S.; Hutton, J. Identification of the type 2 proinsulin processing endopeptidase as PC2, a member of the eukaryote subtilisin family. *J. Biol. Chem.* **1992**, *267*, 15229–15236. [CrossRef]
75. Guest, P.C.; Ravazzola, M.; Davidson, H.W.; Orci, L.; Hutton, J.C. Molecular heterogeneity and cellular localization of carboxypeptidase H in the islets of Langerhans. *Endocrinology* **1991**, *129*, 734–740. [CrossRef] [PubMed]
76. Kim, T.; Tao-Cheng, J.-H.; Eiden, L.E.; Loh, Y.P. Chromogranin A, an "on/off" switch controlling dense-core secretory granule biogenesis. *Cell* **2001**, *106*, 499–509. [CrossRef]
77. Portela-Gomes, G.; Gayen, J.; Grimelius, L.; Stridsberg, M.; Mahata, S. The importance of chromogranin A in the development and function of endocrine pancreas. *Regul. Pept.* **2008**, *151*, 19–25. [CrossRef] [PubMed]
78. Udupi, V.; Lee, H.-M.; Kurosky, A.; Greeley, G.H., Jr. Prohormone convertase-1 is essential for conversion of chromogranin A to pancreastatin. *Regul. Pept.* **1999**, *83*, 123–127. [CrossRef]
79. Schmid, G.M.; Meda, P.; Caille, D.; Wargent, E.; O'Dowd, J.; Hochstrasser, D.F.; Cawthorne, M.A.; Sanchez, J.-C. Inhibition of insulin secretion by betagranin, an N-terminal chromogranin A fragment. *J. Biol. Chem.* **2007**, *282*, 12717–12724. [CrossRef]
80. Hutton, J.C. Insulin secretory granule biogenesis and the proinsulin-processing endopeptidases. *Diabetologia* **1994**, *37*, S48–S56. [CrossRef]
81. Davidson, H.W.; Hutton, J.C. The insulin-secretory-granule carboxypeptidase H. Purification and demonstration of involvement in proinsulin processing. *Biochem. J.* **1987**, *245*, 575–582. [CrossRef]
82. Sandberg, M.; Borg, L.H. Intracellular degradation of insulin and crinophagy are maintained by nitric oxide and cyclo-oxygenase 2 activity in isolated pancreatic islets. *Biol. Cell* **2006**, *98*, 307–315. [CrossRef] [PubMed]
83. Klumperman, J.; Kuliawat, R.; Griffith, J.M.; Geuze, H.J.; Arvan, P. Mannose 6–phosphate receptors are sorted from immature secretory granules via adaptor protein AP-1, clathrin, and syntaxin 6–positive vesicles. *J. Cell Biol.* **1998**, *141*, 359–371. [CrossRef]
84. Arvan, P.; Castle, D. Sorting and storage during secretory granule biogenesis: Looking backward and looking forward. *Biochem. J.* **1998**, *332*, 593–610. [CrossRef] [PubMed]
85. Lin, R.C.; Scheller, R.H. Mechanisms of synaptic vesicle exocytosis. *Annu. Rev. Cell Dev. Biol.* **2000**, *16*, 19–49. [CrossRef] [PubMed]
86. Lam, P.P.; Ohno, M.; Dolai, S.; He, Y.; Qin, T.; Liang, T.; Zhu, D.; Kang, Y.; Liu, Y.; Kauppi, M. Munc18b is a major mediator of insulin exocytosis in rat pancreatic β-cells. *Diabetes* **2013**, *62*, 2416–2428. [CrossRef] [PubMed]
87. Iorio, V.; Festa, M.; Rosati, A.; Hahne, M.; Tiberti, C.; Capunzo, M.; De Laurenzi, V.; Turco, M. BAG3 regulates formation of the SNARE complex and insulin secretion. *Cell Death Dis.* **2015**, *6*, e1684. [CrossRef]
88. Gaisano, H.Y. Recent new insights into the role of SNARE and associated proteins in insulin granule exocytosis. *Diabetes Obes. Metab.* **2017**, *19* (Suppl. 1), 115–123. [CrossRef] [PubMed]
89. Qin, T.; Liang, T.; Zhu, D.; Kang, Y.; Xie, L.; Dolai, S.; Sugita, S.; Takahashi, N.; Ostenson, C.-G.; Banks, K. Munc18b increases insulin granule fusion, restoring deficient insulin secretion in type-2 diabetes human and Goto-Kakizaki rat islets with improvement in glucose homeostasis. *EBioMedicine* **2017**, *16*, 262–274. [CrossRef] [PubMed]
90. Regazzi, R.; Wollheim, C.; Lang, J.; Theler, J.; Rossetto, O.; Montecucco, C.; Sadoul, K.; Weller, U.; Palmer, M.; Thorens, B. VAMP-2 and cellubrevin are expressed in pancreatic beta-cells and are essential for Ca(2+)-but not for GTP gamma S-induced insulin secretion. *EMBO J.* **1995**, *14*, 2723–2730. [CrossRef]
91. Martinez, O.; Goud, B. Rab proteins. *Biochim. Biophys. Acta (BBA) Mol. Cell Res.* **1998**, *1404*, 101–112. [CrossRef]
92. Cheviet, S.V.; Coppola, T.; Haynes, L.P.; Burgoyne, R.D.; Regazzi, R. The Rab-binding protein Noc2 is associated with insulin-containing secretory granules and is essential for pancreatic β-cell exocytosis. *Mol. Endocrinol.* **2004**, *18*, 117–126. [CrossRef]
93. Wang, Z.; Thurmond, D.C. Mechanisms of biphasic insulin-granule exocytosis–roles of the cytoskeleton, small GTPases and SNARE proteins. *J. Cell Sci.* **2009**, *122*, 893–903. [CrossRef] [PubMed]
94. Mizuno, K.; Fujita, T.; Gomi, H.; Izumi, T. Granuphilin exclusively mediates functional granule docking to the plasma membrane. *Sci. Rep.* **2016**, *6*, 1–12. [CrossRef] [PubMed]
95. Hatanaka, M.; Tanabe, K.; Yanai, A.; Ohta, Y.; Kondo, M.; Akiyama, M.; Shinoda, K.; Oka, Y.; Tanizawa, Y. Wolfram syndrome 1 gene (WFS1) product localizes to secretory granules and determines granule acidification in pancreatic β-cells. *Hum. Mol. Genet.* **2011**, *20*, 1274–1284. [CrossRef] [PubMed]
96. Barg, S.; Huang, P.; Eliasson, L.; Nelson, D.J.; Obermüller, S.; Rorsman, P.; Thévenod, F.; Renström, E. Priming of insulin granules for exocytosis by granular Cl^- uptake and acidification. *J. Cell Sci.* **2001**, *114*, 2145–2154. [CrossRef]
97. Gharanei, S.; Zatyka, M.; Astuti, D.; Fenton, J.; Sik, A.; Nagy, Z.; Barrett, T.G. Vacuolar-type H+-ATPase V1A subunit is a molecular partner of Wolfram syndrome 1 (WFS1) protein, which regulates its expression and stability. *Hum. Mol. Genet.* **2013**, *22*, 203–217. [CrossRef]
98. Orci, L.; Ravazzola, M.; Storch, M.-J.; Anderson, R.; Vassalli, J.-D.; Perrelet, A. Proteolytic maturation of insulin is a post-Golgi event which occurs in acidifying clathrin-coated secretory vesicles. *Cell* **1987**, *49*, 865–868. [CrossRef]
99. Heaslip, A.T.; Nelson, S.R.; Lombardo, A.T.; Previs, S.B.; Armstrong, J.; Warshaw, D.M. Cytoskeletal dependence of insulin granule movement dynamics in INS-1 beta-cells in response to glucose. *PLoS ONE* **2014**, *9*, e109082. [CrossRef]

100. Wang, B.; Lin, H.; Li, X.; Lu, W.; Kim, J.B.; Xu, A.; Cheng, K.K. The adaptor protein APPL2 controls glucose-stimulated insulin secretion via F-actin remodeling in pancreatic β-cells. *Proc. Natl. Acad. Sci. USA* **2020**, *117*, 28307–28315. [CrossRef]
101. Ghiasi, S.M.; Dahlby, T.; Andersen, C.H.; Haataja, L.; Petersen, S.; Omar-Hmeadi, M.; Yang, M.; Pihl, C.; Bresson, S.E.; Khilji, M.S. Endoplasmic reticulum chaperone glucose-regulated protein 94 is essential for proinsulin handling. *Diabetes* **2019**, *68*, 747–760. [CrossRef]
102. Varghese, R.; Wong, C.K.; Deol, H.; Wagner, G.F.; DiMattia, G.E. Comparative analysis of mammalian stanniocalcin genes. *Endocrinology* **1998**, *139*, 4714–4725. [CrossRef]
103. Ellard, J.P.; McCudden, C.R.; Tanega, C.; James, K.A.; Ratkovic, S.; Staples, J.F.; Wagner, G.F. The respiratory effects of stanniocalcin-1 (STC-1) on intact mitochondria and cells: STC-1 uncouples oxidative phosphorylation and its actions are modulated by nucleotide triphosphates. *Mol. Cell. Endocrinol.* **2007**, *264*, 90–101. [CrossRef] [PubMed]
104. Iversen, P.; Sorensen, D.; Benestad, H. Inhibitors of angiogenesis selectively reduce the malignant cell load in rodent models of human myeloid leukemias. *Leukemia* **2002**, *16*, 376–381. [CrossRef] [PubMed]
105. Zaidi, D.; Turner, J.K.; Durst, M.A.; Wagner, G.F. Stanniocalcin-1 co-localizes with insulin in the pancreatic islets. *Int. Sch. Res. Not.* **2012**, *2012*, 834359. [CrossRef] [PubMed]
106. Vakilian, M.; Tahamtani, Y.; Ghaedi, K. A review on insulin trafficking and exocytosis. *Gene* **2019**, *706*, 52–61. [CrossRef] [PubMed]
107. Olofsson, C.S.; Gopel, S.O.; Barg, S.; Galvanovskis, J.; Ma, X.; Salehi, A.; Rorsman, P.; Eliasson, L. Fast insulin secretion reflects exocytosis of docked granules in mouse pancreatic B-cells. *Pflugers Arch.* **2002**, *444*, 43–51. [CrossRef]
108. Olofsson, C.S.; Salehi, A.; Holm, C.; Rorsman, P. Palmitate increases L-type Ca^{2+} currents and the size of the readily releasable granule pool in mouse pancreatic β-cells. *J. Physiol.* **2004**, *557*, 935–948. [CrossRef]
109. Pottekat, A.; Becker, S.; Spencer, K.R.; Yates III, J.R.; Manning, G.; Itkin-Ansari, P.; Balch, W.E. Insulin biosynthetic interaction network component, TMEM24, facilitates insulin reserve pool release. *Cell Rep.* **2013**, *4*, 921–930. [CrossRef]
110. Gomi, H.; Mizutani, S.; Kasai, K.; Itohara, S.; Izumi, T. Granuphilin molecularly docks insulin granules to the fusion machinery. *J. Cell Biol.* **2005**, *171*, 99–109. [CrossRef]
111. Gandasi, N.R.; Yin, P.; Omar-Hmeadi, M.; Laakso, E.O.; Vikman, P.; Barg, S. Glucose-dependent granule docking limits insulin secretion and is decreased in human type 2 diabetes. *Cell Metab.* **2018**, *27*, 470–478.e4. [CrossRef]
112. Hao, M.; Li, X.; Rizzo, M.A.; Rocheleau, J.V.; Dawant, B.M.; Piston, D.W. Regulation of two insulin granule populations within the reserve pool by distinct calcium sources. *J. Cell Sci.* **2005**, *118*, 5873–5884. [CrossRef]
113. Gaisano, H.Y. Here come the newcomer granules, better late than never. *Trends Endocrinol. Metab.* **2014**, *25*, 381–388. [CrossRef]
114. Zhu, D.; Koo, E.; Kwan, E.; Kang, Y.; Park, S.; Xie, H.; Sugita, S.; Gaisano, H. Syntaxin-3 regulates newcomer insulin granule exocytosis and compound fusion in pancreatic beta cells. *Diabetologia* **2013**, *56*, 359–369. [CrossRef] [PubMed]
115. Pedersen, M.G.; Sherman, A. Newcomer insulin secretory granules as a highly calcium-sensitive pool. *Proc. Natl. Acad. Sci. USA* **2009**, *106*, 7432–7436. [CrossRef]
116. Kreutzberger, A.J.; Kiessling, V.; Doyle, C.A.; Schenk, N.; Upchurch, C.M.; Elmer-Dixon, M.; Ward, A.E.; Preobraschenski, J.; Hussein, S.S.; Tomaka, W. Distinct insulin granule subpopulations implicated in the secretory pathology of diabetes types 1 and 2. *eLife* **2020**, *9*, e62506. [CrossRef]
117. Hou, N.; Mogami, H.; Kubota-Murata, C.; Sun, M.; Takeuchi, T.; Torii, S. Preferential release of newly synthesized insulin assessed by a multi-label reporter system using pancreatic beta-cell line MIN6. *PLoS ONE* **2012**, *7*, e47921. [CrossRef] [PubMed]
118. Tsuboi, T.; Kitaguchi, T.; Karasawa, S.; Fukuda, M.; Miyawaki, A. Age-dependent preferential dense-core vesicle exocytosis in neuroendocrine cells revealed by newly developed monomeric fluorescent timer protein. *Mol. Biol. Cell* **2010**, *21*, 87–94. [CrossRef]
119. Hays, L.B.; Wicksteed, B.; Wang, Y.; McCuaig, J.F.; Philipson, L.H.; Edwardson, J.M.; Rhodes, C.J. Intragranular targeting of syncollin, but not a syncollinGFP chimera, inhibits regulated insulin exocytosis in pancreatic beta-cells. *J. Endocrinol.* **2005**, *185*, 57–67. [CrossRef]
120. Gauthier, D.J.; Sobota, J.A.; Ferraro, F.; Mains, R.E.; Lazure, C. Flow cytometry-assisted purification and proteomic analysis of the corticotropes dense-core secretory granules. *Proteomics* **2008**, *8*, 3848–3861. [CrossRef] [PubMed]
121. Macklin, A.; Khan, S.; Kislinger, T. Recent advances in mass spectrometry based clinical proteomics: Applications to cancer research. *Clin. Proteom.* **2020**, *17*, 1–25. [CrossRef] [PubMed]
122. Yates, J.R.; Ruse, C.I.; Nakorchevsky, A. Proteomics by mass spectrometry: Approaches, advances, and applications. *Annu. Rev. Biomed. Eng.* **2009**, *11*, 49–79. [CrossRef] [PubMed]
123. Timp, W.; Timp, G. Beyond mass spectrometry, the next step in proteomics. *Sci. Adv.* **2020**, *6*, eaax8978. [CrossRef] [PubMed]
124. Swaminathan, J.; Boulgakov, A.A.; Hernandez, E.T.; Bardo, A.M.; Bachman, J.L.; Marotta, J.; Johnson, A.M.; Anslyn, E.V.; Marcotte, E.M. Highly parallel single-molecule identification of proteins in zeptomole-scale mixtures. *Nat. Biotechnol.* **2018**, *36*, 1076–1082. [CrossRef]
125. Cao, M.; Mao, Z.; Kam, C.; Xiao, N.; Cao, X.; Shen, C.; Cheng, K.K.; Xu, A.; Lee, K.-M.; Jiang, L. PICK1 and ICA69 control insulin granule trafficking and their deficiencies lead to impaired glucose tolerance. *PLoS Biol.* **2013**, *11*, e1001541. [CrossRef] [PubMed]
126. Li, H.; Wei, S.; Cheng, K.; Gounko, N.V.; Ericksen, R.E.; Xu, A.; Hong, W.; Han, W. BIG3 inhibits insulin granule biogenesis and insulin secretion. *EMBO Rep.* **2014**, *15*, 714–722. [CrossRef] [PubMed]
127. Pradhan, A.D.; Manson, J.E.; Meigs, J.B.; Rifai, N.; Buring, J.E.; Liu, S.; Ridker, P.M. Insulin, proinsulin, proinsulin:insulin ratio, and the risk of developing type 2 diabetes mellitus in women. *Am. J. Med.* **2003**, *114*, 438–444. [CrossRef]

128. Brackeva, B.; De Punt, V.; Kramer, G.; Costa, O.; Verhaeghen, K.; Stangé, G.; Sadones, J.; Xavier, C.; Aerts, J.; Gorus, F. Potential of UCHL1 as biomarker for destruction of pancreatic beta cells. *J. Proteom.* **2015**, *117*, 156–167. [CrossRef]
129. Martens, G. Species-related differences in the proteome of rat and human pancreatic beta cells. *J. Diabetes Res.* **2015**, *2015*, 549818. [CrossRef]
130. Brackeva, B.; Kramer, G.; Vissers, J.; Martens, G. Quantitative proteomics of rat and human pancreatic beta cells. *Data Brief* **2015**, *3*, 234–239. [CrossRef]
131. Clardy, S.M.; Mohan, J.F.; Vinegoni, C.; Keliher, E.J.; Iwamoto, Y.; Benoist, C.; Mathis, D.; Weissleder, R. Rapid, high efficiency isolation of pancreatic ss-cells. *Sci. Rep.* **2015**, *5*, 13681. [CrossRef]
132. Jayaraman, S. A novel method for the detection of viable human pancreatic beta cells by flow cytometry using fluorophores that selectively detect labile zinc, mitochondrial membrane potential and protein thiols. *Cytom. A* **2008**, *73*, 615–625. [CrossRef] [PubMed]
133. Lukowiak, B.; Vandewalle, B.; Riachy, R.; Kerr-Conte, J.; Gmyr, V.; Belaich, S.; Lefebvre, J.; Pattou, F. Identification and purification of functional human beta-cells by a new specific zinc-fluorescent probe. *J. Histochem. Cytochem.* **2001**, *49*, 519–528. [CrossRef] [PubMed]
134. Smelt, M.J.; Faas, M.M.; de Haan, B.J.; de Vos, P. Pancreatic beta-cell purification by altering FAD and NAD(P)H metabolism. *Exp. Diabetes Res.* **2008**, *2008*, 165360. [CrossRef] [PubMed]
135. Van De Winkel, M.; Pipeleers, D. Autofluorescence-activated cell sorting of pancreatic islet cells: Purification of insulin-containing B-cells according to glucose-induced changes in cellular redox state. *Biochem. Biophys. Res. Commun.* **1983**, *114*, 835–842. [CrossRef]

metabolites

MDPI

Review

What Is the Metabolic Amplification of Insulin Secretion and Is It (Still) Relevant?

Ingo Rustenbeck [1,*], Torben Schulze [1], Mai Morsi [1,2], Mohammed Alshafei [1] and Uwe Panten [1]

[1] Institute of Pharmacology, Toxicology and Clinical Pharmacy, Technische Universität Braunschweig, D38106 Braunschweig, Germany; to.schulze@tu-bs.de (T.S.); m.morsi@tu-bs.de (M.M.); mohammed.alshafei@tu-bs.de (M.A.); u.panten@tu-bs.de (U.P.)

[2] Department of Pharmacology, Faculty of Pharmacy, Assiut University, Assiut 71526, Egypt

* Correspondence: i.rustenbeck@tu-bs.de; Tel.: +49-(0)53-139-156-70

Abstract: The pancreatic beta-cell transduces the availability of nutrients into the secretion of insulin. While this process is extensively modified by hormones and neurotransmitters, it is the availability of nutrients, above all glucose, which sets the process of insulin synthesis and secretion in motion. The central role of the mitochondria in this process was identified decades ago, but how changes in mitochondrial activity are coupled to the exocytosis of insulin granules is still incompletely understood. The identification of ATP-sensitive K^+-channels provided the link between the level of adenine nucleotides and the electrical activity of the beta cell, but the depolarization-induced Ca^{2+}-influx into the beta cells, although necessary for stimulated secretion, is not sufficient to generate the secretion pattern as produced by glucose and other nutrient secretagogues. The metabolic amplification of insulin secretion is thus the sequence of events that enables the secretory response to a nutrient secretagogue to exceed the secretory response to a purely depolarizing stimulus and is thus of prime importance. Since the cataplerotic export of mitochondrial metabolites is involved in this signaling, an orienting overview on the topic of nutrient secretagogues beyond glucose is included. Their judicious use may help to define better the nature of the signals and their mechanism of action.

Keywords: cytosolic calcium concentration; glucose; insulin secretion; metabolic amplification; mitochondria; nutrient secretagogues

check for updates

Citation: Rustenbeck, I.; Schulze, T.; Morsi, M.; Alshafei, M.; Panten, U. What Is the Metabolic Amplification of Insulin Secretion and Is It (Still) Relevant? *Metabolites* **2021**, *11*, 355. https://doi.org/10.3390/metabo11060355

Academic Editors: Melkam Kebede and Belinda Yau

Received: 7 May 2021
Accepted: 31 May 2021
Published: 2 June 2021

1. The Biphasic Pattern of Insulin Secretion Can Be Produced in Different Experimental Settings

The endocrine pancreas responds to an increase in glucose in the bloodstream with a biphasic pattern of insulin release, represented by a short (5–10 min) first phase, followed by a long, sustained second phase release. This unique feature is best shown in response to a 'square wave' glucose stimulus, which is produced in vivo by the hyperglycaemic clamp [1]. It is obvious that such a stimulation pattern is non-physiological, but the resulting biphasic response is the hallmark of the healthy endocrine pancreas. In type 2 diabetes (T2D) and animal models of this disease, it displays varying degrees of a diminished insulin response, often described to be more prominent during the first phase but also recognizable during the second phase [2,3]. The demonstration that metabolically healthy first-degree relatives of T2D patients have a diminished secretory response during a hyperglycaemic clamp [4] and that insulin-sensitive offspring of T2D patients have diminished beta-cell function [5] suggest that beta-cell dysfunction is an independent pathogenic factor driving the progressive impairment of glucose homeostasis in the course of T2D [6]. The low incidence of T2D during historical periods of food shortage shows that beta-cell dysfunction is rarely sufficient for manifestation in the absence of insulin resistance.

Originally demonstrated in vivo, the biphasic secretion pattern could also be demonstrated in vitro, or more precisely, ex vivo, using the model of the perfused rat [7,8] or mouse pancreas [9] and, after the introduction of the collagenase isolation technique, with the more reductionist model of the batch-perifused isolated islets [10,11]. Theoretically, it is possible that the phenomenologically similar secretion pattern in vivo and ex vivo may result from different mechanisms. In a less apodictic form, namely that the relative contribution of the processes involved in shaping the secretion pattern may vary between the different experimental systems, this argument is worth considering (see below). Just to illustrate this point, the effect of incretins is certainly absent in perifused islets, which may be responsible for their more protracted and less dynamic responses.

Today, a large part of the basic research on beta-cell function is performed using insulin-secreting cell lines, often referred to as clonal beta cells. In addition to the initially perceived advantage, to generate a steady supply of insulin-secreting cells with standardized properties without the hazards and time consumption of collagenase digestion, the ease of transfection has made them popular tools. Thus, cell lines are also used for the study of metabolism-derived signals for secretion (see, e.g., [12,13]). However, it has to be kept in mind that the much larger biosynthetic activity as a precondition for the higher rates of mitosis represents a drain of metabolites. This is one of the reasons why quickly proliferating cell lines, such as RINm5F, have a poor secretory response to metabolic stimulation, whereas highly differentiated cell lines, such as INS-1E, have a better response but proliferate only slowly [14,15]. Interestingly, three-dimensional aggregates of MIN6 cells, also called MIN6 pseudo-islets, show a higher secretory response to glucose than the equivalent number of MIN6 cells in monolayer [16] and can be perifused like primary islets. Still, the kinetics and dynamic range of secretion are inferior to the one of primary islets.

2. The Early Years: Substrate Site Hypothesis versus Receptor Site Hypothesis

Very early investigations, using a number of metabolizable and non-metabolizable carbohydrates, suggested that stimulated insulin secretion required the metabolic breakdown of the glucose [17,18]. For example, glucose and mannose stimulated insulin secretion, whereas 2-deoxyglucose, 3-*O*-methylglucose, and galactose did not. However, it was difficult to verify that increases in the level of potential signal-conveying metabolites occurred during the initial phase of stimulated secretion; thus, the alternative view evolved, namely that glucose stimulates insulin secretion by binding to a plasma membrane receptor [19]. The demonstration that the NAD(P)H-autofluorescence of perifused islets increased, coinciding with the onset of stimulated secretion, suggested that the level of mitochondrial reducing equivalents increased during the onset of secretion [20] and supported the substrate site hypothesis.

Perhaps, some of the earlier controversies were caused by the existence of sweet taste receptors at the beta-cell plasma membrane, the activation of which may lead to a leftward shift of the threshold of glucose stimulation [21]. The receptor concept survived in modified form when the prominent role of the glucokinase, the first glucose-metabolizing enzyme in the beta cells, to regulate the velocity of glucose breakdown was recognized [22–24]. Actually, the biochemical or pharmacological enhancement of the glucokinase activity proved to have insulinotropic consequences [25,26]. Nevertheless, it remained unclear for some time whether proximal events of glucose metabolism could directly affect insulin secretion [27] or whether the complete degradation of the glucose carbon to CO_2 and increased oxidative phosphorylation was necessary for signal generation and not just a complicating by-product of glucose metabolism.

3. What Are Nutrient Secretagogues?

The metabolization of glucose as a precondition for its insulinotropic effect has become textbook knowledge. In consequence, not only the earlier disputes but also, the earlier experimental approaches to characterize the relation between the nutrient and the insulinotropic properties are becoming less widely known. In view of the still unresolved issues, they may not only be of historical interest. The insulinotropic property of non-carbohydrates belongs to this chapter.

In addition to the few insulinotropic carbohydrates, some amino acids and monocarboxylic keto acids are insulinotropic. However, the concentrations needed to elicit insulin secretion are beyond their physiological ranges in the serum of humans and mice. So, they cannot be named physiological stimulators, even if the mechanisms they activate within the beta cell are physiological in the sense that even prolonged stimulation is not detrimental for beta-cell function and viability. This sets them apart from more robust stimuli, such as depolarization, by high extracellular potassium concentration (see below). The feature which made (and still makes) amino acids and keto acids interesting as heuristics tools is that their catabolism and the one of glucose only meet in the mitochondrial matrix, thus reducing the number of possible mechanisms involved in stimulus-secretion coupling.

The amino acid which is able to elicit insulin secretion in the absence of glucose is leucine. Its insulinotropic efficacy is strongly enhanced by combination with glutamine, which by itself is ineffective [28,29]. The series of keto acids are of interest since their catabolism shows differences, as shown by their insulinotropic efficacy. Alpha-ketoisocaproic acid (KIC), alpha-ketocaproate, alpha-ketovalerate, and alpha-keto-beta-methylvalerate are all insulinotropic [30]. KIC has the strongest insulinotropic effect, on a par with glucose, albeit with different kinetics. Alpha-ketoisovalerianic acid (KIV) is not insulinotropic but can enhance the efficacy of other nutrient secretagogues. Actually, the term nutrient secretagogue for such compounds is preferable over "fuels" since they do not only support ATP generation but like glucose, support the biosynthesis of insulin, as has been shown for the combination of leucine and glutamine [31].

All of the above insulinotropic compounds support the generation of reducing equivalents by the Krebs cycle in the mitochondrial matrix, either by being catabolized themselves or by activating the catabolism of endogenous nutrients. Like leucine, BCH (2-amino-bicyclo [2.2.1]heptane-2-carboxylic acid), a non-metabolizable structural analogue of leucine, activates glutamate dehydrogenase and provides the citric acid cycle with alpha-ketoglutarate [32]. Similarly, beta-phenylpyruvate is not metabolizable but serves as transamination partners for glutamate and glutamine [33]. In the course of phenylketonuria, an inborn error of metabolism, phenylpyruvate can reach concentrations that inappropriately stimulate insulin secretion and generate hypoglycaemia [34]. The role as a transamination partner is also relevant for KIC, which is itself metabolized, yielding acetyl-CoA, but also transaminates glutamate and thus provides the Krebs cycle with alpha-ketoglutarate [30]. Inhibition of transamination by aminooxyacetate abolishes the insulinotropic effect of KIC [35].

However, it has to be emphasized that in contrast to glucose, not all of the keto acids are pure nutrient secretagogues. For example, KIC and beta-phenylpyruvate were found to directly inhibit the ATP-sensitive K^+ channel (K_{ATP} channel) [36,37] in addition to inhibiting the channel by the increased ATP/ADP ratio [38,39]. This has to be kept in mind when interpreting secretory responses to keto acid stimulation. In this context, it should be mentioned that distinct from other amino acids, the insulinotropic property of arginine is not caused by metabolism but is exclusively due to the depolarization caused by its electrogenic uptake into the beta cells [40].

4. Which Features of the Mitochondrial Metabolism Are Typical for Beta Cells?

The insulinotropic property of the above amino- and keto-acids play a major role in defining the mitochondria as the central hub of the stimulation of insulin secretion by nutrients (see Figure 1). Several lines of evidence contributed to this progress. First, the demonstration that not only glucose but nutrient secretagogues in general increase NAD(P)H-fluorescence and at the same time decrease FAD-fluorescence [41–43], which points to the mitochondrial matrix. Second, KIC, like glucose and a number of other nutrient secretagogues, increases oxygen consumption supporting a role for oxidative phosphorylation [44–46]. By combining oxygen consumption measurements of perifused islets with secretion measurements, an interesting difference in the kinetics was observed, in that the increase in KIC-stimulated secretion closely parallels the increase in the oxygen consumption rate (OCR), whereas glucose generates a faster increase in the OCR, which then turns into a slowly ascending plateau [47].

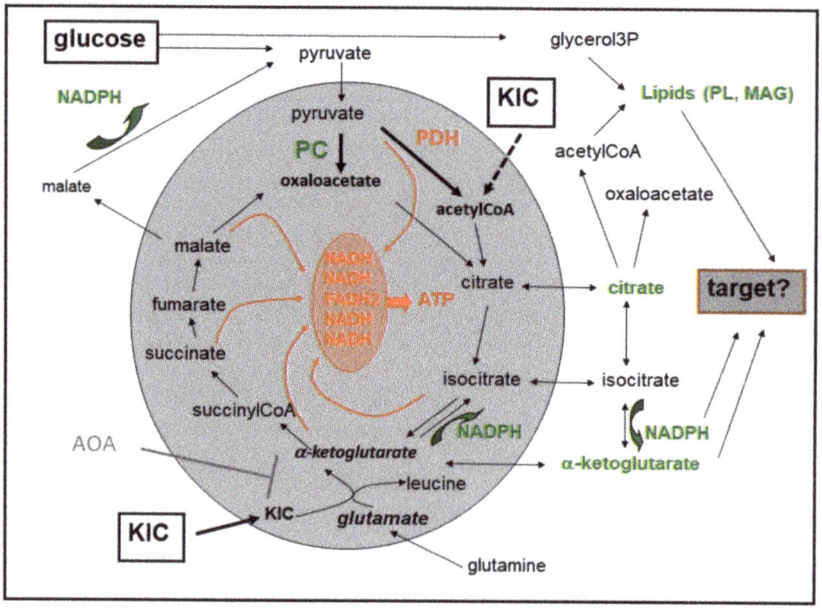

Figure 1. Both glucose and KIC exert an anaplerotic effect on the citric acid cycle in beta cells. Glucose-derived pyruvate is either metabolized by pyruvate dehydrogenase (PDH) to acetyl-CoA or by pyruvate carboxylase (PC) to oxaloacetate. KIC can be degraded, which yields acetyl-CoA. It can also be transaminated with glutamate, which yields alpha-ketoglutarate and leucine. This reaction is essential for the insulinotropic effect since inhibition of transamination by AOA inhibits KIC-induced insulin secretion. The carbon which is fed into the citric acid cycle is derived from glutamate, which, in turn, is generated from endogenous glutamine. The red color marks reactions which support oxidative phosphorylation, the green color marks metabolites and reactions with potential relevance for glucose-dependent amplification. Abbreviations: AOA = aminooxyacetate, KIC = alpha-ketoisocaproic acid, MAG = monoacylglycerol, PL = phospholipids.

It is interesting that islet cells in situ live in a higher oxygen tension than the surrounding exocrine tissue [48], even though mitochondria in beta cells make up only 4% of the intracellular volume, much less than, e.g., in hepatocytes [49]. The higher oxygen tension is likely a consequence of dense vascularization of the islet [50,51]. It is known that three critical dehydrogenases of the Krebs cycle can be activated by increased Ca^{2+} levels [52]. Interestingly, the increase in oxygen consumption, which results from this activation, occurs only when Ca^{2+} is provided by influx via L-type Ca^{2+} channels, not by release from intracellular stores [53].

The role of mitochondrial metabolism for glucose-derived signal generation is emphasized by the fact that the fraction of glucose carbon that is oxidized increases with accelerated glycolysis, even though it is already high at basal levels. Another notable feature is that glucose utilization does not increase with decreasing oxygen levels [54]. The latter feature is explained by the low activity of lactate dehydrogenase in beta cells [55]. Thus, to re-oxidize NADH in the cytosol and to keep the glycolysis running, reducing equivalents have to be transported into the mitochondrial matrix via shuttling mechanisms. This is enabled by a high expression level of the FAD-linked glycerol-3-phosphate dehydrogenase [56]. The use of pyruvate in beta cell mitochondria is characterized by a high activity of pyruvate carboxylase, which is atypical for a cell without gluconeogenic function [57].

Finally, the specific relevance for the beta cell of the close coupling of glucose utilization to mitochondrial metabolism is underlined by the much looser coupling in the glucagon-secreting alpha cells [54]. Given the central position of the mitochondria in the stimulus-secretion coupling, it is not surprising that they are considered as essential contributors to the pathogenesis of type 2 diabetes [58].

5. Which Observations Have Led to the Hypothesis of a Bifurcating Pathway Emanating from the Mitochondria?

The realization that nutrient-stimulated insulin secretion is dependent on mitochondrial activation, or more precisely, the increased oxidative phosphorylation, led to the hypothesis that ATP or perhaps the ATP/ADP-ratio might convey the signal to the electrical activity which had been observed in beta cells [59,60]. Measurements of radioactivity from perifused islets loaded with rubidium (^{86}Rb) identified a decrease in the potassium conductance as the likely initiating event of electrical activity, which was required for the initiation of exocytosis [61]. However, even though it could be shown that raising the glucose concentration increased the ATP content in islets, it did not occur in the insulinotropic concentration range of glucose, and measurements of the ATP/ADP ratio yielded inconclusive results [62,63]. Only when the high background by the ATP content in the insulin granules was diminished by partial degranulation could a convincing correlation between the ATP/ADP ratio and insulin secretion be demonstrated [64].

The missing link between changes in the ATP/ADP ratio and the electrical activity proved to be the K_{ATP} channel, the functional identification of which was a fruit of the newly developed patch clamp technique [65,66]. It took several years to clarify its molecular composition, which consists of an inwardly rectifying K^+ channel (Kir 6.2 = KCNJ11) as the pore-forming unit and a member of the ATP cassette-binding family (SUR1 = ABCC8) as the regulatory subunit [67]. While ATP^{4-} closes the channel by interaction with the channel pore, Mg-ADP binding to the regulatory subunit (also termed sulfonylurea receptor) has an opening effect on the channel and modifies sulfonylurea potency [68,69]. Even though additional metabolism-derived signals, such as H_2O_2 or acyl-CoA, may contribute to the regulation of K_{ATP} channels [70,71], the ATP/ADP ratio as determined in intact islets can be considered as a sufficient measure of the signal by which the beta cell mitochondria initiate the electrical activity of the plasma membrane [72].

After the identification of the K_{ATP} channel as the link between energy metabolism and electrical activity of the beta cell, the view prevailed that the depolarization induced Ca^{2+} influx caused by the closure of the channel would constitute the final common pathway of insulin secretion. This view was regarded as the consensus theory of glucose-induced insulin secretion. A test of this theory was to close the K_{ATP} channel pharmacologically and then raise glucose from a basal to a stimulatory level. If Ca^{2+} influx by K_{ATP} channel closure was the only regulator of secretion, no further increase in secretion was to be expected. However, glucose caused an increase in secretion beyond the level established by the maximally effective concentration of sulfonylurea [73]. The consensus theory was further challenged by another experimental approach, where K_{ATP} channels were opened by diazoxide to prohibit any effect of the energy metabolism on the membrane potential. Depolarization was induced by a high extracellular potassium concentration. Because of the large potassium conductance under this condition, the membrane potential closely follows the Nernst potential. Again, raising glucose to a stimulatory level increased the secretion of isolated islets beyond the level established by depolarization alone [74].

The latter approach has become the experimental standard, in part because of suspected insulinotropic effects of sulfonylureas in addition to the closure of plasma membrane K_{ATP} channels [75,76]. However, interference by intracellular sites of action also limits the value of diazoxide as a pharmacological tool since it was shown to exert direct effects on beta-cell mitochondria [77]. An argument in favor of pharmacologically blocking the K_{ATP} channels is that the depolarization thus produced corresponds to the depolarization strength and pattern of nutrient stimulation, whereas depolarization by high potassium can be much stronger, depending on the actual concentration, and lacks the typical pattern of action potential spiking [78,79].

Initially often named "K_{ATP} channel-independent pathway" [80], the name "amplifying pathway", as suggested by Henquin [81], has become widely accepted. This choice of name was based on the evidence that insulin secretion is not increased if the plasma membrane is not depolarized, which under physiological conditions requires the closure of the K_{ATP} channels. For this reason, the pathway is not "independent" in the proper sense. The pathway leading to the K_{ATP} channel closure and Ca^{2+} influx, in contrast, was termed "triggering pathway" [81]. Another nomenclature tries to avoid a functional definition by simply naming the stimulus-secretion coupling via K_{ATP} channel closure the "canonical pathway" and leaving the role of those signals undefined, which affect secretion without causing changes in the electrical activity [82]. What is important, though, is not to confuse the metabolic amplification, an inherent property of nutrient secretagogues, with a receptor-mediated enhancement of secretion by neurotransmitters or hormones, such as GLP-1.

6. Models of the Biphasic Kinetics of Secretion

A certain weakness in the definition of the amplifying pathway is that it relies on the use of pharmacological agents prior to the nutrient stimulation. This, by necessity, influences the secretion kinetics of the nutrient. So, to discuss whether the metabolic amplification primarily affects early or late periods of the biphasic secretion pattern, we have to briefly touch on the topic of which mechanisms underlie the biphasic kinetics of secretion. In principle, two different points of view can be distinguished: one considers the limited number of secretion-ready granules as the main responsible factor, whereas the other considers the evolving pattern of metabolism-derived signals as the underlying cause.

The more popular hypothesis proposes that the biphasic secretion pattern results from the existence of two different pools of granules. Its origins can be traced back to the two-compartment model of secretion, proposing that insulin is contained in a stable compartment and in a much smaller compartment that is labile to stimulation [83]. Based on the combination of ultrastructural and electrophysiological data, the correlate of the latter compartment was suggested to be a limited number of secretion-ready granules that are firmly attached to the plasma membrane ("docked") and fully prepared ("primed") for fusion [84–86]. The high fusion rate produced by depolarization-induced Ca^{2+} influx depletes this pool of secretion-ready granules causing the transient decrease in the rate of insulin secretion, observed after 10–15 min of continuous stimulation (somewhat longer than in vivo). The progressively faster replenishment of this pool by metabolism-dependent recruitment of a distant reserve pool causes the subsequent recovery in the insulin release rate, taking shape as the second phase [87,88]. Since the accelerated rate of replenishment now matches the rate of exocytosis, the secretion rate of the second phase can be maintained for several hours until a general desensitization towards stimulation sets in, which reduces the secretion rate to about one-third [89].

The alternative view can be named the metabolic control hypothesis. This hypothesis proposes that the glucose metabolism in the beta cell generates stimulatory as well as inhibitory signals [90,91]. The initial stimulatory effect is followed by a more slowly evolving inhibitory signal which reduces the secretion rate. This process, which leads to the nadir of the secretion rate, was named time-dependent inhibition. The hypothetical inhibitory signal is then succeeded or overruled by a third signal evolving from the glucose metabolism. This even more retarded reaction was named time-dependent potentiation. The time-dependent inhibition and potentiation do not only differ in their kinetics but also their dependency on the glucose concentration [92]. The main obstacle for this hypothesis to gain a wider acceptance is that no clear mechanism for the inhibitory signal could be demonstrated.

On the other hand, the hypothesis of the time-dependent potentiation bears a clear resemblance to the phenomenon of metabolic amplification. A comparison of the storage- and the signal-limited model of insulin secretion gave slightly better fits for the simulation of the signal-limited model [93]. Further progress in characterizing the fate of the insulin granules may eventually show that both hypotheses are not mutually exclusive but rather that the earlier model (limited storage) is related to the latter.

In this context, it is worth pointing out that a fully developed first phase with a transient decrease in the secretion rate requires the "non-physiological" square-wave stimulus (Figure 2). A ramp-like stimulation generates a continuously ascending rate of secretion, where the initial phase is only different from the following phase by a more steeply ascending rate of secretion [94]. In perifused islets, such a pattern also results when a maximally effective glucose stimulus is applied after a period of nutrient deprivation [47,95]. This suggests that the velocity with which metabolites are generated may not only determine the initial increase in the secretion rate but also the extent of the transient decrease.

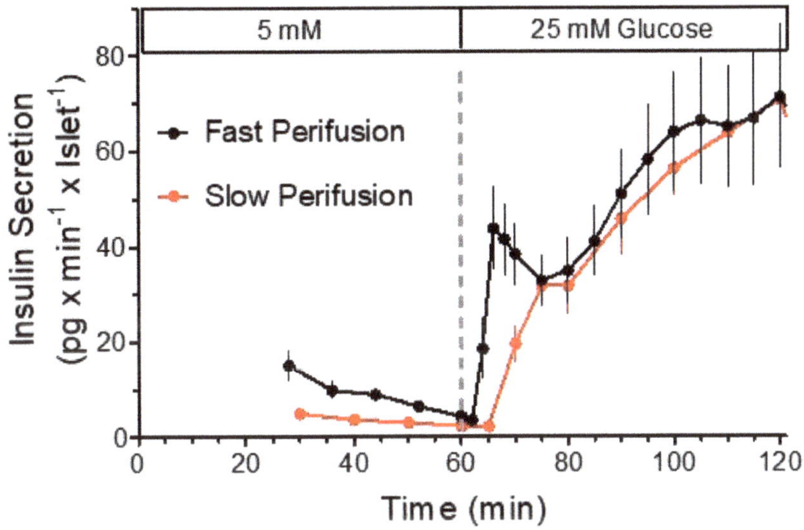

Figure 2. The kinetics of nutrient-induced insulin secretion are strongly dependent on the velocity of nutrient increase. The only difference between the experimental conditions of the secretory responses shown in this figure is the pump rate of the perifusion and, by consequence, the steepness of the glucose gradient. The final concentration was reached after 5 min or 35 min, respectively.

7. Relation of the Triggering and Amplifying Pathways to the Biphasic Secretion Kinetics

The hypothesis that the imbalance between the fusion rate of the fully prepared granules and the velocity of granule replenishment is responsible for the first phase entails that the relevance of the mitochondrial metabolism for secretion becomes visible during the build-up of the second phase. Consequently, the triggering signal was often held to be responsible for the first phase and the amplifying signal for the second phase of glucose-stimulated secretion, see, e.g., [96].

This view was supported by observations on INS1 cells with inactivated mtDNA, which did not secrete insulin in response to glucose but did so in response to KCl depolarization [97]. However, the short-term treatment of perifused islets with mitochondrial inhibitors virtually abolished not only the insulin secretion elicited by glucose but also the one by KCl or sulfonylureas [98]. MIN6 cells devoid of mitochondrial DNA showed a markedly reduced secretion in response to the sulfonylurea, glibenclamide [99]. So, the extent to which the triggering signal alone elicits secretion is less well defined than often realized. In this context, it is worth mentioning that Ca^{2+} influx does not simply constitute the signal for granule fusion but that it may affect granule transport and thus the availability of fusion-ready granules. In fact, the biphasic increase in the cytosolic Ca^{2+} concentration, often seen in intact islets [100,101], suggests a role for the Ca^{2+} signal beyond triggering.

As mentioned above, potassium depolarization has no upper limit within the physiological working range of the beta cell, in contrast to K_{ATP} channel closure. The closure of the K_{ATP} channels results in a plateau depolarization of about 20 mV by a still ill-defined inward leak current. The superimposed action potentials result from the phasic influx of Ca^{2+} via L-type channels, as can be seen by the use of blockers of these channels [102]. So, for the experimental triggering signal to mimic the consequence of K_{ATP} channel closure by nutrient stimulation, a depolarization by about 20 mV has to be produced to open voltage-dependent Ca^{2+} channels. This is the depolarizing strength of 15 mM KCl, which only elicits a small transient increase in secretion far below the typical range of the first phase [102]. Of note, the presence of 15 mM KCl is not without effect. It enhances the secretion elicited by glucose and by sulfonylureas [102,103].

These observations suggest that a strong depolarization (e.g., by 40 mM KCl and more) can produce a first phase-like monophasic secretion, but that the weaker depolarization produced by nutrient stimulation generates the first phase because the metabolic amplifying is already effective. This conclusion concurs with the one drawn from observations on the changing relation between first and second phase secretion in dependence on the prestimulatory conditions [104]. In short, metabolic amplification is likely effective from the beginning of nutrient-stimulated secretion.

Finally, it has to be mentioned that the live-cell imaging of fluorescently labelled granules by TIRF microscopy have led to a much more detailed view on the insulin granules and the events preceding the fusion process, see, e.g., [105–107]. However, these data have not yet had a major impact on the hypothesis of triggering and amplifying pathways. Certainly, the quantitative description of granule genesis, transport, and fusion under triggering and amplifying conditions will be of major help to understand how the amplifying signals enhance the secretion rate.

8. Cataplerosis and Putative Amplification Signals

In contrast to the triggering pathway, a broadly accepted model of the amplifying pathway has not yet emerged. A consensus exists in that the export of metabolites from the mitochondrial matrix into the cytosol (cataplerosis) is of importance [108]. Cataplerosis is any process by which carbon leaves the citric acid cycle in a form different from carbon dioxide. Thereby the need for refilling the citric acid cycle (anaplerosis) is generated. So, it can be hypothesized that a compound qualifies for the role of a nutrient secretagogue when it increases the rate of oxidative phosphorylation and at the same time supports cataplerosis by refilling the citric acid cycle [57].

According to this theory, glucose is insulinotropic not only because it generates NADH and $FADH_2$ in the matrix space but also because it refills the pool of oxaloacetate by the action of pyruvate carboxylase. Actually, only 60% of the glucose-derived pyruvate entering the mitochondria is metabolized by pyruvate dehydrogenase to acetyl-CoA; the other 40% is built up to oxaloacetate. This compensates for the loss of cataplerotic metabolites and permits the constant running of the citric acid cycle [109]. This scenario would apply when mitochondria release glutamate, which could be generated from alpha-ketoglutarate by the glutamate dehydrogenase reaction. Glutamate was the first in the line of candidate signal-conveying compounds [110], also comprising alpha-ketoglutarate [111] or short-chain acyl-CoA [112].

Some authors view the cataplerosis as part of one or more cycles of export and re-import of metabolites to convey reducing equivalents out of the mitochondria to ultimately increase the cytosolic levels of NADPH [113,114]. NADPH is discussed as a regulator of exocytosis in beta cells via the deSUMOylation of SNARE proteins by SENP1 protein [115,116]. It has been shown repeatedly that glucose stimulation increases the level of

NADPH (not to be confused with NAD(P)H, which designates the sum of both NADH and NADPH in conventional fluorometry). But such an increase was not found for the amplifying effect of KIC [117]. Thus, NADPH is unlikely to be an indispensable signal for metabolic amplification. From a wider perspective, it can be postulated that candidate signalling compounds, such as monoacylglycerol [118], which can result from the metabolism of glucose, but not of nutrient secretagogues in general, may contribute to the metabolic amplification but are unlikely to be indispensable for this process.

As mentioned above, KIC has two different sites of entry into the citric acid cycle; in addition to acetyl-CoA, it can generate alpha-ketoglutarate as long as the transamination partner glutamate is available. The comparison of the metabolic amplification by KIC and by glucose has led to an unexpected result. To obtain a metabolic situation of the beta cells, where KIC metabolism was only minimally affected by the preceding glucose metabolism, a 60 min nutrient-free perifusion preceded the application of exogenous nutrient. The triggering signal was provided by blocking the K_{ATP} channels with a maximally effective concentration of sulfonylurea. The insulinotropic effect of KIC under this condition was about as strong as in the absence of the sulfonylurea (KIC at >5 mmol/L is a K_{ATP} channel blocker in its own right), whereas the insulinotropic effect of glucose under the same condition was virtually abolished and did not recover within the next hour. Or, in other words, the metabolic amplification of glucose, but not by KIC, had been blocked by the exposure to sulfonylurea throughout the experiment [119]. Later, it was confirmed that the moderate depolarization by 15 mM KCl could substitute for the action of the sulfonylurea [47].

It can thus be hypothesized that prolonged depolarization in the absence of nutrients leads to a critically low concentration of a metabolite that cannot be replenished by glucose alone. This view was supported by the observation that the secretion can be restarted by adding KIV [120] or by glutamine, the effect of which could be enhanced by BCH [47]. Remarkably, the insulinotropic effect of glucose, once restarted, remained elevated after removal of the anaplerotic starters, demonstrating that the regained rate of anaplerosis translates into the rate of secretion (Figures 3 and 4). Taken together, the above findings and considerations are consistent with the view that mitochondrial export of citrate and acetoacetate, both sources of cytosolic acetyl-CoA, in insulin-secreting cells, amplifies insulin secretion by enhancing the cytosolic supply of acetyl-CoA [120–122].

Recently, evidence on the role of cAMP as a signal-conveying compound of glucose-stimulated insulin secretion has been presented [123], reflecting the earlier suggestion of cAMP as the mediator of the time-dependent potentiation [124]. While the enhancing function of receptor-mediated cAMP increase on insulin granule exocytosis is beyond doubt [125], it is difficult to relate cAMP generation to the cataplerotic export of citrate cycle metabolites. As a consequence, this hypothesis reduces the role of cataplerosis in stimulus-secretion coupling to maintaining the biosynthesis of insulin in response to nutrient stimulation. A more complete view of how nutrient-stimulated granule biogenesis, transport, and degradation contribute to secretion kinetics is needed to consider this possibility.

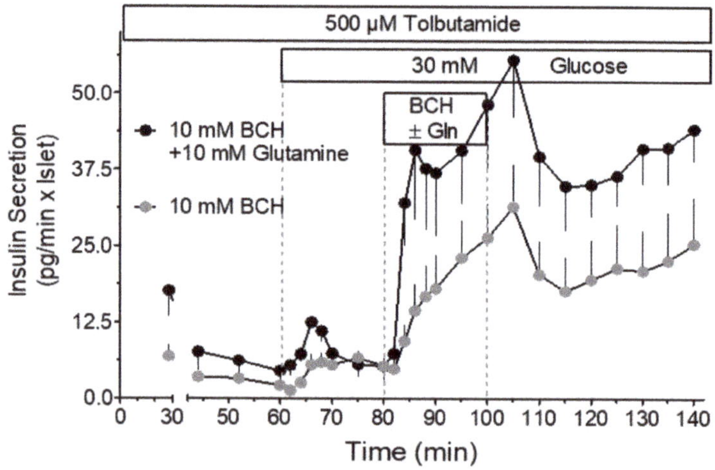

Figure 3. The rescue of the metabolic amplification of glucose, abolished by a period of depolarization in the absence of nutrients. The amplification was restarted by perifusion with BCH and continued after wash-out of BCH. The concomitant presence of glutamine and BCH resulted in a virtual jump-start. Again, the elevated secretion level continued after wash-out of the starters. Since tolbutamide depolarizes beta cells even in the absence of glucose, the triggering signal is continuously present. For the underlying mechanisms, see Figure 4. Adapted from Reference [47].

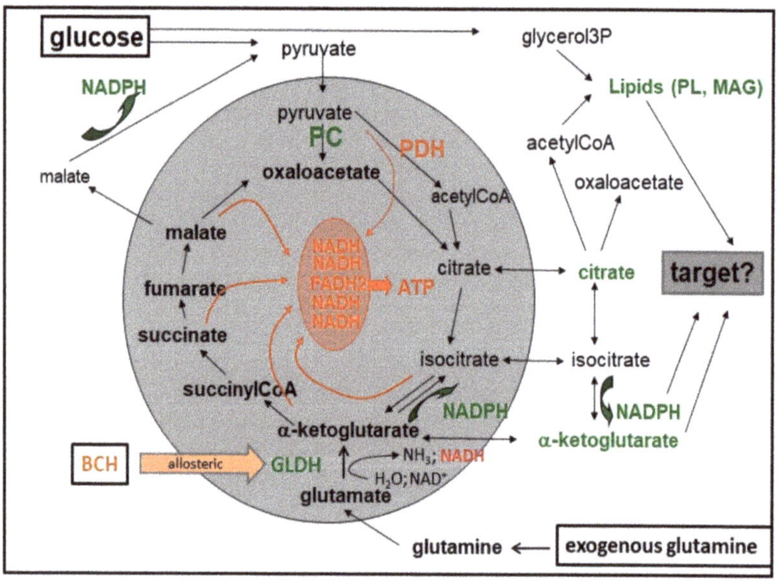

Figure 4. The provision of the citric acid cycle with alpha-ketoglutarate may not only run via transamination but also via glutamate dehydrogenase. This enzyme is allosterically activated by leucine and by (±) BCH, a non-metabolizable leucine analog. In principle, the reaction can run in either direction, but normally the level of endogenous glutamate is sufficient to support the generation of alpha-ketoglutarate. The deaminating, anaplerotic reaction is further enhanced by the exogenous offer of glutamine. The cataplerotic export of acetoacetate, which depends on sufficient levels of succinate, is omitted for clarity. Abbreviations: BCH = 2-amino-bicyclo [2.2.1]heptane-2-carboxylic acid, GlDH = glutamate dehydrogenase.

9. Concluding Remarks

In this review, we have tried to outline the evolution of the ideas on how nutrient secretagogues stimulate insulin secretion. The central position of the mitochondrial metabolism in this process can be regarded as firmly established. Its specific features in the beta-cell suggest that cataplerosis is a major contributor to nutrient-induced signaling and likely produces the phenomenon of metabolic amplification. A broad consensus on the specific nature of the signaling compound(s) and on its (their) mechanism(s) of action has so far failed to emerge. For a review dealing specifically with the shortcomings of current models of stimulus-secretion coupling, the reader is referred to [126]. Here, we have placed an emphasis on the alternative experimental protocol and on the use of amino acids and keto acids as experimental tools to delineate the mechanisms operative during the metabolic amplification (for an encompassing overview, see [127]). These observations, which were obtained with freshly isolated islets, support the view that in addition to the ATP/ADP ratio, the second branch of signaling originates from the beta-cell mitochondria. Whatever its precise nature, it will impact the provision of granules to the sites of exocytosis. It is conceivable that multiple sites of action are involved in this process, which makes the concept of reversible lysine acetylation by increased levels of cytosolic acetyl-CoA attractive [128]. Based on the modeling of clinical and experimental data, defective metabolic amplification has been suggested to underlie the beta-cell dysfunction during type 2 diabetes [129]. Thus, metabolic amplification continues to be a relevant topic in diabetes research, and its mechanisms warrant further investigation.

Author Contributions: Conceptualization, I.R. and U.P.; writing—original draft preparation I.R.; writing—review and editing I.R., T.S., M.M., M.A. All authors have read and agreed to the published version of the manuscript.

Funding: The authors' research mentioned in this review was supported by the Deutsche Forschungsgemeinschaft (Ru 368/5-1 to 5-4) and by grants from the Deutsche Diabetes Gesellschaft (to I.R. and T.S.).

Institutional Review Board Statement: Not applicable.

Informed Consent Statement: Not applicable.

Data Availability Statement: The figures are original work of the authors.

Acknowledgments: Skillful technical assistance by Claudia Bütefisch, Angela Hahlbohm, Verena Lier-Glaubitz and Sabine Warmbold is gratefully acknowledged.

Conflicts of Interest: The authors declare no conflict of interest.

References

1. Cerasi, E.; Luft, R. Plasma-insulin response to sustained hyperglycemia induced by glucose infusion in human subjects. *Lancet* **1963**, *2*, 1359–1361. [CrossRef]
2. Gerich, J.E. Is Reduced First-Phase Insulin Release the Earliest Detectable Abnormality in Individuals Destined to Develop Type 2 Diabetes? *Diabetes* **2002**, *51*, S117–S121. [CrossRef] [PubMed]
3. Del Prato, S. Loss of early insulin secretion leads to postprandial hyperglycaemia. *Diabetologia* **2003**, *46* (Suppl. 1), M2–M8. [CrossRef] [PubMed]
4. Pimenta, W.; Korytkowski, M.; Mitrakou, A.; Jenssen, T.; Yki-Jarvinen, H.; Evron, W.; Dailey, G.; Gerich, J. Pancreatic beta-cell dysfunction as the primary genetic lesion in NIDDM. Evidence from studies in normal glucose-tolerant individuals with a first-degree NIDDM relative. *JAMA* **1995**, *273*, 1855–1861. [CrossRef]
5. Stadler, M.; Pacini, G.; Petrie, J.; Luger, A.; Anderwald, C. Beta cell (dys)function in non-diabetic offspring of diabetic patients. *Diabetologia* **2009**, *52*, 2435–2444. [CrossRef]
6. Wood, A.R.; Jonsson, A.; Jackson, A.U.; Wang, N.; van Leewen, N.; Palmer, N.D.; Kobes, S.; Deelen, J. A Genome-Wide Association Study of IVGTT-Based Measures of First-Phase Insulin Secretion Refines the Underlying Physiology of Type 2 Diabetes Variants. *Diabetes* **2017**, *66*, 2296–2309. [CrossRef]
7. Curry, D.L.; Bennett, L.L.; Grodsky, G.M. Dynamics of Insulin Secretion by the Perfused Rat Pancreas. *Endocrinology* **1968**, *83*, 572–584. [CrossRef] [PubMed]
8. Nesher, R.; Waldman, L.; Cerasi, E. Time-dependent inhibition of insulin release: Glucose-arginine interactions in the perfused rat pancreas. *Diabetologia* **1984**, *26*, 146–149. [CrossRef]

9. Lenzen, S. Insulin secretion by isolated perfused rat and mouse pancreas. *Am. J. Physiol. Metab.* **1979**, *236*, E391. [CrossRef]
10. Panten, U.; Kriegstein, E.; Poser, W.; Schönborn, J.; Hasselblatt, A. Effects of L-leucine and α-ketoisocaproic acid upon insulin secretion and metabolism of isolated pancreatic islets. *FEBS Lett.* **1972**, *20*, 225–228. [CrossRef]
11. Henquin, J.-C. Relative Importance of Extracellular and Intracellular Calcium for the Two Phases of Glucose-Stimulated Insulin Release: Studies with Theophylline*. *Endocrinology* **1978**, *102*, 723–730. [CrossRef] [PubMed]
12. Minami, K.; Yano, H.; Miki, T.; Nagashima, K.; Wang, C.-Z.; Tanaka, H.; Miyazaki, J.-I.; Seino, S. Insulin secretion and differential gene expression in glucose-responsive and -unresponsive MIN6 sublines. *Am. J. Physiol. Metab.* **2000**, *279*, E773–E781. [CrossRef] [PubMed]
13. McClenaghan, N.H.; Barnett, C.R.; O'Harte, F.P.; Flatt, P.R. Mechanisms of amino acid-induced insulin secretion from the glu-cose-responsive BRIN-BD11 pancreatic B-cell line. *J. Endocrinol.* **1996**, *151*, 349–357. [CrossRef]
14. Halban, P.A.; Praz, G.A.; Wollheim, C. Abnormal glucose metabolism accompanies failure of glucose to stimulate insulin release from a rat pancreatic cell line (RINm5F). *Biochem. J.* **1983**, *212*, 439–443. [CrossRef]
15. Merglen, A.; Theander, S.; Rubi, B.; Chaffard, G.; Wollheim, C.B.; Maechler, P. Glucose Sensitivity and Metabolism-Secretion Coupling Studied during Two-Year Continuous Culture in INS-1E Insulinoma Cells. *Endocrinology* **2004**, *145*, 667–678. [CrossRef]
16. Luther, M.J.; Hauge-Evans, A.; Souza, K.L.; Jörns, A.; Lenzen, S.; Persaud, S.J.; Jones, P.M. MIN6 beta-cell-beta-cell interactions influ-ence insulin secretory responses to nutrients and non-nutrients. *Biochem. Biophys. Res. Commun.* **2006**, *343*, 99–104. [CrossRef]
17. Grodsky, G.M.; Batts, A.A.; Bennett, L.L.; Vcella, C.; McWilliams, N.B.; Smith, D.F. Effects of carbohydrates on secretion of insulin from isolated rat pancreas. *Am. J. Physiol. Content* **1963**, *205*, 638–644. [CrossRef]
18. Coore, H.; Randle, P. Regulation of insulin secretion studied with pieces of rabbit pancreas incubated in vitro. *Biochem. J.* **1964**, *93*, 66–78. [CrossRef]
19. Matschinsky, F.M.; Landgraf, R.; Ellerman, J.; Kotler-Brajtburg, J. Glucoreceptor mechanisms in islets of Langerhans. *Diabetes* **1972**, *21*, 555–569. [CrossRef]
20. Panten, U.; Christians, J.; von Kriegstein, E.; Poser, W.; Hasselblatt, A. Effect of carbo-hydrates upon fluorescence of reduced pyridine nucleotides from perifused isolated pancreatic islets. *Diabetologia* **1973**, *9*, 477–482. [CrossRef]
21. Sanchez-Andres, J.V.; Malaisse, W.J.; Kojima, I. Electrophysiology of the pancreatic islet β-cell sweet taste receptor TIR3. *Pflügers Arch. Eur. J. Physiol.* **2018**, *471*, 647–654. [CrossRef]
22. Randle, P.J. Glucokinase and candidate genes for Type 2 (non-insulin-dependent) diabetes mellitus. *Diabetologia* **1993**, *36*, 269–275. [CrossRef]
23. Matschinsky, F.M. Evolution of the glucokinase glucose sensor paradigm for pancreatic beta cells. *Diabetologia* **1993**, *36*, 1215–1217. [CrossRef]
24. Malaisse, W.J. Is glucokinase the beta-cell glucoreceptor? *Diabetologia* **1994**, *37*, 442.
25. Baltrusch, S.; Langer, S.; Massa, L.; Tiedge, M.; Lenzen, S. Improved Metabolic Stimulus for Glucose-Induced Insulin Secretion through GK and PFK-2/FBPase-2 Coexpression in Insulin-Producing RINm5F Cells. *Endocrinology* **2006**, *147*, 5768–5776. [CrossRef]
26. Doliba, N.M.; Fenner, D.; Zelent, B.; Bass, J.; Sarabu, R.; Matschinsky, F.M. Repair of diverse diabetic defects of β-cells in man and mouse by pharmacological glucokinase activation. *Diabetes, Obes. Metab.* **2012**, *14*, 109–119. [CrossRef]
27. MacDonald, M.J. Elusive proximal signals of beta-cells for insulin secretion. *Diabetes* **1990**, *39*, 1461–1466. [CrossRef]
28. Malaisse, W.J.; Sener, A.; Malaisse-Lagae, F.; Welsh, M.; Matthews, D.E.; Bier, D.M.; Hellerström, C. The stimulus-secretion coupling of amino acid-induced insulin release. Metabolic response of pancreatic islets of L-glutamine and L-leucine. *J. Biol. Chem.* **1982**, *257*, 8731–8737. [CrossRef]
29. Malaisse, W.J. Branched-chain amino and keto acid metabolism in pancreatic islets. *Adv. Enzym. Regul.* **1986**, *25*, 203–217. [CrossRef]
30. Lenzen, S.; Schmidt, W.; Rustenbeck, I.; Panten, U. Transamination of neutral amino acids and 2-keto acids in pancreatic B-cell mitochondria. *Biosci. Rep.* **1986**, *6*, 163–169. [CrossRef]
31. Uchizono, Y.; Alarcón, C.; Wicksteed, B.L.; Marsh, B.J.; Rhodes, C.J. The balance between proinsulin biosynthesis and insulin secre-tion: Where can imbalance lead? *Diabetes Obes. Metab.* **2007**, *9* (Suppl. 2), 56–66. [CrossRef] [PubMed]
32. Gylfe, E. Comparison of the effects of leucines, non-metabolizable leucine analogues and other insulin secretagogues on the activity of glutamate dehydrogenase. *Acta Diabetol.* **1976**, *13*, 20–24. [CrossRef] [PubMed]
33. Lenzen, S.; Rustenbeck, I.; Panten, U. Transamination of 3-phenylpyruvate in pancreatic B-cell mitochondria. *J. Biol. Chem.* **1984**, *259*, 2043–2046. [CrossRef]
34. Gazit, V.; Ben-Abraham, R.; Rudin, M.; Katz, Y. Glucose-lowering effect of beta-phenylpyruvate in neonatal mice: A possible mechanism for phenylketonuria-related neurodegenerative changes. *Brain Res. Dev. Brain Res.* **2003**, *141*, 137–140. [CrossRef]
35. Welsh, M.; Sener, A.; Malaisse-Lagae, F.; Malaisse, W.J. The stimulus-secretion coupling of amino acid-induced insulin release. Inhibition of islet respiration and insulin release by aminooxyacetate. *Mol. Cell. Biochem.* **1984**, *63*, 33–37. [CrossRef]
36. Bränström, R.; Efendić, S.; Berggren, P.O.; Larsson, O. Direct inhibition of the pancreatic beta-cell ATP-regulated potassium channel by alpha-ketoisocaproate. *J. Biol. Chem.* **1998**, *273*, 14113–14118. [CrossRef]
37. Heissig, H.; Urban, K.A.; Hastedt, K.; Zünkler, B.J.; Panten, U. Mechanism of the Insulin-Releasing Action of α-Ketoisocaproate and Related α-Keto Acid Anions. *Mol. Pharmacol.* **2005**, *68*, 1097–1105. [CrossRef]

38. Ashcroft, F.M.; Ashcroft, S.J.; Harrison, D.E. Effects of 2-ketoisocaproate on insulin release and single potassium channel activity in dispersed rat pancreatic beta-cells. *J. Physiol.* **1987**, *385*, 517–529. [CrossRef]
39. Duchen, M.R.; Smith, P.A.; Ashcroft, F.M. Substrate-dependent changes in mitochondrial function, intracellular free calcium con-centration and membrane channels in pancreatic beta-cells. *Biochem. J.* **1993**, *294*, 35–42. [CrossRef]
40. Smith, P.A.; Sakura, H.; Coles, B.; Gummerson, N.; Proks, P.; Ashcroft, F.M. Electrogenic arginine transport mediates stimu-lus-secretion coupling in mouse pancreatic beta-cells. *J. Physiol.* **1997**, *499*, 625–635. [CrossRef]
41. Panten, U.; Ishida, H. Fluorescence of oxidized flavoproteins from perifused isolated pancreatic islets. *Diabetologia* **1975**, *11*, 569–573. [CrossRef]
42. Panten, U. Effects of alpha-ketomonocarboxylic acids upon insulin secretion and metabolism of isolated pancreatic islets. *Naunyn-Schmiedeberg's Arch. Pharmacol.* **1975**, *291*, 405–420. [CrossRef]
43. Rocheleau, J.V.; Head, W.S.; Piston, D.W. Quantitative NAD(P)H/Flavoprotein Autofluorescence Imaging Reveals Metabolic Mechanisms of Pancreatic Islet Pyruvate Response. *J. Biol. Chem.* **2004**, *279*, 31780–31787. [CrossRef]
44. Hellerström, C. Effects of Carbohydrates on the Oxygen Consumption of Isolated Pancreatic Islets of Mice. *Endocrinology* **1967**, *81*, 105–112. [CrossRef]
45. Panten, U.; Klein, H. O2Consumption by Isolated Pancreatic Islets, as Measured in a Microincubation System with a Clark-Type Electrode*. *Endocrinology* **1982**, *111*, 1595–1600. [CrossRef]
46. Doliba, N.M.; Wehrli, S.L.; Vatamaniuk, M.Z.; Qin, W.; Buettger, C.W.; Collins, H.W.; Matschinsky, F.M. Metabolic and ionic coupling factors in amino acid-stimulated insulin release in pancreatic beta-HC9 cells. *Am. J. Physiol. Endocrinol. Metab.* **2007**, *292*, E1507–E1519. [CrossRef]
47. Schulze, T.; Morsi, M.; Reckers, K.; Brüning, D.; Seemann, N.; Panten, U.; Rustenbeck, I. Metabolic amplification of insulin secretion is differentially desensitized by depolarization in the absence of exogenous fuels. *Metabolism* **2017**, *67*, 1–13. [CrossRef]
48. Bergsten, P.; Westerlund, J.; Liss, P.; Carlsson, P.-O. Primary in vivo oscillations of metabolism in the pancreas. *Diabetes* **2002**, *51*, 699–703. [CrossRef]
49. Dean, P.M. Ultrastructural morphometry of the pancreatic beta-cell. *Diabetologia* **1973**, *9*, 115–119. [CrossRef]
50. Bonner-Weir, S.; Orci, L. New perspectives on the microvasculature of the islets of Langerhans in the rat. *Diabetes* **1982**, *31*, 883–889. [CrossRef]
51. Nyman, L.R.; Ford, E.; Powers, A.C.; Piston, D.W. Glucose-dependent blood flow dynamics in murine pancreatic islets in vivo. *Am. J. Physiol. Metab.* **2010**, *298*, E807–E814. [CrossRef]
52. Denton, R.M. Regulation of mitochondrial dehydrogenases by calcium ions. *Biochim. Biophys. Acta.* **2009**, *1787*, 1309–1316. [CrossRef]
53. Gilbert, M.; Jung, S.R.; Reed, B.J.; Sweet, I.R. Islet oxygen consumption and insulin secretion tightly coupled to calcium derived from L-type calcium channels but not from the endoplasmic reticulum. *J. Biol. Chem.* **2008**, *283*, 24334–24342. [CrossRef]
54. Schuit, F.; De Vos, A.; Farfari, S.; Moens, K.; Pipeleers, D.; Brun, T.; Prentki, M. Metabolic Fate of Glucose in Purified Islet Cells. *J. Biol. Chem.* **1997**, *272*, 18572–18579. [CrossRef] [PubMed]
55. Tamarit-Rodriguez, J.; Idahl, L.A.; Giné, E.; Alcazar, O.; Sehlin, J. Lactate production in pancreatic islets. *Diabetes* **1998**, *47*, 1219–1223. [CrossRef] [PubMed]
56. Sekine, N.; Cirulli, V.; Regazzi, R.; Brown, L.J.; Gine, E.; Tamarit-Rodriguez, J.; Girotti, M.; Marie, S.; MacDonald, M.J.; Wollheim, C.B.; et al. Low lactate dehydrogenase and high mitochondrial glycerol phosphate dehydrogenase in pancreatic beta-cells. Potential role in nutrient sensing. *J. Biol. Chem.* **1994**, *269*, 4895–4902. [CrossRef]
57. Jitrapakdee, S.; Wutthisathapornchai, A.; Wallace, J.C.; Macdonald, M.J. Regulation of insulin secretion: Role of mitochondrial signalling. *Diabetologia* **2010**, *53*, 1019–1032. [CrossRef]
58. Fex, M.; Nicholas, L.M.; Vishnu, N.; Medina, A.; Sharoyko, V.V.; Nicholls, D.G.; Spégel, P.; Mulder, H. The pathogenetic role of β-cell mitochondria in type 2 diabetes. *J. Endocrinol.* **2018**, *236*, R145–R159. [CrossRef]
59. Dean, P.M.; Matthews, E.K. Electrical activity in pancreatic islet cells. *Nature* **1968**, *219*, 389–390. [CrossRef]
60. Meissner, H.P. Electrical characteristics of the beta-cells in pancreatic islets. *J. Physiol.* **1976**, *72*, 757–767.
61. Henquin, J.C. D-Glucose inhibits potassium efflux from pancreatic islet cells. *Nat. Cell Biol.* **1978**, *271*, 271–273. [CrossRef]
62. Ashcroft, S.J.H.; Weerasinghe, L.C.C.; Randle, P.J. Interrelationship of islet metabolism, adenosine triphosphate content and insulin release. *Biochem. J.* **1973**, *132*, 223–231. [CrossRef]
63. Erecińska, M.; Bryła, J.; Michalik, M.; Meglasson, M.D.; Nelson, D. Energy metabolism in islets of Langerhans. *Biochim. Biophys. Acta* **1992**, *1101*, 273–295. [CrossRef]
64. Detimary, P.; Jonas, J.C.; Henquin, J.C. Possible links between glucose-induced changes in the energy state of pancreatic B cells and insulin release. Unmasking by decreasing a stable pool of adenine nucleotides in mouse islets. *J. Clin. Investig.* **1995**, *96*, 1738–1745. [CrossRef]
65. Cook, D.L.; Hales, C.N. Intracellular ATP directly blocks K+ channels in pancreatic B-cells. *Nature* **1984**, *311*, 271–273. [CrossRef]
66. Ashcroft, F.M.; Rorsman, P. Electrophysiology of the pancreatic beta-cell. *Prog. Biophys. Mol. Biol.* **1989**, *54*, 87–143. [CrossRef]
67. Inagaki, N.; Gonoi, T.; Clement, J.P.; Namba, N.; Inazawa, J.; Gonzalez, G.; Aguilar-Bryan, L.; Seino, S.; Bryan, J. Reconstitution of IKATP: An inward rectifier subunit plus the sulfonylurea receptor. *Science* **1995**, *270*, 1166–1170. [CrossRef]
68. Zünkler, B.; Lins, S.; Ohno-Shosaku, T.; Trube, G.; Panten, U. Cytosolic ADP enhances the sensitivity to tolbutamide of ATP-dependent K+channels from pancreatic B-cells. *FEBS Lett.* **1988**, *239*, 241–244. [CrossRef]

69. Nichols, C.G. KATP channels as molecular sensors of cellular metabolism. *Nat. Cell Biol.* **2006**, *440*, 470–476. [CrossRef]
70. Plecitá-Hlavatá, L.; Jaburek, M.; Holendova, B.; Tauber, J.; Pavluch, V.; Berková, Z.; Cahová, M.; Schröder, K.; Brandes, R.P.; Siemen, D.; et al. Glucose-Stimulated Insulin Secretion Fundamentally Requires H2O2 Signaling by NADPH Oxidase 4. *Diabetes* **2020**, *69*, 1341–1354. [CrossRef]
71. Bränström, R.; Aspinwall, C.A.; Välimäki, S.; Ostensson, C.G.; Tibell, A.; Eckhard, M.; Brandhorst, H.; Corkey, B.E.; Berggren, P.O.; Lars-son, O. Long-chain CoA esters activate human pancreatic beta-cell KATP channels: Potential role in Type 2 diabetes. *Diabetologia* **2004**, *47*, 277–283. [CrossRef] [PubMed]
72. Rorsman, P.; Ashcroft, F.M. Pancreatic β-Cell Electrical Activity and Insulin Secretion: Of Mice and Men. *Physiol. Rev.* **2018**, *98*, 117–214. [CrossRef] [PubMed]
73. Panten, U.; Schwanstecher, M.; Wallasch, A.; Lenzen, S. Glucose both inhibits and stimulates insulin secretion from isolated pan-creatic islets exposed to maximally effective concentrations of sulfonylureas. *Naunyn Schmiedebergs Arch. Pharmacol.* **1988**, *338*, 459–462. [CrossRef] [PubMed]
74. Gembal, M.; Gilon, P.; Henquin, J.C. Evidence that glucose can control insulin release independently from its action on ATP-sensitive K+ channels in mouse B cells. *J. Clin. Investig.* **1992**, *89*, 1288–1295. [CrossRef]
75. Eliasson, L.; Renström, E.; Ammälä, C.; Berggren, P.O.; Bertorello, A.M.; Bokvist, K.; Chibalin, A.; Deeney, J.T.; Flatt, P.R.; Gäbel, J.; et al. PKC-dependent stimulation of exocytosis by sulfonylureas in pancreatic beta cells. *Science* **1996**, *271*, 813–815. [CrossRef]
76. Geng, X.; Li, L.; Bottino, R.; Balamurugan, A.N.; Bertera, S.; Densmore, E.; Su, A.; Chang, Y.; Trucco, M.; Drain, P. Antidiabetic sulfonyl-urea stimulates insulin secretion independently of plasma membrane KATP channels. *Am. J. Physiol. Endocrinol. Metab.* **2007**, *293*, E293–E301. [CrossRef]
77. Grimmsmann, T.; Rustenbeck, I. Direct effects of diazoxide on mitochondria in pancreatic B-cells and on isolated liver mitochondria. *Br. J. Pharmacol.* **1998**, *123*, 781–788. [CrossRef]
78. Willenborg, M.; Hatlapatka, K.; Ghaly, H.; Belz, M.; Panten, U.; Rustenbeck, I. Studies of first phase insulin secretion using imposed plasma membrane depolarization. *Front. Biosci.* **2011**, *3*, 662–679.
79. Willenborg, M.; Belz, M.; Schumacher, K.; Paufler, A.; Hatlapatka, K.; Rustenbeck, I. Ca^{2+}-dependent desensitization of insulin secretion by strong potassium depolarization. *Am. J. Physiol. Metab.* **2012**, *303*, E223–E233. [CrossRef]
80. Komatsu, M.; Sato, Y.; Aizawa, T.; Hashizume, K. KATP channel-independent glucose action: An elusive pathway in stimu-lus-secretion coupling of pancreatic beta-cell. *Endocr. J.* **2001**, *48*, 275–288. [CrossRef]
81. Henquin, J.-C. Triggering and amplifying pathways of regulation of insulin secretion by glucose. *Diabetes* **2000**, *49*, 1751–1760. [CrossRef]
82. Nicholls, D.G. The Pancreatic β-Cell: A Bioenergetic Perspective. *Physiol. Rev.* **2016**, *96*, 1385–1447. [CrossRef]
83. Grodsky, G.M. A threshold distribution hypothesis for packet storage of insulin and its mathematical modeling. *J. Clin. Investig.* **1972**, *51*, 2047–2059. [CrossRef]
84. Eliasson, L.; Renström, E.; Ding, W.-G.; Proks, P.; Rorsman, P. Rapid ATP-Dependent Priming of Secretory Granules Precedes Ca2+-Induced Exocytosis in Mouse Pancreatic B-Cells. *J. Physiol.* **1997**, *503*, 399–412. [CrossRef]
85. Barg, S.; Eliasson, L.; Renström, E.; Rorsman, P. A subset of 50 secretory granules in close contact with L-type Ca2+ channels ac-counts for first-phase insulin secretion in mouse beta-cells. *Diabetes* **2002**, *51* (Suppl. 1), S74–S82. [CrossRef]
86. Rorsman, P.; Renström, E. Insulin granule dynamics in pancreatic beta cells. *Diabetologia* **2003**, *46*, 1029–1045. [CrossRef]
87. Daniel, S.; Noda, M.; Straub, S.G.; Sharp, G.W. Identification of the docked granule pool responsible for the first phase of glu-cose-stimulated insulin secretion. *Diabetes* **1999**, *48*, 1686–1690. [CrossRef]
88. Straub, S.G.; Shanmugam, G.; Sharp, G.W. Stimulation of insulin release by glucose is associated with an increase in the number of docked granules in the beta-cells of rat pancreatic islets. *Diabetes* **2004**, *53*, 3179–3183. [CrossRef]
89. Grodsky, G.M. A new phase of insulin secretion. How will it contribute to our understanding of beta-cell function? *Diabetes* **1989**, *38*, 673–678. [CrossRef]
90. Nesher, R.; Cerasi, E. Modeling Phasic Insulin Release: Immediate and Time-Dependent Effects of Glucose. *Diabetes* **2002**, *51*, S53–S59. [CrossRef]
91. Nesher, R.; Cerasi, E. Biphasic Insulin Release as the Expression of Combined Inhibitory and Potentiating Effects of Glucose*. *Endocrinology* **1987**, *121*, 1017–1024. [CrossRef]
92. Grill, V.; Adamson, U.; Cerasi, E. Immediate and Time-Dependent Effects of Glucose on Insulin Release from Rat Pancreatic Tissue. *J. Clin. Investig.* **1978**, *61*, 1034–1043. [CrossRef]
93. O'Connor, M.D.; Landahl, H.; Grodsky, G.M. Comparison of storage- and signal-limited models of pancreatic insulin secretion. *Am. J. Physiol.* **1980**, *238*, R378–R389. [CrossRef]
94. O'Connor, M.D.L.; Landahl, H.D.; Grodsky, G.M. Role of Rate of Change of Glucose Concentration as a Signal for Insulin Release. *Endocrinology* **1977**, *101*, 85–88. [CrossRef]
95. Morsi, M.; Schulze, T.; Früh, E.; Brüning, D.; Panten, U.; Rustenbeck, I. Fresh and cultured mouse islets differ in their response to nutrient stimulation. *Endocr. Connect.* **2020**, *9*, 769–782. [CrossRef]
96. Straub, S.G.; Sharp, G.W.G. Hypothesis: One rate-limiting step controls the magnitude of both phases of glucose-stimulated insulin secretion. *Am. J. Physiol. Physiol.* **2004**, *287*, C565–C571. [CrossRef]
97. Kennedy, E.D.; Maechler, P.; Wollheim, C.B. Effects of depletion of mitochondrial DNA in metabolism secretion coupling in INS-1 cells. *Diabetes* **1998**, *47*, 374–380. [CrossRef]

98. Rustenbeck, I.; Herrmann, C.; Grimmsmann, T. Energetic requirement of insulin secretion distal to calcium influx. *Diabetes* **1997**, *46*, 1305–1311. [CrossRef]
99. Tsuruzoe, K.; Araki, E.; Furukawa, N.; Shirotani, T.; Matsumoto, K.; Kaneko, K.; Motoshima, H.; Yoshizato, K.; Shirakami, A.; Kishikawa, H.; et al. Creation and characterization of a mitochondrial DNA-depleted pancreatic beta-cell line: Impaired insulin secretion induced by glucose, leucine, and sulfonylureas. *Diabetes* **1998**, *47*, 621–631. [CrossRef]
100. Dolenšek, J.; Stožer, A.; Skelin Klemen, M.; Miller, E.W.; Slak Rupnik, M. The relationship between membrane potential and calci-um dynamics in glucose-stimulated beta cell syncytium in acute mouse pancreas tissue slices. *PLoS ONE* **2013**, *8*, e82374. [CrossRef]
101. Schulze, T.; Mattern, K.; Erfle, P.; Brüning, D.; Scherneck, S.; Dietzel, A.; Rustenbeck, I. A Parallel Perifusion Slide from Glass for the Functional and Morphological Analysis of Pancreatic Islets. *Front. Bioeng. Biotechnol.* **2021**, *9*, 615639. [CrossRef] [PubMed]
102. Hatlapatka, K.; Willenborg, M.; Rustenbeck, I. Plasma membrane depolarization as a determinant of the first phase of insulin secretion. *Am. J. Physiol. Metab.* **2009**, *297*, E315–E322. [CrossRef] [PubMed]
103. Belz, M.; Willenborg, M.; Görgler, N.; Hamada, A.; Schumacher, K.; Rustenbeck, I. Insulinotropic effect of high potassium concen-tration beyond plasma membrane depolarization. *Am. J. Physiol. Endocrinol. Metab.* **2014**, *306*, E697–E706. [CrossRef] [PubMed]
104. Henquin, J.C.; Nenquin, M.; Stiernet, P.; Ahren, B. In vivo and in vitro glucose-induced biphasic insulin secretion in the mouse: Pattern and role of cytoplasmic Ca2+ and amplification signals in beta-cells. *Diabetes* **2006**, *55*, 441–451. [CrossRef]
105. Müller, A.; Mziaut, H.; Neukam, M.; Knoch, K.-P.; Solimena, M. A 4D view on insulin secretory granule turnover in the β-cell. *Diabetes Obes. Metab.* **2017**, *19*, 107–114. [CrossRef]
106. Brüning, D.; Reckers, K.; Drain, P.; Rustenbeck, I. Glucose but not KCl diminishes submembrane granule turnover in mouse be-ta-cells. *J. Mol. Endocrinol.* **2017**, *59*, 311–324. [CrossRef]
107. Gandasi, N.; Yin, P.; Omar-Hmeadi, M.; Laakso, E.O.; Vikman, P.; Barg, S. Glucose-Dependent Granule Docking Limits Insulin Secretion and Is Decreased in Human Type 2 Diabetes. *Cell Metab.* **2018**, *27*, 470–478.e4. [CrossRef]
108. Farfari, S.; Schulz, V.; Corkey, B.; Prentki, M. Glucose-regulated anaplerosis and cataplerosis in pancreatic beta-cells: Possible im-plication of a pyruvate/citrate shuttle in insulin secretion. *Diabetes* **2000**, *49*, 718–726. [CrossRef]
109. Khan, A.; Ling, Z.C.; Landau, B.R. Quantifying the Carboxylation of Pyruvate in Pancreatic Islets. *J. Biol. Chem.* **1996**, *271*, 2539–2542. [CrossRef]
110. Maechler, P.; Wollheim, C.B. Mitochondrial glutamate acts as a messenger in glucose-induced insulin exocytosis. *Nat. Cell Biol.* **1999**, *402*, 685–689. [CrossRef]
111. Rabaglia, M.E.; Gray-Keller, M.P.; Frey, B.L.; Shortreed, M.R.; Smith, L.M.; Attie, A.D. Alpha-Ketoisocaproate-induced hy-persecretion of insulin by islets from diabetes susceptible mice. *Am. J. Physiol. Endocrinol. Metab.* **2005**, *289*, E218–E224. [CrossRef]
112. MacDonald, M.J.; Smith, A.D.; Hasan, N.M.; Sabat, G.; Fahien, L.A. Feasibility of pathways for transfer of acyl groups from mi-tochondria to the cytosol to form short chain acyl-CoAs in the pancreatic beta cell. *J. Biol. Chem.* **2007**, *282*, 30596–30606. [CrossRef]
113. Macdonald, M.J. Feasibility of a mitochondrial pyruvate malate shuttle in pancreatic islets. Further implication of cytosolic NADPH in insulin secretion. *J. Biol. Chem.* **1995**, *270*, 20051–20058. [CrossRef]
114. Jensen, M.V.; Joseph, J.W.; Ronnebaum, S.M.; Burgess, S.C.; Sherry, A.D.; Newgard, C.B. Metabolic cycling in control of glu-cose-stimulated insulin secretion. *Am. J. Physiol. Endocrinol. Metab.* **2008**, *295*, E1287–E1297. [CrossRef]
115. Vergari, E.; Plummer, G.; Dai, X.; Macdonald, P.E. DeSUMOylation Controls Insulin Exocytosis in Response to Metabolic Signals. *Biomolecules* **2012**, *2*, 269–281. [CrossRef] [PubMed]
116. Ferdaoussi, M.; Dai, X.; Jensen, M.V.; Wang, R.; Peterson, B.S.; Huang, C.; Ilkayeva, O.; Smith, N.; Miller, N.; Hajmrle, C.; et al. Iso-citrate-to-SENP1 signaling amplifies insulin secretion and rescues dysfunctional β cells. *J. Clin. Investig.* **2015**, *125*, 3847–3860. [CrossRef]
117. Panten, U.; Rustenbeck, I. Fuel-induced amplification of insulin secretion in mouse pancreatic islets exposed to a high sulfonylurea concentration: Role of the NADPH/NADP+ ratio. *Diabetologia* **2007**, *51*, 101–109. [CrossRef]
118. Prentki, M.; Corkey, B.E.; Madiraju, S.R.M. Lipid-associated metabolic signalling networks in pancreatic beta cell function. *Diabetologia* **2019**, *63*, 10–20. [CrossRef]
119. Urban, K.A.; Panten, U. Selective loss of glucose-induced amplification of insulin secretion in mouse pancreatic islets pretreated with sulfonylurea in the absence of fuels. *Diabetologia* **2005**, *48*, 2563–2566. [CrossRef] [PubMed]
120. Panten, U.; Früh, E.; Reckers, K.; Rustenbeck, I. Acute metabolic amplification of insulin secretion in mouse islets: Role of cytosolic acetyl-CoA. *Metabolism* **2016**, *65*, 1225–1229. [CrossRef] [PubMed]
121. Panten, U.; Willenborg, M.; Schumacher, K.; Hamada, A.; Ghaly, H.; Rustenbeck, I. Acute metabolic amplification of insulin secretion in mouse islets is mediated by mitochondrial export of metabolites, but not by mitochondrial energy generation. *Metabolism* **2013**, *62*, 1375–1386. [CrossRef]
122. MacDonald, M.J.; Hasan, N.M.; Longacre, M.J. Studies with leucine, β-hydroxybutyrate and ATP citrate lyase-deficient beta cells support the acetoacetate pathway of insulin secretion. *Biochim. Biophys. Acta (BBA) Gen. Subj.* **2008**, *1780*, 966–972. [CrossRef]
123. Tengholm, A.; Gylfe, E. cAMP signalling in insulin and glucagon secretion. *Diabetes Obes. Metab.* **2017**, *19*, 42–53. [CrossRef]

124. Rabinovitch, A.; Grill, V.; Renold, A.E.; Cerasi, E. Insulin release and cyclic AMP accumulation in response to glucose in pancre-atic islets of fed and starved rats. *J. Clin. Investig.* **1976**, *58*, 1209–1216. [CrossRef]
125. Dyachok, O.; Isakov, Y.; Sågetorp, J.; Tengholm, A. Oscillations of cyclic AMP in hormone-stimulated insulin-secreting beta-cells. *Nature* **2006**, *439*, 349–352. [CrossRef]
126. Henquin, J.C.; Nenquin, M.; Ravier, M.A.; Szollosi, A. Shortcomings of current models of glucose-induced insulin secretion. *Diabetes Obes. Metab.* **2009**, *11*, 168–179. [CrossRef]
127. Kalwat, M.A.; Cobb, M.H. Mechanisms of the amplifying pathway of insulin secretion in the β cell. *Pharmacol. Ther.* **2017**, *179*, 17–30. [CrossRef]
128. Choudhary, C.; Weinert, B.T.; Nishida, Y.; Verdin, E.; Mann, M. The growing landscape of lysine acetylation links metabolism and cell signalling. *Nat. Rev. Mol. Cell Biol.* **2014**, *15*, 536–550. [CrossRef]
129. Grespan, E.; Giorgino, T.; Arslanian, S.; Natali, A.; Ferrannini, E.; Mari, A. Defective Amplifying Pathway of β-Cell Secretory Re-sponse to Glucose in Type 2 Diabetes: Integrated Modeling of In Vitro and In Vivo Evidence. *Diabetes* **2018**, *67*, 496–506. [CrossRef]

metabolites

Review

Understanding the Long-Lasting Effects of Fetal Nutrient Restriction versus Exposure to an Obesogenic Diet on Islet-Cell Mass and Function

Stephanie E. O'Hara, **Kelly M. Gembus** and **Lisa M. Nicholas** *

The Robinson Research Institute, Adelaide Medical School, University of Adelaide, Adelaide, SA 5005, Australia; stephanie.ohara@adelaide.edu.au (S.E.O.); kelly.gembus@adelaide.edu.au (K.M.G.)
* Correspondence: lisa.nicholas@adelaide.edu.au

Abstract: Early life represents a window of phenotypic plasticity. Thus, exposure of the developing fetus to a compromised nutritional environment can have long term consequences for their health. Indeed, undernutrition or maternal intake of an obesogenic diet during pregnancy leads to a heightened risk of type 2 diabetes (T2D) and obesity in her offspring in adult life. Given that abnormalities in beta-cell function are crucial in delineating the risk of T2D, studies have investigated the impact of these exposures on islet morphology and beta-cell function in the offspring in a bid to understand why they are more at risk of T2D. Interestingly, despite the contrasting maternal metabolic phenotype and, therefore, intrauterine environment associated with undernutrition versus high-fat feeding, there are a number of similarities in the genes/biological pathways that are disrupted in offspring islets leading to changes in function. Looking to the future, it will be important to define the exact mechanisms involved in mediating changes in the gene expression landscape in islet cells to determine whether the road to T2D development is the same or different in those exposed to different ends of the nutritional spectrum.

Keywords: pancreatic beta-cells; intrauterine growth restriction; maternal obesity; developmental programming

Citation: O'Hara, S.E.; Gembus, K.M.; Nicholas, L.M. Understanding the Long-Lasting Effects of Fetal Nutrient Restriction versus Exposure to an Obesogenic Diet on Islet-Cell Mass and Function. *Metabolites* **2021**, *11*, 514. https://doi.org/10.3390/metabo11080514

Academic Editors: Melkam Kebede and Belinda Yau

Received: 13 July 2021
Accepted: 29 July 2021
Published: 4 August 2021

1. Introduction

Before the 'developmental origins of adult health and disease' paradigm emerged, it was widely accepted that type 2 diabetes (T2D) was wholly genetically determined [1]. However, in 1991, upon linking birth weight records to glucose tolerance in men, a relationship was observed between lower weight at birth and a higher risk of T2D in adult life; men with the lowest weight at birth were six times more likely to develop T2D when compared to the heaviest [2]. This association has since been replicated in a number of epidemiological studies [3], is present in both men and women [4] and persists even after adjusting for ethnicity, childhood socioeconomic status and adult lifestyle factors [5].

In 1992, the 'thrifty phenotype' hypothesis was put forward by Hales and Barker to explain the association between poor fetal growth and increased T2D risk in adult life. This hypothesis posits that when faced with poor nutrition, the fetus adapts to their environment by selectively protecting brain growth at the detriment of other tissues and organs. This includes impaired development of the endocrine pancreas, leading to abnormal pancreatic beta-cell mass/function and reduced capacity for insulin secretion, which persists into adult life [1]. Indeed, offspring who were exposed to periods of famine whilst in utero are at a higher risk of T2D in adult life (reviewed in [6]).

In 1998, Hattersley and Tooke put forth the 'fetal insulin' hypothesis as an alternative explanation for the association between low birth weight and risk of T2D. This hypothesis proposes that both phenotypes are the result of a genetic predisposition for reduced insulin secretion and action leading to decreased growth in fetal life and a decreased capacity to secrete and respond to insulin and, therefore, higher T2D risk [7,8]. Support for the 'fetal insulin' hypothesis emerged from genome-wide association studies showing that a number of T2D risk loci are associated with lower birthweight. Furthermore, the strongest associations are at loci that primarily affect beta-cell function (reviewed in [8]). However, it has since been argued that genetic susceptibility does not solely account for the relationship between low birth weight and T2D; in a study of monozygotic twins discordant for T2D, the diabetic twin had a lower birth weight compared to their genetically identical non-diabetic co-twin [9].

Over the last two decades, the increasing number of studies reporting incidence of high birth weight (especially given the higher prevalence of maternal obesity/gestational diabetes in more contemporary cohorts) have provided additional clarity regarding the relationship between low versus birth weight and T2D risk; systematic review and meta-analysis have shown that babies born with either low (<2.5 kg) or high (>4.5 kg) birth weight are most at risk for T2D in later life [3,10]. This is particularly worrying given that in many populations, nearly half of women of child-bearing age are either overweight or obese [11]; not only are these women twice more likely to have a baby with high birth weight [12], adult children of overweight and obese mothers also have a 1.4- and 3.5-fold higher incidence of T2D, respectively, compared to children of normal weight mothers [13].

Similar to fetal undernutrition, it remains unresolved whether genetic predisposition [14,15] or the intrauterine/early life environment [13] have a larger part to play in driving the relationship between maternal obesity and offspring T2D risk. However, given that defective pancreatic beta-cells account for most (if not all) forms of diabetes, a large number of studies have been undertaken to elucidate the impact of intrauterine growth restriction (IUGR) as well as sub-optimal maternal nutrition (in the form of undernutrition or exposure to foods that are high in fat and sugar with low nutritional value) on islet morphology and beta-cell function in the offspring.

In this review, we consolidate the evidence across various small and large animal models of restricted fetal growth and maternal caloric excess and discuss both the differences and similarities in islet-specific phenotypes that arise in offspring exposed to these contrasting maternal environments. We also evaluate the proposed mechanisms driving beta-cell dysfunction to determine whether these early life insults result in specific vulnerabilities within beta-cells, leading to an increased risk of T2D. Finally, we share some ideas that we believe are important areas of future research in the field.

2. An Overview of Animal Models of Restricted Fetal Growth and Exposure to Maternal Caloric Excess

Animal models have been indispensable for understanding the potential causative mechanisms underlying the relationship between a compromised early life environment and future risk of T2D. The use of genetically identical rodent strains has enabled researchers to dissect the relative contributions of specific components of the maternal (and hence intrauterine) milieu on offspring phenotypes. Furthermore, given that it is not possible to obtain longitudinal samples from human offspring and that the nutritional history of human islet donors are unknown, mechanistic understanding of altered islet function and mass in affected offspring have solely relied on animal models.

2.1. Animal Models of Restricted Fetal Growth

Inadequate nutrition during fetal development, usually the result of maternal undernutrition and placental insufficiency, leads to IUGR and low birth weight. Given that amino acid supply is disrupted in growth restricted babies [16,17] and that amino acids are critically important for insulin production, Hales and Barker postulated that defects in the development of fetal pancreatic beta-cells are due to maternal amino acid deficiency [1]. Thus, a number of studies have investigated the impact of IUGR on offspring islets by feeding rodents an isocaloric diet with approximately 60% less protein content compared to controls [18–26], which leads to ~10% reduction in offspring birth weight followed by a period of catch-up growth [21,27–29]. This has been a core model used in the field.

It should be noted, however, that not all models of maternal protein restriction result in an IUGR phenotype in the offspring [20] for reasons that remain unknown. Another well-characterized rodent model of IUGR that has been used to study its effect on islet development is uterine artery ligation. This model induces uteroplacental insufficiency, therefore, limiting the supply of nutrients and oxygen to the fetus leading to a 10–20% decrease in body weight [30–37].

A similar effect has also been achieved in a sheep model of IUGR by surgically removing most of the sites of placentation prior to pregnancy [38,39]. In contrast, a much more severe effect on offspring birth weight (40% reduction) is observed when pregnant sheep are exposed to elevated ambient temperatures in mid to late gestation [40–44]. Despite differences in the method used to induce IUGR and the timing and duration of insult, impaired glucose tolerance and T2D is observed in the offspring across most models, which is in line with findings in human cohorts.

2.2. *Animal Models of Maternal Caloric Excess*

Given the rising rate of overweight and obesity amongst women of child-bearing age, a number of animal models have been developed involving excess maternal caloric intake during pregnancy and lactation. Japanese Macaques fed a high-fat diet has been used as a model that most closely mimics maternal obesity in humans [45–48] given that dams typically give birth to a single offspring at a time and macaque colonies are genetically heterogeneous. Islets from the Japanese Macaque are also similar to humans in terms of structure, insulin secretion and the expression of endocrine hormones and key transcription factors [49]. Another large animal that has been used to model maternal obesity is sheep. In addition to having only one to three offspring per pregnancy, islet development in sheep is also similar to humans. However, rather than feeding ewes a diet high in fat, obesity is achieved using the same diet as controls but ewes are fed at 150% of their energy requirements [50,51].

Most rodent models of maternal caloric excess involve consumption of a diet that is either high in fat or high in both fat and sugar (with the latter being more reflective of an obesogenic Western diet in humans). Fat content varies across studies ranging from 25 to 50% kcal and is mainly composed of saturated fats. In general, these models result in a maternal phenotype of obesity and impaired glucose tolerance during pregnancy in studies where mice are fed the obesogenic diet for 6–10 weeks before mating (reviewed in [52]). To elucidate the specific maternal characteristics associated with an obese pregnancy in mediating altered islet phenotype in the offspring, one study used the agouti viable yellow (Avy) mouse to model maternal obesity without impaired glucose tolerance [53]. The agouti signaling molecule antagonizes satiety signaling. Thus, Avy/a mice develop obesity when provided ad libitum access to standard chow due to hyperphagia. In contrast, mice carrying a silent allele are lean and metabolically healthy [54–56].

In contrast to animal models of restricted fetal growth where metabolic dysfunction is observed in the offspring across most models, the differences in diet composition as well as the duration and timing of maternal over-feeding has led to metabolic phenotypes that are not always entirely consistent between studies (reviewed in [52]). Despite this, it is clear from the numerous studies published across a number of different species that exposure to an obesogenic intrauterine environment predominantly results in metabolic dysfunction including obesity and impaired glucose tolerance in the offspring in adult life (reviewed in [57,58]).

3. The Impact of a Compromised Early Life Environment on Beta-Cell Morphology and Function

3.1. *The Impact of Restricted Nutrition versus Caloric Excess on Islet Morphology*

Compromised nutrition during development, irrespective of whether in the form of maternal nutrient restriction (low-protein diet) or exposure to an obesogenic environment, results in reduced beta-cell mass at birth in rodents [18,59,60] (Figure 1). This is despite a divergent birth weight phenotype between the models; a maternal low-protein diet results in reduced birth weight whereas exposure to maternal obesity has no impact on offspring birth weight [59]. In the case of offspring exposed to a maternal low-protein diet, diminished beta-cell mass is due to reduced proliferation and increased apoptosis of beta-cells in fetal life associated with reduced levels of growth factors [60–63]. Whether the same occurs in offspring exposed to maternal obesity has not been investigated.

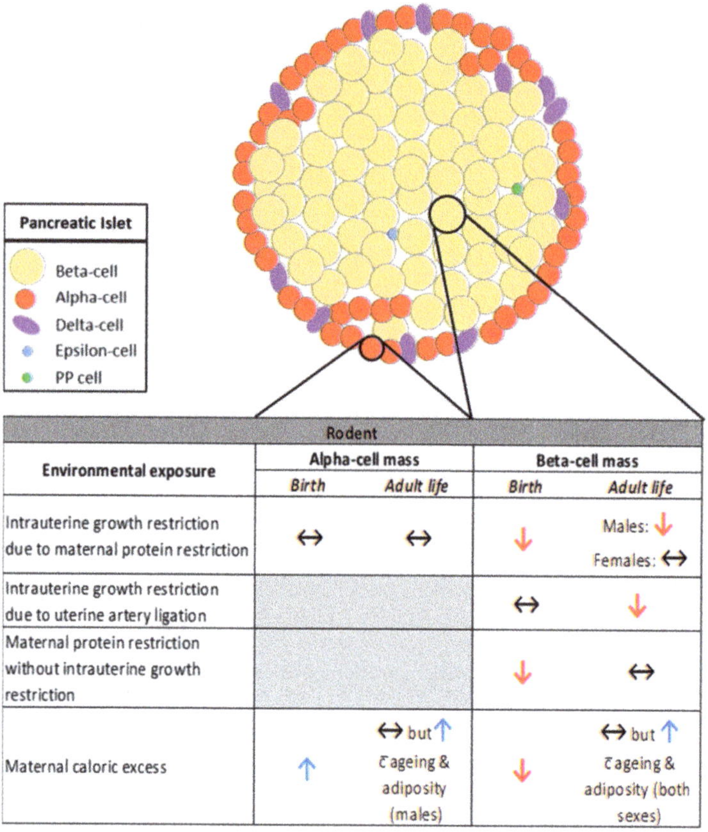

Rodent				
Environmental exposure	**Alpha-cell mass**		**Beta-cell mass**	
	Birth	*Adult life*	*Birth*	*Adult life*
Intrauterine growth restriction due to maternal protein restriction	↔	↔	↓	Males: ↓ Females: ↔
Intrauterine growth restriction due to uterine artery ligation			↔	↓
Maternal protein restriction without intrauterine growth restriction			↓	↔
Maternal caloric excess	↑	↔ but ↑ c̄ ageing & adiposity (males)	↓	↔ but ↑ c̄ ageing & adiposity (both sexes)

Sheep				
Environmental exposure	**Alpha-cell mass**		**Beta-cell mass**	
	Fetal life	*Neonates*	*Fetal life*	*Neonates*
Intrauterine growth restriction	↓	↔	↓	↔
Maternal caloric excess			↓	

Non-human primate				
Environmental exposure	**Alpha-cell mass**		**Beta-cell mass**	
	Fetal life	*Early childhood*	*Fetal life*	*Early childhood*
Maternal caloric excess	↓	↓	↔	↔

Figure 1. The impact of exposure to restricted nutrition versus caloric excess in rodent, sheep and non-human primate on offspring alpha- and beta-cell mass. *Red arrow: Decreased; Blue arrow: Increased; Black arrow: No change; Gray shading indicates data is not available.*

In adult life, beta-cell mass remains reduced in the offspring of protein-restricted dams [18,63]. Importantly, the relationship between lower birth weight (<3 kg versus >3 kg) and reduced beta-cell area in adult life is also observed in humans [64]. In the case of maternal obesity, offspring beta-cells have the ability, to a certain extent, to compensate for changes in body weight and adiposity; beta-cell mass is unchanged in lean, metabolically healthy male and female adult offspring of obese dams [65] but increases with increasing age and adiposity in both sexes [59,66,67] (Figure 1).

In contrast to a maternal low-protein diet, IUGR in rodents induced by uterine artery ligation (which leads to an acute decrease in fetal nutrients and oxygen near term) does not impact on offspring beta-cell mass at birth despite a ~15% reduction in birth weight [68]. There is, however, an eventual decline in beta-cell mass from seven weeks of age into adult life that occurs before the onset of hyperglycemia and is not associated with increased beta-cell apoptosis [68]. This suggests that irrespective of whether fetal exposure to restricted nutrition is chronic or acute, beta-cells are particularly sensitive to IUGR and, importantly, these changes are permanent; there is no recovery of beta-cell mass in offspring despite the strong replicative potential of newborn beta-cells and the normalization of body weight [18,63]. Findings from Theys and colleagues suggest that perhaps male offspring are particularly susceptible; beta-cell mass was unaffected in adult female offspring exposed to maternal protein restriction [21]. Interestingly, when maternal protein restriction is not accompanied by IUGR, whilst the fraction of beta-cells is still reduced in the newborn this is recovered by the time both male and female offspring reach adulthood [20]. Taken together, these findings highlight the complexities involved in understanding how compromised nutrient availability during discrete windows of development can lead to longer term changes in beta-cell mass.

Similar to what is observed in rodents, beta-cell mass is reduced in fetal sheep in late (90% of) gestation in response to both IUGR [44] and maternal obesity [51] with the former due to decreased proliferation and the latter increased apoptosis. In contrast to rodents, however, beta-cell mass is normalized in growth restricted lambs by two weeks of age [69] (Figure 1) suggesting that sheep beta-cells have increased plasticity despite nutrient restriction in utero. While the longer term impact of exposure to maternal obesity on beta-cell mass in sheep offspring has not been investigated, acute insulin response to glucose is decreased in these offspring during a glucose tolerance test in adult life [70].

In a non-human primate model, beta-cell mass is comparable between offspring exposed to either a maternal Western-style or control diet during pregnancy and lactation. This was observed in both fetal life [46] and early childhood (three years of age) [45] (Figure 1). We speculate that one of the reasons beta-cell mass is not impacted in these young offspring is that maternal metabolic dysfunction in response to an obesogenic diet is less pronounced in Japanese Macaques compared to rodents. This is the case even for those with a higher sensitivity to weight gain and insulin resistance in response to a high-fat diet [71].

In contrast to beta-cell mass, alpha-cell mass is unchanged in growth-restricted rodent offspring in late gestation [63] and at birth [72]. There is, however, a decrease in alpha-cell mass one week after birth but this is normalized by two weeks of age and maintained into adult life [72] (Figure 1). Alpha-cell mass is differentially impacted by IUGR in sheep offspring; it is reduced in late gestation due to decreased pancreas weight (the percentage of glucagon-positive area is not different between groups) [44]. Like beta-cell mass, alpha-cell mass is normalized in growth restricted lambs by two weeks of age given that pancreas weight is no longer different between the groups [69] (Figure 1). A recent study in humans also found no association between birthweight and alpha-cell mass in adult life [64].

Maternal high-fat feeding throughout pregnancy leads to increased alpha-cell volume in one-day-old rodent offspring [73]. This increase is not maintained in lean, metabolically healthy adult offspring of obese dams [65] but similar to beta-cells, alpha-cell mass is increased in response to increasing adiposity and age in male [59,74] but not female [67] offspring (Figure 1). Alpha-cell mass in sheep offspring exposed to maternal obesity has not been investigated. The alpha-cell phenotype observed in rodent models of maternal obesity/high-fat feeding are in contrast to findings in Japanese Macaques where a maternal Western-style diet leads to decreased alpha-cell mass in fetal life [46] (despite minimal effect on the overall physiological phenotype of the fetus), which still persists at three years of age in both male and female offspring [45] (Figure 1). Whether these changes impact on offspring T2D risk is unclear given that there is no consensus on how alpha-cell mass is impacted in response to T2D in both rodents and humans [75–77]. It should also be noted that whether the population of other endocrine-cell types in offspring islets is affected by exposure to either restricted nutrition or caloric excess has not been investigated.

3.2. The Impact of Restricted Nutrition versus Caloric Excess on Beta-Cell Function

Insulin producing beta-cells play a central role in the pathogenesis of T2D; the inability of beta-cells to adapt and compensate for increased insulin demand, for example, which develops as a consequence of insulin resistance in ageing or obesity, leads to T2D [78]. In contrast, sustained beta-cell adaptation is capable of preventing T2D, even in the face of severe insulin resistance [79]. In addition to having more beta-cells, plasticity of beta-cell function also contributes to the compensatory increase in insulin output. Moreover, given that impaired beta-cell function is an early feature of T2D pathogenesis [80], it is important to determine whether beta-cell function is impacted in offspring exposed to restricted nutrition versus caloric excess. This will enable us to better understand the etiology of T2D in response to compromised nutrition during development. Unfortunately, in contrast to the number of studies that have determined the impact on beta-cell mass, only a handful of studies have directly investigated beta-cell function in the offspring.

In rodent models, both IUGR (maternal protein restriction, uterine artery ligation) and exposure to maternal diet-induced obesity is associated with impaired glucose-stimulated insulin secretion in islets of adult offspring [60,63,68,81]. This is perhaps not surprising given that in most studies, islet function was assessed in offspring with increased adiposity and impaired glucose tolerance. Furthermore, it may also explain why glucose intolerance is present in offspring exposed to maternal high-fat feeding despite having greater beta-cell mass [82]. Overall, severity of the insulin secretion phenotype varies across models with the most pronounced effect seen in adult offspring that were growth restricted in utero; the insulin secretory response to glucose is virtually absent in these offspring [63,68]. Surprisingly, islets from offspring exposed to maternal protein restriction but who were not growth restricted at birth also displayed impaired glucose-stimulated insulin secretion and reduced islet insulin content [20]. Similar to the observations made in maternal obesity-exposed mice, these offspring have impaired glucose tolerance despite increased beta-cell mass [20]. It should be noted that these studies did not compare effects between male and female offspring.

In Japanese Macaques, although the effects of a maternal high-fat diet on offspring islet morphology are subtle, both first- and second-phase insulin secretion was increased in response to elevated glucose levels when compared to controls. This was found in very young (three years old), metabolically healthy offspring [45], however, it is unknown whether this was present in only one or both sexes.

Only a small number of studies have compared the sex-specific impact of compromised nutrition during development on offspring beta-cell function. These studies found that in response to either IUGR by maternal protein restriction or to maternal diet-induced obesity, islets from female offspring secrete more insulin for a given glucose load compared to male offspring. This occurs in the absence of increased insulin demand i.e., impaired glucose tolerance and increased adiposity in both male and female offspring [21,65,66]. Furthermore, in mice, only adult male offspring exposed to maternal obesity develop T2D (at six months of age) despite increased adiposity in both sexes [83]. Therefore, we speculate that at least in the case of exposure to an obesogenic diet during pregnancy and lactation, beta-cells from female offspring are primed to better cope with increasing insulin demand in adult life. However, there may be a limit to this ability especially in the face of both ageing and obesity [81]. Curiously, when maternal obesity is modelled using Avy mice that are normoglycemic during pregnancy, female offspring are more vulnerable to metabolic disease with age compared to male offspring. Aged females developed glucose intolerance and had reduced glucose-stimulated insulin secretion following nearly nine months of high-fat feeding in postnatal life [53]. This is in contrast to another study, which found that the latent predisposition to metabolic disease in offspring of Avy dams was more prominent in males who developed glucose intolerance and insulin resistance after only three weeks on a high-fat diet [55]. The reason(s) for the contrasting findings between studies is unknown.

Findings are less consistent in relation to the impact of maternal obesity on basal insulin secretion. Furthermore, there is variability in what constitutes 'basal' glucose levels with studies using either lower (1, 2, and 2.8 mM glucose) or slightly higher (5.5 mM glucose) concentrations. We have previously shown that islets from metabolically healthy female offspring of obese dams have a slight but significant increase in insulin secretion compared to controls when incubated in 2.8 mM glucose [65]. In studies performed on offspring in which weight gain, increased adiposity or glucose intolerance is observed, whilst one study found that islets from male offspring exposed to maternal obesity secreted almost double the amount of insulin in response to 1 and 5.5 mM glucose [84], others found no impact on basal insulin secretion (in both males and females) [60,63,81]. Meanwhile, Zambrano and colleagues found that insulin secretion in response to 5.5mM glucose was decreased in both male and female offspring [67]. These inconsistencies in study outcomes suggest that basal insulin secretion may be particularly sensitive to even slight differences (which are likely to be present between studies) in either the pre- or postnatal environment or their interaction.

4. Similarities and Differences in the Mechanisms Identified in Offspring Exposed to Restricted Nutrition versus Caloric Excess

Thus far, studies have taken a targeted approach to identifying mechanisms that are contributing to compromised islet function and mass in offspring exposed to malnutrition, focusing on genes and pathways with known effects on beta-cell function. Importantly, the impact is observed even in very young offspring and persists as they age suggesting that these mechanisms are programmed by the intrauterine environment and may, therefore, be the central driver of T2D risk in affected offspring.

4.1. The impact of Restricted Nutrition versus Caloric Excess on Pdx1 Expression

Pancreatic and duodenal homeobox 1 (Pdx1) is a transcription factor that is critical for the regulation of pancreatic development and beta-cell differentiation. Pdx1 levels are reduced in human islets from T2D donors [85] and compromised Pdx1 expression in beta-cells is tightly linked to hyperglycemia and loss of cellular identity [86]. Consequently, most studies have investigated the impact of either nutrient restriction or maternal high-fat feeding on Pdx1 expression in the offspring. In rodent studies, both lead to compromised expression in the pancreas/islets of offspring (Table 1).

Both uterine artery ligation and exposure to a maternal low-protein diet lead to a permanent suppression of Pdx1 in the offspring [26,33] irrespective of whether restricted nutrition leads to IUGR [20]. Interestingly, in those exposed to uterine artery ligation, beta-cell mass is unchanged at birth despite a 50% decrease in Pdx1 expression in fetal life [33]. Importantly, this permanent suppression of Pdx1 across the life course may offer an explanation for why offspring exposed to restricted nutrition are unable to expand their beta-cell mass in response to a gain in adiposity and insulin resistance in later life. It will be important to determine whether Pdx1 expression is different between male and female offspring exposed to restricted nutrition and whether this mediates sex-specific differences in T2D susceptibility.

Table 1. Similarities and differences in the mechanisms identified in offspring exposed to restricted nutrition versus caloric excess. *Red arrow: Decreased; Blue arrow: Increased; Black arrow: No change. Gray shading indicates data is not available.*

Environmental Exposure	Pdx1 Expression		Mitochondrial Metabolism +	Oxidative Stress →	Functional Outcome	mTOR Abundance/Activity	
	Fetal Life	*Adult Life*	*Adult Life*			*Birth*	*Adult Life*
Intrauterine growth restriction (due to maternal protein restriction or uterine artery ligation)	↓ *(26, 33)*	↓ *(26, 33)*	↓ glucose-stimulated ATP *(87)* ↓ expression of mitochondrial-encoded genes *(87)* Males and females: ↓ glucose-stimulated ATP *(21)* Females only: ↓ expression of Tfam *(21)* ↔ expression of mitochondrial-encoded genes *(21)*	↑ oxidative stress *(87)* Males only: ↑ reactive oxygen species *(21)*	↓ glucose and mitochondrial-fuel stimulated insulin secretion *(87)* Males and females: ↑ glucose-stimulated insulin secretion *(21)*	↔ mTOR protein *(18)*	↓ mTOR protein *(18)*
Maternal protein restriction without intrauterine growth restriction		↓ *(20)*				mTOR ↓ activity *(20)*	mTOR ↓ activity *(20)*
Maternal caloric excess		Males: ↓ *(65,73,81)* Females: ↔ *(65)*	<u>*Exposure to a high-fat, high-sugar diet from before and throughout pregnancy and lactation (64):*</u>				
			Females: ↑ mitochondrial respiration ↑ expression of mitochondrial-encoded genes ↑ mitochondrial density Males: ↓ mitochondrial respiration ↔ expression of mitochondrial-encoded genes ↔ mitochondrial density	Males and females: ↑ reactive oxygen species	Females: ↑ glucose and mitochondrial-fuel stimulated insulin secretion Males: ↔ glucose and mitochondrial-fuel stimulated insulin		
			<u>*Exposure to a high-fat diet from conception and throughout pregnancy and lactation (65):*</u>				
				Males but not females: ↑ oxidative stress ↑ expression of genes involved in superoxide production			
			<u>*Exposure to a high-fat diet from conception and throughout pregnancy and lactation (91):*</u>				
			↓ glucose-stimulated ATP ↓ basal ATP content ↔ mitochondrial DNA content ↔ expression of Tfam ↔ expression of mitochondrial-encoded genes		↓ glucose stimulated insulin secretion		

In the case of exposure to maternal high-fat feeding throughout pregnancy and lactation, Pdx1 expression is reduced in islets of adult male offspring [66,74,82]. This effect also persists in the following (F2) generation [74]. Interestingly, in contrast to nutrient restricted offspring, beta-cell mass is increased in these offspring despite lower levels of Pdx1 [74,82]. Given that these studies investigated Pdx1 in offspring that were already metabolically compromised, it remains unclear whether compromised expression is present in offspring islets before the onset of metabolic dysfunction and is, therefore, programmed by maternal caloric excess or whether it is merely a consequence of declining islet function in these offspring. Whilst evidence for the former is present in the non-human primate model of maternal high-fat feeding; Pdx1 expression is not altered in fetal pancreas at gestational day 130 (early third trimester) [46], whether the same occurs in rodents is yet to be determined.

Thus far, only one study has compared Pdx1 expression in male versus female offspring. The authors found that in mice exposed to a maternal high-fat diet, Pdx1 expression was decreased in males but unchanged in females at four months of age [66]. The presence of this sex-specific difference in expression is likely because only male offspring displayed increased adiposity and were glucose intolerant at this age. In contrast, female offspring remained metabolically healthy [66].

4.2. The Impact of Restricted Nutrition versus Caloric Excess on Mitochondrial Metabolism and Oxidative Stress

Mitochondrial metabolism plays a central role in regulating insulin release from beta-cells. Mitochondria are the main source of ATP (energy) which together with other mitochondrial factors accomplish the coupling of glucose metabolism to insulin secretion [87]. Importantly, it has been shown that islet mitochondrial function in the offspring is sensitive to programming by the intrauterine and early life nutritional milieu (Table 1). Islets from offspring exposed to uterine artery ligation produce less ATP in response to 16.7mM glucose; an effect that is already present at one week of age and declines further with age [88]. A similar pattern is also observed for the activities of complex I and III of the mitochondrial electron transport chain and expression of mitochondrial-encoded genes resulting in decreased glucose- and leucine (mitochondrial fuel)-stimulated insulin secretion [88].

Mitochondrial dysfunction is frequently associated with oxidative stress, a phenomenon caused by an imbalance between the production and clearance of reactive oxygen species (ROS) in cells. In beta-cells, the production of ROS such as superoxide anions and hydrogen peroxide is coupled to glycolytic and respiratory metabolism [89]. These cells, however, are particularly vulnerable to oxidative stress due to their low levels of antioxidant enzymes, which is the first line of defense against excessive ROS levels [90]. Indeed, islets from offspring exposed to uterine artery ligation also show signs of oxidative stress (inferred from the presence of 4-Hydroxynonenal-protein adducts) from one week of age [88]. This is despite having higher levels of the mitochondrial antioxidant enzyme, manganese superoxide dismutase compared to controls [88], suggesting that whilst islets are able to mount an antioxidant defense, the system is likely overwhelmed by ROS.

T2D develops more rapidly in offspring growth restricted by uterine artery ligation compared to those exposed to maternal protein restriction; the former are diabetic at three months of age whereas at this same age, the latter have normal glucose tolerance. Despite this difference, protein-restricted offspring have compromised ATP production in response to glucose; an effect which is present in both males and females [21]. Surprisingly, glucose-stimulated insulin secretion was preserved in islets with female offspring secreting even more insulin than males in response to 16.7mM glucose. Expression of the mitochondrial transcription factor, Tfam, which is essential for mitochondrial DNA transcription and maintenance was decreased in female but not male islets. However, this did not have an overall effect of reducing the expression of mitochondrial-encoded genes [21]. In contrast, islets from male offspring exposed to a maternal low-protein diet produced more ROS in response to both low and high glucose levels [21], highlighting the complexities involved in interpreting the effects of compromised nutrition and growth on the mechanisms regulating islet function.

Exposure to a maternal obesogenic diet also leads to compromised mitochondrial function in offspring islets. However, timing of this nutritional insult and how severely the metabolic health of dams are compromised during pregnancy appears to determine whether or not there are sex-specific differences in susceptibility. We recently investigated mitochondrial function in islets from metabolically healthy adult mouse offspring (eight weeks of age) exposed to maternal obesity from before and throughout pregnancy and lactation. In this model, dams are heavier, fatter and glucose intolerant during pregnancy [91]. Islets from female offspring displayed increased mitochondrial respiration and density as well as increased expression of both mitochondrial- and nuclear-encoded components of the electron transport chain [65]. Not surprisingly, glucose- and mitochondrial fuel-stimulated insulin secretion is also increased in these offspring. Male offspring, however, have compromised mitochondrial respiration characterized by decreased ATP synthesis-driven respiration and increased "uncoupled" respiration but unlike in females, insulin secretion, mitochondrial density and the expression of genes encoding components of the electron transport chain are unaffected [65]. Upon stimulation with glucose, islets from both male and female offspring of obese dams produced more ROS compared to controls, however, only females had increased expression of antioxidant enzymes suggesting that islets from male offspring may be more vulnerable to oxidative stress [65]. A similar phenotype is also observed in the study by Yokomizo et al. where maternal high-fat feeding commences from conception. In this model, dams have increased body weight and higher fasting blood glucose levels during pregnancy. At four months of age, only islets from male offspring have signs of oxidative stress (measured by intensity of 8-hydroxy-2′-deoxyguanosine) and increased expression of Nox4 and gp91phox, which are involved in superoxide production [66]. It should be noted that at this age, male (and not female) offspring of high fat diet-fed dams have increased adiposity and impaired glucose tolerance compared to controls, however, the impact on glucose-stimulated insulin secretion in islets was not investigated.

In contrast, in the model used by Theys and colleagues, rats fed an obesogenic diet from conception have comparable body weight to controls throughout pregnancy except just before delivery when high fat diet-fed dams were lighter than controls [92]. In this study, dams were fed a diet with a much lower fat content compared to the study by Yokomizo et al. (23% versus 62% in [66]). Interestingly, at three months of age, islet function is already severely compromised in these offspring; both male and female offspring showed a loss of response to glucose, i.e., insulin secretion was not different between low and high glucose conditions [92]. In line this with finding, there was also no increase in glucose-stimulated ATP production. Basal ATP content was also reduced in offspring islets. Given that mitochondrial DNA content and the expression of Tfam and mitochondrial-encoded genes are unaffected in these islets, the mechanisms driving lowered ATP production remains to be determined.

4.3. *Maternal Protein Restriction Leads to Reduced mTOR Signaling in Offspring Islets*

Mammalian target of rapamycin (mTOR) is a nutrient-responsive kinase that exists as two functionally and structurally distinct complexes: mTOR complex 1 (mTORC1) and mTOR complex 2 (mTORC2). mTOR signaling plays a key role in integrating hormonal and nutritional stimuli to regulate cellular metabolism, survival and growth. mTOR is, therefore, integral for the maintenance of beta-cell development and function (reviewed in [93]). Given its critical role in nutrient-sensing and beta-cell homeostasis, studies have investigated whether the nutritional milieu experienced during fetal life leads to programming of altered mTOR signaling in the longer term and its effect on islet function. This has been investigated in the context of exposure to maternal protein restriction, which showed a persistent decrease in mTOR protein and/or mTOR activity, irrespective of whether offspring were growth restricted at birth or whether beta-cell mass is altered [18,20] (Table 1). Specifically, maternal protein restriction without IUGR leads to reduced mTORC1 activity (measured by phosphorylation of ribosomal protein S6 at Ser240) at birth when offspring also have a reduced beta-cell/pancreas ratio (beta-cell fraction). This effect persists in the offspring at three months of age despite a normalization of beta-cell mass. Importantly, transient activation of mTORC1 in beta-cells of protein-restricted offspring in the last week of pregnancy is able to rescue glucose intolerance in adult life and the defect in beta-cell fraction at birth by normalizing beta-cell proliferation in these mice [20]. In contrast, although IUGR offspring also have reduced beta-cells at birth, mTOR protein abundance is unchanged until after weaning with levels decreased at both 30 and 130 days of age (along with decreased beta-cell mass). However, whether mTOR activity is also altered in these offspring is unknown [18].

Islet mTOR protein and/or activity is also sensitive to the timing of protein restriction; when this is confined only to the last week of pregnancy, there is no impact on both mTOR protein abundance and phosphorylation of ribosomal protein S6 at Ser240 in the islets of three-month-old adult offspring. Whilst it is not known whether beta-cell function and mass are preserved at this age, glucose-stimulated insulin secretion is impaired and beta-cell mass increased in male offspring at 12 months of age [19].

4.4. Epigenetic Changes Underlying the Developmental Origins of Pancreatic Islet Dysfunction

Epigenetics involves a complex array of mechanisms including DNA methylation, histone modifications and microRNAs (miRNAs) that act to interpret the genome in a cell type-specific way. The rapid rise in T2D incidence over the past decades has highlighted the role of epigenetics (in addition to genetics) in contributing to disease risk. Thus far, only a small number of studies have focused on the role of epigenetics in mediating changes in gene expression and, therefore, function that is observed in islets of offspring exposed to compromised early life nutrition. However, we anticipate this to be an area of research with rapid growth in the coming years.

4.5. The Impact of Restricted Nutrition on DNA Methylation and Histone Modifications

DNA methylation is a major component of the mammalian epigenome and the only one for which inheritance through multiple cell divisions has been demonstrated. It, therefore, serves an essential function for stabilizing gene expression patterns throughout life [94]. DNA methylation can also be altered in response to environmental factors present in both pre- and postnatal life leading to cellular dysfunction and disease. For example, DNA methylation and chromatin accessibility of regulatory regions is altered in pancreatic islets from T2D donors with corresponding changes in gene expression and islet function [95–98]. Importantly, environmentally driven changes in DNA methylation can be maintained throughout life. Thus, aberrant epigenetic changes that result from in utero exposure to poor maternal nutritional states, e.g., undernutrition or obesity can have long-lasting effects on offspring health.

To date, our knowledge of such changes has been limited to the impact of IUGR on DNA methylation and histone modifications in islets. In particular, exposure to uterine artery ligation in fetal life leads to genome-wide changes in DNA methylation at intergenic (and not promoter) regions in islets from seven-week-old offspring [35]. At this age, fasting hyperglycemia has not yet emerged, however, islets already have compromised function [88]. Whole-genome histone modification maps of islets from two and 10-week-old offspring (using the same model as above) to define active promoters, transcriptionally silent chromatin and active enhancers also identified that the majority of differential histone marks were located in intergenic regions [32]. To better understand how DNA methylation contributes to the mechanisms underpinning T2D, it will be important to assign these sites of gene regulation, e.g., enhancers to their target gene(s). The latter is particularly challenging given that enhancers are often separated from the target genes they regulate by hundreds of thousands of base pairs.

In addition to a genome-wide approach, studies have also focused on identifying epigenetic dysregulation at specific candidate sites. Using the same model as above, Park and colleagues showed progressive changes in epigenetic regulation of Pdx1. This is observed through a gain in promoter methylation, which is present in offspring islets at six months of age but not at two and seven weeks. Additionally, there is also early and progressive changes in the acetylation and methylation of histones at the Pdx1 promoter. Interestingly, this is present in fetal life and with larger differences at two weeks and six months of age [34]. Consequently, there is a decline in Pdx1 expression in the islets of IUGR offspring with age, resulting in a complete absence of Pdx1 in three-month-old rats [33].

IUGR also leads to epigenetic changes that alter the expression of hepatocyte nuclear factor 4-α (HNF4α) in islets from three and 15-month-old rat offspring [99]. HNF4α is a transcription factor required for beta-cell differentiation and glucose homeostasis, and consequently has been linked to development of a number of forms of T2D. Specifically, maternal protein restriction during pregnancy and lactation leads to epigenetic silencing at the HNF4α enhancer region, which weakens the interaction with its promoter, resulting in a permanent reduction in gene expression. Furthermore, progressive epigenetic silencing of the entire HNF4α locus in islets in response to ageing is more pronounced in rats exposed to poor maternal diet. Importantly, these epigenetic changes precede T2D development in offspring, which occurs at 17 months of age and, therefore, represents a mechanistic basis for the cellular "memory" linking maternal diet to the development of T2D in the offspring [99]. It will be interesting to determine whether, like Pdx1, these changes are already present in fetal life.

4.6. The Impact of Restricted Nutrition on miRNA Expression

In addition to DNA methylation and histone modifications, miRNAs also have a central role to play in the development and function of beta-cells (reviewed in [100]). MiRNAs are small (~19–23 nucleotides) non-coding RNAs involved in post-transcriptional gene regulation and are, therefore, fundamental for cellular identity and function. So far, three studies have investigated the impact of maternal protein restriction on miRNA expression in offspring islets using an unbiased PCR-based array [20,26,60]. Alejandro et al. identified 14 miRNAs that were differentially expressed between offspring groups; two increased and 12 reduced. These measures were performed in three-month-old offspring who displayed glucose intolerance and impaired glucose-stimulated insulin secretion. Importantly, maternal protein restriction did not lead to IUGR in this model [20]. It is, therefore, not surprising that none of these same miRNAs were differentially expressed in islets from adult offspring who were growth restricted in utero; Su and colleagues showed that a different suite of 20 miRNAs were dysregulated in response to maternal protein restriction-induced IUGR [60]. Follow-up experiments focused on miR-15b, which is highly conserved amongst vertebrates and is up-regulated in IUGR islets, identified reduced expression of its target genes cyclin D1 and D2 at the protein level. Furthermore, inhibition of miR-15b in vitro was able to rescue the defective beta-cell proliferation and insulin secretion observed in offspring islets [60].

In contrast to these studies, which investigated miRNAs in the islets of adult offspring, Zhang et al. determined miRNA expression in whole pancreas from IUGR offspring in mid- and late-gestation and identified 31 miRNAs with increased and 54 miRNAs with decreased expression compared to controls [26]. Curiously, differences in miRNA expression are not maintained across gestation; most miRNAs were differentially expressed only at one fetal time point. Furthermore, given differences in the IUGR phenotype, tissue type and time points across the models, it is not surprising that there were no differentially expressed miRNAs in common between the studies. The expression of two miRNAs (miR-542-3p and miR-342-5p), however, are altered in response to IUGR. miR-542-3p is up-regulated in the fetal pancreas in mid- and late-gestation [26] but is decreased in the islets of adult offspring [60]. The expression of miR-342-5p is higher both in the pancreas of IUGR fetuses at embryonic day 19 [26] and in islets of three-month-old non-IUGR offspring [20]. It will be interesting to determine whether the presence of inflammatory markers is also increased in these offspring given that the expression of this miRNA is up-regulated in human islets in response to cytokine exposure [101]. It is important to note that whilst the studies by Alejandro et al. [20] and Zhang et al. [26] used comparable fold-change cut-offs for analysis ($\geq$1.4 for up-regulated miRNAs and $\leq$0.55 for down-regulated miRNAs), the study by Su et al. showed only a modest change in up-regulated expression ranging from 1.1 to 1.9 fold [60].

Studies have also taken a candidate approach when screening for differentially expressed miRNAs. For example, Dumortier et al. measured the expression of miR-375, an islet-abundant miRNA with multiple roles in beta-cell function and found its expression increased in the pancreas of IUGR fetuses at 21 days gestation. This change persists in offspring islets at three months of age. Expression of pyruvate dehydrogenase kinase 1, a target of miR-375, was reduced at the mRNA and protein level [63]. Furthermore, in addition to the miRNAs identified by PCR-array, Alejandro and colleagues also measured the expression of miR-7 and miR-199a-3p, which regulate mTOR given that, as discussed above, exposure to maternal protein restriction leads to decreased insulin secretion that is due, in part, to reduced mTOR signaling. Expression of both miRNAs were increased in the islets of adult offspring [20]. The expression of miR-199a-3p has also been shown to be increased in islets from diabetic db/db mice but its expression is not yet altered in these mice in the pre-diabetic state [102].

Whilst some progress has been made to delineate epigenetic mechanisms driving islet dysfunction and T2D risk in offspring exposed to a nutrient-restricted intrauterine environment, there is still plenty of room for growth in this field especially in defining epigenetic networks that are driving sex-specific differences in islet function and which are also cell type-specific. Moreover, given that different environmental stimuli such as glucose, fatty acids and cytokines can regulate DNA methylation and miRNA expression [102–107], it is key that we understand which changes are programmed by the in utero environment and maintained throughout life, irrespective of the postnatal environment versus those that manifest as a result of declining metabolic health. This knowledge is equally important in the context of exposure to maternal caloric excess especially given that the impact of maternal obesity on the epigenetic landscape of offspring islets is still unknown.

5. Conclusions and Future Directions

It is clear from studies highlighted in this review that fetal malnutrition (encompassing exposure to either restricted nutrition or to an obesogenic diet) has long-lasting effects on islet-cell mass and function. Importantly, the decline in islet function and mass is not solely in response to pre-existing obesity and insulin resistance in adult offspring. Rather, there is inherent vulnerability in these islets resulting in their inability to successfully compensate when faced with increased functional demand.

Given this knowledge, it is key that we define the mechanisms mediating compromised islet function to better understand how T2D develops in those affected by malnutrition during fetal development. Whilst some progress has been made in this area, looking to the future, we believe that the field will greatly benefit from studies that move beyond a candidate approach to epigenetic and gene expression analyses toward genome-wide investigations. Importantly, the availability of next-generation sequencing technologies will enable the mapping of omics data to metabolic traits and will also allow for the characterization of novel mechanisms, for example, by combining data of transcription factor networks, chromatin state and gene transcription to generate an integrated map of gene regulation in islet cells. Furthermore, given that pancreatic islets consist of several endocrine cell types, the cell-specific pathways that are contributing to the functional phenotype should also be defined.

Finally, it is important to state that underpinning all of this research in animal models is the significant need to ensure that the mechanisms identified are also relevant to islet dysfunction and T2D pathogenesis in humans. This includes defining the outcomes relating to the interaction between an underlying genetic predisposition (to T2D) and exposure to in utero environmental drivers.

Author Contributions: Conceptualization, S.E.O., K.M.G., L.M.N.; writing—original draft preparation, S.E.O.; writing—review and editing, K.M.G., L.M.N. All authors have read and agreed to the published version of the manuscript.

Funding: L.M.N. is supported by a C J Martin Fellowship from the National Health and Medical Research Council (GNT1092158).

Conflicts of Interest: The authors declare no conflict of interest.

References

1. Hales, C.N.; Barker, D.J. Type 2 (non-insulin-dependent) diabetes mellitus: The thrifty phenotype hypothesis. *Diabetologia* **1992**, *35*, 595–601. [CrossRef] [PubMed]
2. Hales, C.N.; Barker, D.J.; Clark, P.M.; Cox, L.J.; Fall, C.; Osmond, C.; Winter, P.D. Fetal and infant growth and impaired glucose tolerance at age 64. *BMJ* **1991**, *303*, 1019–1022. [CrossRef] [PubMed]
3. Harder, T.; Rodekamp, E.; Schellong, K.; Dudenhausen, J.W.; Plagemann, A. Birth weight and subsequent risk of type 2 diabetes: A meta-analysis. *Am. J. Epidemiol.* **2007**, *165*, 849–857. [CrossRef]
4. Phipps, K.; Barker, D.J.; Hales, C.N.; Fall, C.H.; Osmond, C.; Clark, P.M. Fetal growth and impaired glucose tolerance in men and women. *Diabetologia* **1993**, *36*, 225–228. [CrossRef] [PubMed]

5. Rich-Edwards, J.W.; Colditz, G.A.; Stampfer, M.J.; Willett, W.C.; Gillman, M.W.; Hennekens, C.H.; Speizer, F.E.; Manson, J.E. Birthweight and the risk for type 2 diabetes mellitus in adult women. *Ann. Intern. Med.* **1999**, *130*, 278–284. [CrossRef]
6. Stein, A.D.; Obrutu, O.E.; Behere, R.V.; Yajnik, C.S. Developmental undernutrition, offspring obesity and type 2 diabetes. *Diabetologia* **2019**, *62*, 1773–1778. [CrossRef]
7. Hattersley, A.T.; Tooke, J.E. The fetal insulin hypothesis: An alternative explanation of the association of low birthweight with diabetes and vascular disease. *Lancet* **1999**, *353*, 1789–1792. [CrossRef]
8. Hughes, A.E.; Hattersley, A.T.; Flanagan, S.E.; Freathy, R.M. Two decades since the fetal insulin hypothesis: What have we learned from genetics? *Diabetologia* **2021**, *64*, 717–726. [CrossRef]
9. Poulsen, P.; Vaag, A.A.; Kyvik, K.O.; Moller Jensen, D.; Beck-Nielsen, H. Low birth weight is associated with niddm in discordant monozygotic and dizygotic twin pairs. *Diabetologia* **1997**, *40*, 439–446. [CrossRef]
10. Knop, M.R.; Geng, T.T.; Gorny, A.W.; Ding, R.; Li, C.; Ley, S.H.; Huang, T. Birth weight and risk of type 2 diabetes mellitus, cardiovascular disease, and hypertension in adults: A meta-analysis of 7 646 267 participants from 135 studies. *J. Am. Heart Assoc.* **2018**, *7*, e008870. [CrossRef]
11. Branum, A.M.; Kirmeyer, S.E.; Gregory, E.C. Prepregnancy body mass index by maternal characteristics and state: Data from the birth certificate, 2014. *Natl. Vital Stat. Rep.* **2016**, *65*, 1–11.
12. Gaudet, L.; Ferraro, Z.M.; Wen, S.W.; Walker, M. Maternal obesity and occurrence of fetal macrosomia: A systematic review and meta-analysis. *Biomed Res. Int.* **2014**, *2014*, 640291. [CrossRef] [PubMed]
13. Lahti-Pulkkinen, M.; Bhattacharya, S.; Wild, S.H.; Lindsay, R.S.; Raikkonen, K.; Norman, J.E.; Bhattacharya, S.; Reynolds, R.M. Consequences of being overweight or obese during pregnancy on diabetes in the offspring: A record linkage study in aberdeen, scotland. *Diabetologia* **2019**, *62*, 1412–1419. [CrossRef] [PubMed]
14. Tyrrell, J.; Richmond, R.C.; Palmer, T.M.; Feenstra, B.; Rangarajan, J.; Metrustry, S.; Cavadino, A.; Paternoster, L.; Armstrong, L.L.; De Silva, N.M.G.; et al. Genetic evidence for causal relationships between maternal obesity-related traits and birth weight. *JAMA* **2016**, *315*, 1129–1140. [CrossRef] [PubMed]
15. Richmond, R.C.; Timpson, N.J.; Felix, J.F.; Palmer, T.; Gaillard, R.; McMahon, G.; Davey Smith, G.; Jaddoe, V.W.; Lawlor, D.A. Using genetic variation to explore the causal effect of maternal pregnancy adiposity on future offspring adiposity: A mendelian randomisation study. *PLOS Med.* **2017**, *14*, e1002221. [CrossRef]
16. Cetin, I.; Marconi, A.M.; Bozzetti, P.; Sereni, L.P.; Corbetta, C.; Pardi, G.; Battaglia, F.C. Umbilical amino acid concentrations in appropriate and small for gestational age infants: A biochemical difference present in utero. *Am. J. Obstet. Gynecol.* **1988**, *158*, 120–126. [CrossRef]
17. Dicke, J.M.; Henderson, G.I. Placental amino acid uptake in normal and complicated pregnancies. *Am. J. Med Sci.* **1988**, *295*, 223–227. [CrossRef] [PubMed]
18. King, R.; Hill, J.L.; Saha, B.; Tong, Y.; Strutt, B.J.; Russell, M.A.; Morgan, N.G.; Richardson, S.J.; Hill, D.J. Offspring of mice exposed to a low-protein diet in utero demonstrate changes in mtor signaling in pancreatic islets of langerhans, associated with altered glucagon and insulin expression and a lower β-cell mass. *Nutrients* **2019**, *11*, 605. [CrossRef] [PubMed]
19. Alejandro, E.U.; Jo, S.; Akhaphong, B.; Llacer, P.R.; Gianchandani, M.; Gregg, B.; Parlee, S.D.; MacDougald, O.A.; Bernal-Mizrachi, E. Maternal low-protein diet on the last week of pregnancy contributes to insulin resistance and β-cell dysfunction in the mouse offspring. *Am. J. Physiol. Regul. Integr. Comp. Physiol.* **2020**, *319*, R485–R496. [CrossRef] [PubMed]
20. Alejandro, E.U.; Gregg, B.; Wallen, T.; Kumusoglu, D.; Meister, D.; Chen, A.; Merrins, M.J.; Satin, L.S.; Liu, M.; Arvan, P.; et al. Maternal diet-induced micrornas and mtor underlie β cell dysfunction in offspring. *J. Clin. Investig.* **2014**, *124*, 4395–4410. [CrossRef]
21. Theys, N.; Bouckenooghe, T.; Ahn, M.T.; Remacle, C.; Reusens, B. Maternal low-protein diet alters pancreatic islet mitochondrial function in a sex-specific manner in the adult rat. *Am. J. Physiol. Regul. Integr. Comp. Physiol.* **2009**, *297*, R1516–R1525. [CrossRef]
22. Tarry-Adkins, J.L.; Chen, J.H.; Smith, N.S.; Jones, R.H.; Cherif, H.; Ozanne, S.E. Poor maternal nutrition followed by accelerated postnatal growth leads to telomere shortening and increased markers of cell senescence in rat islets. *FASEB J.* **2009**, *23*, 1521–1528. [CrossRef]
23. Rodríguez-Trejo, A.; Ortiz-López, M.G.; Zambrano, E.; Granados-Silvestre Mde, L.; Méndez, C.; Blondeau, B.; Bréant, B.; Nathanielsz, P.W.; Menjivar, M. Developmental programming of neonatal pancreatic β-cells by a maternal low-protein diet in rats involves a switch from proliferation to differentiation. *Am. J. Physiol. Endocrinol. Metab.* **2012**, *302*, E1431–E1439. [CrossRef]
24. Fernandez-Twinn, D.S.; Wayman, A.; Ekizoglou, S.; Martin, M.S.; Hales, C.N.; Ozanne, S.E. Maternal protein restriction leads to hyperinsulinemia and reduced insulin-signaling protein expression in 21-mo-old female rat offspring. *Am. J. Physiol. Regul. Integr. Comp. Physiol.* **2005**, *288*, R368–R373. [CrossRef]
25. Petry, C.J.; Dorling, M.W.; Pawlak, D.B.; Ozanne, S.E.; Hales, C.N. Diabetes in old male offspring of rat dams fed a reduced protein diet. *Int. J. Exp. Diabetes Res.* **2001**, *2*, 139–143. [CrossRef] [PubMed]
26. Zhang, L.; Chen, W.; Dai, Y.; Zhu, Z.; Liu, Q. Detection of expressional changes induced by intrauterine growth restriction in the developing rat pancreas. *Exp. Biol. Med.* **2016**, *241*, 1446–1456. [CrossRef] [PubMed]
27. Martin-Gronert, M.S.; Tarry-Adkins, J.L.; Cripps, R.L.; Chen, J.H.; Ozanne, S.E. Maternal protein restriction leads to early life alterations in the expression of key molecules involved in the aging process in rat offspring. *Am. J. Physiol. Integr. Comp. Physiol.* **2008**, *294*, R494–R500. [CrossRef]

28. Tarry-Adkins, J.L.; Fernandez-Twinn, D.S.; Chen, J.H.; Hargreaves, I.P.; Neergheen, V.; Aiken, C.E.; Ozanne, S.E. Poor maternal nutrition and accelerated postnatal growth induces an accelerated aging phenotype and oxidative stress in skeletal muscle of male rats. *Dis. Model. Mech.* **2016**, *9*, 1221–1229. [CrossRef]
29. Chen, J.H.; Martin-Gronert, M.S.; Tarry-Adkins, J.; Ozanne, S.E. Maternal protein restriction affects postnatal growth and the expression of key proteins involved in lifespan regulation in mice. *PLoS ONE* **2009**, *4*, e4950. [CrossRef] [PubMed]
30. Rashid, C.S.; Lien, Y.C.; Bansal, A.; Jaeckle-Santos, L.J.; Li, C.; Won, K.J.; Simmons, R.A. Transcriptomic analysis reveals novel mechanisms mediating islet dysfunction in the intrauterine growth-restricted rat. *Endocrinology* **2018**, *159*, 1035–1049. [CrossRef]
31. Akhaphong, B.; Lockridge, A.; Jo, S.; Mohan, R.; Wilcox, J.A.; Wing, C.R.; Regal, J.F.; Alejandro, E.U. Reduced uterine perfusion pressure causes loss of pancreatic β-cell area but normal function in fetal rat offspring. *Am. J. Physiol. Regul. Integr. Comp. Physiol.* **2018**, *315*, R1220–R1231. [CrossRef]
32. Lien, Y.C.; Wang, P.Z.; Lu, X.M.; Simmons, R.A. Altered transcription factor binding and gene bivalency in islets of intrauterine growth retarded rats. *Cells* **2020**, *9*, 1435. [CrossRef]
33. Stoffers, D.A.; Desai, B.M.; DeLeon, D.D.; Simmons, R.A. Neonatal exendin-4 prevents the development of diabetes in the intrauterine growth retarded rat. *Diabetes* **2003**, *52*, 734–740. [CrossRef]
34. Park, J.H.; Stoffers, D.A.; Nicholls, R.D.; Simmons, R.A. Development of type 2 diabetes following intrauterine growth retardation in rats is associated with progressive epigenetic silencing of pdx1. *J. Clin. Investig.* **2008**, *118*, 2316–2324. [CrossRef]
35. Thompson, R.F.; Fazzari, M.J.; Niu, H.; Barzilai, N.; Simmons, R.A.; Greally, J.M. Experimental intrauterine growth restriction induces alterations in DNA methylation and gene expression in pancreatic islets of rats. *J. Biol. Chem.* **2010**, *285*, 15111–15118. [CrossRef] [PubMed]
36. Delghingaro-Augusto, V.; Madad, L.; Chandra, A.; Simeonovic, C.J.; Dahlstrom, J.E.; Nolan, C.J. Islet inflammation, hemosiderosis, and fibrosis in intrauterine growth-restricted and high fat-fed sprague-dawley rats. *Am. J. Pathol.* **2014**, *184*, 1446–1457. [CrossRef] [PubMed]
37. Siebel, A.L.; Gallo, L.A.; Guan, T.C.; Owens, J.A.; Wlodek, M.E. Cross-fostering and improved lactation ameliorates deficits in endocrine pancreatic morphology in growth-restricted adult male rat offspring. *J. Dev. Orig. Health Dis.* **2010**, *1*, 234–244. [CrossRef]
38. Robinson, J.; Kingston, E.; Jones, C.; Thorburn, G. Studies on experimental growth retardation in sheep. The effect of removal of a endometrial caruncles on fetal size and metabolism. *J. Dev. Physiol.* **1979**, *1*, 379–398.
39. Owens, J.A.; Gatford, K.L.; De Blasio, M.J.; Edwards, L.J.; McMillen, I.C.; Fowden, A.L. Restriction of placental growth in sheep impairs insulin secretion but not sensitivity before birth. *J. Physiol.* **2007**, *584*, 935–949. [CrossRef] [PubMed]
40. Leos, R.A.; Anderson, M.J.; Chen, X.; Pugmire, J.; Anderson, K.A.; Limesand, S.W. Chronic exposure to elevated norepinephrine suppresses insulin secretion in fetal sheep with placental insufficiency and intrauterine growth restriction. *Am. J. Physiol. Endocrinol. Metab.* **2010**, *298*, E770–E778. [CrossRef] [PubMed]
41. Rozance, P.J.; Limesand, S.W.; Barry, J.S.; Brown, L.D.; Hay, W.W. Glucose replacement to euglycemia causes hypoxia, acidosis, and decreased insulin secretion in fetal sheep with intrauterine growth restriction. *Pediatr. Res.* **2009**, *65*, 72–78. [CrossRef] [PubMed]
42. Kelly, A.C.; Bidwell, C.A.; McCarthy, F.M.; Taska, D.J.; Anderson, M.J.; Camacho, L.E.; Limesand, S.W. Rna sequencing exposes adaptive and immune responses to intrauterine growth restriction in fetal sheep islets. *Endocrinology* **2017**, *158*, 743–755. [CrossRef] [PubMed]
43. Boehmer, B.H.; Brown, L.D.; Wesolowski, S.R.; Hay, W.W.; Rozance, P.J. Pulsatile hyperglycemia increases insulin secretion but not pancreatic β-cell mass in intrauterine growth-restricted fetal sheep. *J. Dev. Orig. Health Dis.* **2018**, *9*, 492–499. [CrossRef] [PubMed]
44. Limesand, S.W.; Jensen, J.; Hutton, J.C.; Hay, W.W., Jr. Diminished beta-cell replication contributes to reduced beta-cell mass in fetal sheep with intrauterine growth restriction. *Am. J. Physiol. Regul. Integr. Comp. Physiol.* **2005**, *288*, R1297–R1305. [CrossRef]
45. Elsakr, J.M.; Dunn, J.C.; Tennant, K.; Zhao, S.K.; Kroeten, K.; Pasek, R.C.; Takahashi, D.L.; Dean, T.A.; Velez Edwards, D.R.; McCurdy, C.E.; et al. Maternal western-style diet affects offspring islet composition and function in a non-human primate model of maternal over-nutrition. *Mol. Metab.* **2019**, *25*, 73–82. [CrossRef]
46. Comstock, S.M.; Pound, L.D.; Bishop, J.M.; Takahashi, D.L.; Kostrba, A.M.; Smith, M.S.; Grove, K.L. High-fat diet consumption during pregnancy and the early post-natal period leads to decreased α cell plasticity in the nonhuman primate. *Mol. Metab.* **2012**, *2*, 10–22. [CrossRef] [PubMed]
47. Nicol, L.E.; Grant, W.F.; Comstock, S.M.; Nguyen, M.L.; Smith, M.S.; Grove, K.L.; Marks, D.L. Pancreatic inflammation and increased islet macrophages in insulin-resistant juvenile primates. *J. Endocrinol.* **2013**, *217*, 207–213. [CrossRef] [PubMed]
48. Pound, L.D.; Comstock, S.M.; Grove, K.L. Consumption of a western-style diet during pregnancy impairs offspring islet vascularization in a japanese macaque model. *Am. J. Physiol. Endocrinol. Metab.* **2014**, *307*, E115–E123. [CrossRef]
49. Conrad, E.; Dai, C.; Spaeth, J.; Guo, M.; Cyphert, H.A.; Scoville, D.; Carroll, J.; Yu, W.M.; Goodrich, L.V.; Harlan, D.M.; et al. The mafb transcription factor impacts islet α-cell function in rodents and represents a unique signature of primate islet β-cells. *Am. J. Physiol. Endocrinol. Metab.* **2016**, *310*, E91–E102. [CrossRef]
50. Ford, S.P.; Zhang, L.; Zhu, M.; Miller, M.M.; Smith, D.T.; Hess, B.W.; Moss, G.E.; Nathanielsz, P.W.; Nijland, M.J. Maternal obesity accelerates fetal pancreatic beta-cell but not alpha-cell development in sheep: Prenatal consequences. *Am. J. Physiol. Regul. Integr. Comp. Physiol.* **2009**, *297*, R835–R843. [CrossRef]

51. Zhang, L.; Long, N.M.; Hein, S.M.; Ma, Y.; Nathanielsz, P.W.; Ford, S.P. Maternal obesity in ewes results in reduced fetal pancreatic β-cell numbers in late gestation and decreased circulating insulin concentration at term. *Domest. Anim. Endocrinol.* **2011**, *40*, 30–39. [CrossRef]
52. Nicholas, L.M.; Ozanne, S.E. Early life programming in mice by maternal overnutrition: Mechanistic insights and interventional approaches. *Philos. Trans. R. Soc. Lond. Ser. B Biol. Sci.* **2019**, *374*, 20180116. [CrossRef] [PubMed]
53. Han, J.; Xu, J.; Epstein, P.N.; Liu, Y.Q. Long-term effect of maternal obesity on pancreatic beta cells of offspring: Reduced beta cell adaptation to high glucose and high-fat diet challenges in adult female mouse offspring. *Diabetologia* **2005**, *48*, 1810–1818. [CrossRef] [PubMed]
54. Duhl, D.M.; Vrieling, H.; Miller, K.A.; Wolff, G.L.; Barsh, G.S. Neomorphic agouti mutations in obese yellow mice. *Nat. Genet.* **1994**, *8*, 59–65. [CrossRef]
55. Li, C.C.; Young, P.E.; Maloney, C.A.; Eaton, S.A.; Cowley, M.J.; Buckland, M.E.; Preiss, T.; Henstridge, D.C.; Cooney, G.J.; Febbraio, M.A.; et al. Maternal obesity and diabetes induces latent metabolic defects and widespread epigenetic changes in isogenic mice. *Epigenetics* **2013**, *8*, 602–611. [CrossRef]
56. Waterland, R.A.; Travisano, M.; Tahiliani, K.G.; Rached, M.T.; Mirza, S. Methyl donor supplementation prevents transgenerational amplification of obesity. *Int. J. Obes.* **2008**, *32*, 1373–1379. [CrossRef] [PubMed]
57. Duque-Guimaraes, D.E.; Ozanne, S.E. Nutritional programming of insulin resistance: Causes and consequences. *Trends Endocrinol. Metab.* **2013**, *24*, 525–535. [CrossRef] [PubMed]
58. Nicholas, L.M.; Morrison, J.L.; Rattanatray, L.; Zhang, S.; Ozanne, S.E.; McMillen, I.C. The early origins of obesity and insulin resistance: Timing, programming and mechanisms. *Int. J. Obes.* **2016**, *40*, 229–238. [CrossRef]
59. Bringhenti, I.; Moraes-Teixeira, J.A.; Cunha, M.R.; Ornellas, F.; Mandarim-de-Lacerda, C.A.; Aguila, M.B. Maternal obesity during the preconception and early life periods alters pancreatic development in early and adult life in male mouse offspring. *PLoS ONE* **2013**, *8*, e55711. [CrossRef]
60. Su, Y.; Jiang, X.; Li, Y.; Li, F.; Cheng, Y.; Peng, Y.; Song, D.; Hong, J.; Ning, G.; Cao, Y.; et al. Maternal low protein isocaloric diet suppresses pancreatic β-cell proliferation in mouse offspring via mir-15b. *Endocrinology* **2016**, *157*, 4782–4793. [CrossRef]
61. Petrik, J.; Reusens, B.; Arany, E.; Remacle, C.; Coelho, C.; Hoet, J.J.; Hill, D.J. A low protein diet alters the balance of islet cell replication and apoptosis in the fetal and neonatal rat and is associated with a reduced pancreatic expression of insulin-like growth factor-ii. *Endocrinology* **1999**, *140*, 4861–4873. [CrossRef]
62. Reusens, B.; Remacle, C. Programming of the endocrine pancreas by the early nutritional environment. *Int. J. Biochem. Cell Biol.* **2006**, *38*, 913–922. [CrossRef] [PubMed]
63. Dumortier, O.; Hinault, C.; Gautier, N.; Patouraux, S.; Casamento, V.; Van Obberghen, E. Maternal protein restriction leads to pancreatic failure in offspring: Role of misexpressed microrna-375. *Diabetes* **2014**, *63*, 3416. [CrossRef] [PubMed]
64. Sasaki, H.; Saisho, Y.; Inaishi, J.; Watanabe, Y.; Tsuchiya, T.; Makio, M.; Sato, M.; Kitago, M.; Yamada, T.; Itoh, H. Associations of birthweight and history of childhood obesity with beta cell mass in japanese adults. *Diabetologia* **2020**, *63*, 1199–1210. [CrossRef] [PubMed]
65. Nicholas, L.M.; Nagao, M.; Kusinski, L.C.; Fernandez-Twinn, D.S.; Eliasson, L.; Ozanne, S.E. Exposure to maternal obesity programs sex differences in pancreatic islets of the offspring in mice. *Diabetologia* **2020**, *63*, 324–337. [CrossRef]
66. Yokomizo, H.; Inoguchi, T.; Sonoda, N.; Sakaki, Y.; Maeda, Y.; Inoue, T.; Hirata, E.; Takei, R.; Ikeda, N.; Fujii, M.; et al. Maternal high-fat diet induces insulin resistance and deterioration of pancreatic beta-cell function in adult offspring with sex differences in mice. *Am. J. Physiol. Endocrinol. Metab.* **2014**, *306*, E1163–E1175. [CrossRef]
67. Zambrano, E.; Sosa-Larios, T.; Calzada, L.; Ibanez, C.A.; Mendoza-Rodriguez, C.A.; Morales, A.; Morimoto, S. Decreased basal insulin secretion from pancreatic islets of pups in a rat model of maternal obesity. *J. Endocrinol.* **2016**, *231*, 49–57. [CrossRef]
68. Simmons, R.A.; Templeton, L.J.; Gertz, S.J. Intrauterine growth retardation leads to the development of type 2 diabetes in the rat. *Diabetes* **2001**, *50*, 2279. [CrossRef]
69. Camacho, L.E.; Chen, X.; Hay, W.W., Jr.; Limesand, S.W. Enhanced insulin secretion and insulin sensitivity in young lambs with placental insufficiency-induced intrauterine growth restriction. *Am. J. Physiol. Regul. Integr. Comp. Physiol.* **2017**, *313*, R101–R109. [CrossRef] [PubMed]
70. Long, N.M.; George, L.A.; Uthlaut, A.B.; Smith, D.T.; Nijland, M.J.; Nathanielsz, P.W.; Ford, S.P. Maternal obesity and increased nutrient intake before and during gestation in the ewe results in altered growth, adiposity, and glucose tolerance in adult offspring1. *J. Anim. Sci.* **2010**, *88*, 3546–3553. [CrossRef]
71. McCurdy, C.E.; Bishop, J.M.; Williams, S.M.; Grayson, B.E.; Smith, M.S.; Friedman, J.E.; Grove, K.L. Maternal high-fat diet triggers lipotoxicity in the fetal livers of nonhuman primates. *J. Clin. Investig.* **2009**, *119*, 323–335. [CrossRef] [PubMed]
72. Cox, A.R.; Gottheil, S.K.; Arany, E.J.; Hill, D.J. The effects of low protein during gestation on mouse pancreatic development and beta cell regeneration. *Pediatr. Res.* **2010**, *68*, 16–22. [CrossRef] [PubMed]
73. Cerf, M.E.; Williams, K.; Nkomo, X.I.; Muller, C.J.; Du Toit, D.F.; Louw, J.; Wolfe-Coote, S.A. Islet cell response in the neonatal rat after exposure to a high-fat diet during pregnancy. *Am. J. Physiol. Regul. Integr. Comp. Physiol.* **2005**, *288*, R1122–R1128. [CrossRef] [PubMed]
74. Graus-Nunes, F.; Dalla Corte Frantz, E.; Lannes, W.R.; da Silva Menezes, M.C.; Mandarim-de-Lacerda, C.A.; Souza-Mello, V. Pregestational maternal obesity impairs endocrine pancreas in male f1 and f2 progeny. *Nutrition* **2015**, *31*, 380–387. [CrossRef]

75. Yoon, K.H.; Ko, S.H.; Cho, J.H.; Lee, J.M.; Ahn, Y.B.; Song, K.H.; Yoo, S.J.; Kang, M.I.; Cha, B.Y.; Lee, K.W.; et al. Selective beta-cell loss and alpha-cell expansion in patients with type 2 diabetes mellitus in korea. *J. Clin. Endocrinol. Metab.* **2003**, *88*, 2300–2308. [CrossRef] [PubMed]
76. Henquin, J.C.; Rahier, J. Pancreatic alpha cell mass in european subjects with type 2 diabetes. *Diabetologia* **2011**, *54*, 1720–1725. [CrossRef]
77. Meier, J.J.; Ueberberg, S.; Korbas, S.; Schneider, S. Diminished glucagon suppression after β-cell reduction is due to impaired α-cell function rather than an expansion of α-cell mass. *Endocrinol. Metab.* **2011**, *300*, E717–E723. [CrossRef] [PubMed]
78. Kahn, S.E.; Hull, R.L.; Utzschneider, K.M. Mechanisms linking obesity to insulin resistance and type 2 diabetes. *Nature* **2006**, *444*, 840–846. [CrossRef]
79. Wortham, M.; Sander, M. Mechanisms of beta-cell functional adaptation to changes in workload. *Diabetes Obes. Metab.* **2016**, *18* (Suppl. 1), 78–86. [CrossRef]
80. Chen, C.; Cohrs, C.M.; Stertmann, J.; Bozsak, R.; Speier, S. Human beta cell mass and function in diabetes: Recent advances in knowledge and technologies to understand disease pathogenesis. *Mol. Metab.* **2017**, *6*, 943–957. [CrossRef]
81. Taylor, P.D.; McConnell, J.; Khan, I.Y.; Holemans, K.; Lawrence, K.M.; Asare-Anane, H.; Persaud, S.J.; Jones, P.M.; Petrie, L.; Hanson, M.A.; et al. Impaired glucose homeostasis and mitochondrial abnormalities in offspring of rats fed a fat-rich diet in pregnancy. *Am. J. Physiol. Regul. Integr. Comp. Physiol.* **2005**, *288*, R134–R139. [CrossRef] [PubMed]
82. Bringhenti, I.; Ornellas, F.; Mandarim-de-Lacerda, C.A.; Aguila, M.B. The insulin-signaling pathway of the pancreatic islet is impaired in adult mice offspring of mothers fed a high-fat diet. *Nutrition* **2016**, *32*, 1138–1143. [CrossRef] [PubMed]
83. Samuelsson, A.M.; Matthews, P.A.; Argenton, M.; Christie, M.R.; McConnell, J.M.; Jansen, E.H.J.M.; Piersma, A.H.; Ozanne, S.E.; Twinn, D.F.; Remacle, C.; et al. Diet-induced obesity in female mice leads to offspring hyperphagia, adiposity, hypertension, and insulin resistance: A novel murine model of developmental programming. *Hypertension* **2008**, *51*, 383–392. [CrossRef] [PubMed]
84. Srinivasan, M.; Katewa, S.D.; Palaniyappan, A.; Pandya, J.D.; Patel, M.S. Maternal high-fat diet consumption results in fetal malprogramming predisposing to the onset of metabolic syndrome-like phenotype in adulthood. *Am. J. Physiol. Endocrinol. Metab.* **2006**, *291*, E792–E799. [CrossRef]
85. Guo, S.; Dai, C.; Guo, M.; Taylor, B.; Harmon, J.S.; Sander, M.; Robertson, R.P.; Powers, A.C.; Stein, R. Inactivation of specific β cell transcription factors in type 2 diabetes. *J. Clin. Investig.* **2013**, *123*, 3305–3316. [CrossRef] [PubMed]
86. Gao, T.; McKenna, B.; Li, C.; Reichert, M.; Nguyen, J.; Singh, T.; Yang, C.; Pannikar, A.; Doliba, N.; Zhang, T. Pdx1 maintains β cell identity and function by repressing an α cell program. *Cell Metab.* **2014**, *19*, 259–271. [CrossRef]
87. Maechler, P.; Wollheim, C.B. Mitochondrial function in normal and diabetic β-cells. *Nature* **2001**, *414*, 807–812. [CrossRef]
88. Simmons, R.A.; Suponitsky-Kroyter, I.; Selak, M.A. Progressive accumulation of mitochondrial DNA mutations and decline in mitochondrial function lead to β-cell failure. *J. Biol. Chem.* **2005**, *280*, 28785–28791. [CrossRef]
89. Pi, J.; Bai, Y.; Zhang, Q.; Wong, V.; Floering, L.M.; Daniel, K.; Reece, J.M.; Deeney, J.T.; Andersen, M.E.; Corkey, B.E.; et al. Reactive oxygen species as a signal in glucose-stimulated insulin secretion. *Diabetes* **2007**, *56*, 1783–1791. [CrossRef]
90. Lei, X.G.; Vatamaniuk, M.Z. Two tales of antioxidant enzymes on beta cells and diabetes. *Antioxid. Redox Signal.* **2011**, *14*, 489–503. [CrossRef]
91. Fernandez-Twinn, D.S.; Gascoin, G.; Musial, B.; Carr, S.; Duque-Guimaraes, D.; Blackmore, H.L.; Alfaradhi, M.Z.; Loche, E.; Sferruzzi-Perri, A.N.; Fowden, A.L.; et al. Exercise rescues obese mothers' insulin sensitivity, placental hypoxia and male offspring insulin sensitivity. *Sci. Rep.* **2017**, *7*, 44650. [CrossRef]
92. Theys, N.; Ahn, M.T.; Bouckenooghe, T.; Reusens, B.; Remacle, C. Maternal malnutrition programs pancreatic islet mitochondrial dysfunction in the adult offspring. *J. Nutr. Biochem.* **2011**, *22*, 985–994. [CrossRef] [PubMed]
93. Ardestani, A.; Lupse, B.; Kido, Y.; Leibowitz, G.; Maedler, K. Mtorc1 signaling: A double-edged sword in diabetic β cells. *Cell Metab.* **2018**, *27*, 314–331. [CrossRef] [PubMed]
94. Avrahami, D.; Kaestner, K.H. The dynamic methylome of islets in health and disease. *Mol. Metab.* **2019**, *27s*, S25–S32. [CrossRef] [PubMed]
95. Dayeh, T.; Volkov, P.; Salo, S.; Hall, E.; Nilsson, E.; Olsson, A.H.; Kirkpatrick, C.L.; Wollheim, C.B.; Eliasson, L.; Ronn, T.; et al. Genome-wide DNA methylation analysis of human pancreatic islets from type 2 diabetic and non-diabetic donors identifies candidate genes that influence insulin secretion. *PLoS Genet.* **2014**, *10*, e1004160. [CrossRef] [PubMed]
96. Ling, C.; Del Guerra, S.; Lupi, R.; Rönn, T.; Granhall, C.; Luthman, H.; Masiello, P.; Marchetti, P.; Groop, L.; Del Prato, S. Epigenetic regulation of ppargc1a in human type 2 diabetic islets and effect on insulin secretion. *Diabetologia* **2008**, *51*, 615–622. [CrossRef]
97. Volkov, P.; Bacos, K.; Ofori, J.K.; Esguerra, J.L.; Eliasson, L.; Ronn, T.; Ling, C. Whole-genome bisulfite sequencing of human pancreatic islets reveals novel differentially methylated regions in type 2 diabetes pathogenesis. *Diabetes* **2017**, *66*, 1074–1085. [CrossRef]
98. Yang, B.T.; Dayeh, T.A.; Volkov, P.A.; Kirkpatrick, C.L.; Malmgren, S.; Jing, X.; Renström, E.; Wollheim, C.B.; Nitert, M.D.; Ling, C. Increased DNA methylation and decreased expression of pdx-1 in pancreatic islets from patients with type 2 diabetes. *Mol. Endocrinol.* **2012**, *26*, 1203–1212. [CrossRef] [PubMed]
99. Sandovici, I.; Smith, N.H.; Nitert, M.D.; Ackers-Johnson, M.; Uribe-Lewis, S.; Ito, Y.; Jones, R.H.; Marquez, V.E.; Cairns, W.; Tadayyon, M.; et al. Maternal diet and aging alter the epigenetic control of a promoter–enhancer interaction at the Hnf4a gene in rat pancreatic islets. *Proc. Natl. Acad. Sci. USA* **2011**, *108*, 5449. [CrossRef]

100. Eliasson, L.; Esguerra, J.L.S. Microrna networks in pancreatic islet cells: Normal function and type 2 diabetes. *Diabetes* **2020**, *69*, 804. [CrossRef] [PubMed]
101. Grieco, F.A.; Sebastiani, G.; Juan-Mateu, J.; Villate, O.; Marroqui, L.; Ladrière, L.; Tugay, K.; Regazzi, R.; Bugliani, M.; Marchetti, P.; et al. Micrornas mir-23a-3p, mir-23b-3p, and mir-149-5p regulate the expression of proapoptotic bh3-only proteins dp5 and puma in human pancreatic β-cells. *Diabetes* **2017**, *66*, 100–112. [CrossRef]
102. Nesca, V.; Guay, C.; Jacovetti, C.; Menoud, V.; Peyot, M.-L.; Laybutt, D.R.; Prentki, M.; Regazzi, R. Identification of particular groups of micrornas that positively or negatively impact on beta cell function in obese models of type 2 diabetes. *Diabetologia* **2013**, *56*, 2203–2212. [CrossRef] [PubMed]
103. Hall, E.; Dekker Nitert, M.; Volkov, P.; Malmgren, S.; Mulder, H.; Bacos, K.; Ling, C. The effects of high glucose exposure on global gene expression and DNA methylation in human pancreatic islets. *Mol. Cell. Endocrinol.* **2018**, *472*, 57–67. [CrossRef]
104. Hall, E.; Jönsson, J.; Ofori, J.K.; Volkov, P.; Perfilyev, A.; Dekker Nitert, M.; Eliasson, L.; Ling, C.; Bacos, K. Glucolipotoxicity alters insulin secretion via epigenetic changes in human islets. *Diabetes* **2019**, *68*, 1965–1974. [CrossRef] [PubMed]
105. Hall, E.; Volkov, P.; Dayeh, T.; Bacos, K.; Rönn, T.; Nitert, M.D.; Ling, C. Effects of palmitate on genome-wide mrna expression and DNA methylation patterns in human pancreatic islets. *BMC Med.* **2014**, *12*, 103. [CrossRef] [PubMed]
106. Roggli, E.; Britan, A.; Gattesco, S.; Lin-Marq, N.; Abderrahmani, A.; Meda, P.; Regazzi, R. Involvement of micrornas in the cytotoxic effects exerted by proinflammatory cytokines on pancreatic β-cells. *Diabetes* **2010**, *59*, 978–986. [CrossRef]
107. Esguerra, J.L.S.; Bolmeson, C.; Cilio, C.M.; Eliasson, L. Differential glucose-regulation of micrornas in pancreatic islets of non-obese type 2 diabetes model goto-kakizaki rat. *PLoS ONE* **2011**, *6*, e18613. [CrossRef]

metabolites

Review

FFAR from the Gut Microbiome Crowd: SCFA Receptors in T1D Pathology

Medha Priyadarshini [1], Kristen Lednovich [1], Kai Xu [1], Sophie Gough [1], Barton Wicksteed [1] and Brian T. Layden [1,2,*]

[1] Department of Medicine, Division of Endocrinology, Diabetes, and Metabolism, University of Illinois at Chicago, Chicago, IL 60612, USA; mpriya2@uic.edu (M.P.); kledno2@uic.edu (K.L.); kxu12@uic.edu (K.X.); sgough2@uic.edu (S.G.); bartonw@uic.edu (B.W.)
[2] Jesse Brown Veterans Affair Medical Center, Chicago, IL 60612, USA
* Correspondence: blaydel@uic.edu; Tel.: +1-312-996-6060; Fax: +1-312-413-0437

Abstract: The gut microbiome has emerged as a novel determinant of type 1 diabetes (T1D), but the underlying mechanisms are unknown. In this context, major gut microbial metabolites, short-chain fatty acids (SCFAs), are considered to be an important link between the host and gut microbiome. We, along with other laboratories, have explored how SCFAs and their cognate receptors affect various metabolic conditions, including obesity, type 2 diabetes, and metabolic syndrome. Though gut microbiome and SCFA-level changes have been reported in T1D and in mouse models of the disease, the role of SCFA receptors in T1D remains under explored. In this review article, we will highlight the existing and possible roles of these receptors in T1D pathology. We conclude with a discussion of SCFA receptors as therapeutic targets for T1D, exploring an exciting new potential for novel treatments of glucometabolic disorders.

Keywords: free fatty acid receptor (FFA) 2; FFA3; gut microbiome; incretin; insulin secretion; short-chain fatty acids; type 1 diabetes

Citation: Priyadarshini, M.; Lednovich, K.; Xu, K.; Gough, S.; Wicksteed, B.; Layden, B.T. FFAR from the Gut Microbiome Crowd: SCFA Receptors in T1D Pathology. *Metabolites* **2021**, *11*, 302. https://doi.org/10.3390/metabo11050302

Academic Editors: Melkam Kebede and Belinda Yau

Received: 15 April 2021
Accepted: 7 May 2021
Published: 11 May 2021

1. Introduction

T1D is an organ-specific autoimmune disease characterized by the destruction of the β cells. While the etiology of T1D is not fully understood, the prevailing paradigm hypothesizes that an individual's genetic background plays a central role in the risk of disease development, and more than 50 T1D-associated genes have been identified through extensive genetic studies [1]. However, genetic predisposition alone cannot explain recent rises in the rates of T1D worldwide. Environmental factors, including mode of delivery during birth [2], breast versus formula feeding [2], use of antibiotics in early life [3,4], toxicant exposure [5], etc., have lately emerged as factors potentially affecting T1D onset and progression. Of note, most of these factors converge upon the gut microbiome (GM), making this a novel critical factor, the modulation of which can accelerate or offset T1D progression [6,7].

Comprising over 10 trillion cells and outnumbering host cells approximately 10:1, the GM refers to the totality of microorganisms, including bacteria, fungi, archaea, and viruses, that inhabit the intestinal tract [8]. The vast majority of gut microbes play either commensal or mutualistic roles, including nutrient absorption and digestion, the regulation of endocrine functions and signaling to the brain, eliminating toxins, and producing crucial metabolites for the body. Importantly, the GM is the major generator of short-chain fatty acids (SCFAs) through the intestinal fermentation of dietary complex carbohydrates that cannot be metabolized by the host. SCFAs, along with their role as a nutrient, have an expansive repertoire of functional roles, including the regulation of glucose, lipid, and energy metabolism, modulation of gene expression, cell proliferation, and inflammation, and have localized effects on intestinal function [9]. Additionally, SCFAs act as major mediators of crosstalk between the GM and human host, as fluctuating levels of SCFAs are influenced by changes in gut microbial composition and are largely impacted by factors such as diet and physiological state. Bridging the environment–gut microbiome–host axis, it is not surprising that SCFAs have emerged as major factors in the development of metabolic disorders [10]. Their role in T1D has been recognized recently and is discussed elsewhere [11].

The numerous actions of SCFAs are carried out, in part, through interaction with their membrane receptors, which act both as sensors and mediators of their effects [12]. Two of the important SCFA receptors are free fatty acid receptors (FFAR), FFA2 and FFA3, which belong to the G-protein-coupled receptor family. With widespread tissue expression, the roles of FFA2 and FFA3 as major regulators of metabolism and immunity, as well as in the mediation of microbiome–host crosstalk, are emerging [12]. As discussed in subsequent sections, this is accomplished through a variety of mechanisms, including the modulation of insulin secretion in the pancreas, incretin secretion in the intestine, and the modulation of inflammatory responses. Furthermore, there is a growing body of data supporting the involvement of SCFA receptors in the pathogenesis of T1D [13]. In this review, we also explore their involvement in the pathogenesis of T1D by summarizing the current body of data derived from experimental mouse models and identify areas of interest for future research in this direction.

2. Gut Microbiome in T1D

Our understanding of the role of GM in T1D is based on rodent and human studies (summarized in [7,14]). In general, the intestinal microbiota of T1D and healthy subjects are distinctly different in mouse and human studies, with the former showing reduced microbial diversity and a reduced presence of butyrate producers along with proinflammatory dysbiosis [7,15]. For example, in Non-obese Diabetic (NOD) mice (susceptible to developing diabetes in an autoimmune fashion similar to humans) and Non-obese Diabetes Resistant (NOR) mice, the NOD microbiota contained more pathobionts compared to beneficial bacteria present in the NOR. The ileal NOD microbiota was reduced in segmented filamentous bacteria (SFB) and *Lactobacillus*spp., while the ileal NOR microbiota were reduced in *Anaeroplasma*spp. and *Desulfovibrio*spp. The colonic NOD microbiota was reduced in members of the Alphaproteobacteria class, *Ruminococcus gnavus*, and absent in *Bacteroides acidfaciens* when compared to NOR mice. The colonic NOR microbiota were absent in *Prevotella*spp. but detected in NOD mice. Importantly, SFB and *Lactobacillus* are associated with protection against the development of autoimmune diabetes [16,17], while expansion of strains such as *Prevotella* correlates with detrimental changes in the gut mucosal immune system [18]. The transfer of NOD gut microbes promoted pancreatic inflammation in NOR mice [19]. Blocking the interaction between GM and NOD host by genetic knockout of MYD88, an adapter protein responsible for activating receptors for gut microbial signals and lodging appropriate innate immune response, protected against T1D development. Interestingly, this protective effect was lost in absence of the gut microbiome and could be gained back through recolonization with NOD MYD88 knockout gut microbiota [20]. These results suggested that while the interaction of the GM with the innate immune system is a predisposing factor for T1D [20], it is the balance in gut microbial features affecting tolerance versus T1D development that affects the disease pathogenesis.

Gut microbial changes occurring in T1D have also been analyzed in human cohorts. Two studies utilizing the Environmental Determinants of Diabetes in the Young (TEDDY) cohort have yielded a functional profile of the developing gut metagenome, identifying the relationship between the microbiome and islet immunity and T1D, as well as other major childhood events [21,22]. Similar to observations in NOD mice, control children had higher beneficial bacteria such as *Bifidobacterium* and *Lactobacillus* [21,22], while in children with islet autoimmunity and T1D, there was a high prevalence of *Erythisopelotrichaecae* and fewer *Lactococcus*, *Streptococcus*, and *Akkermansia* [22]. Additionally, the GM of control children had more genes associated with fermentation and biosynthesis of SCFAs [21]. An earlier study comparing children prior to T1D onset (defined as presence of at least two β-cell-specific autoantibodies) with autoantibody-negative children also demonstrated that individuals with a greater number of autoantibodies to β cell antigens had a lower abundance of lactate and butyrate-producing gut bacteria [23].

Despite some key differences in rodent and human T1D forms [24], a pattern of GM modulation in the disease has emerged [7]. Crucial in this pattern are the observations that (a) deviation from optimal microbial homeostasis may lead to loss of self-tolerance and rise in proinflammatory signals [19,21,22,25], and (b) functional net effects of these deviations depend on co-occurring microbial communities and have often been manifested as changes in levels of plasma and fecal SCFAs [19,21,26,27]. In the animal form of T1D, the restoration of optimal GM community (either nutritionally or through fecal microbial transfer, FMT) alleviates and/or delays features of severe forms of the disease [15,28,29]. Similarly, GM restoration in human T1D has shown the benefits of an optimal GM community [14,30]. However, the interrelationship between GM changes and T1D, though apparent, has not been proven causal.

3. SCFAs, FFA2, and FFA3

SCFAs, mainly acetate, butyrate, and propionate, are gut microbiota fermentation byproducts of indigestible fiber. The majority of SCFAs are readily absorbed and utilized by the colonocytes as an energy source. Remaining SCFAs are drained into the hepatic and portal venous systems before emerging into the systemic circulation [31]. SCFAs affect host physiology in numerous ways, acting both as metabolic substrates and signaling molecules. Distinct GM and SCFA profiles in T1D versus controls provide compelling evidence for the roles of SCFA receptors in disease pathology. For a better understanding of this role, a review of known functions of FFA2 and FFA3 centering on the endocrine pancreas, immune cells, and gut is presented. Discussion of the roles of these receptors in other tissues has been covered elsewhere [32–34].

3.1. SCFAs as Signaling Molecules

SCFAs act as extracellular signaling molecules by binding to their cognate G-protein-coupled receptors (GPCRs), FFA2 and FFA3, which can bind all the three SCFAs but with discrete efficacies (Table 1) and G-protein coupling profiles. Due to their coexpression in tissues, shared endogenous ligands, and lack of selective synthetic ligands, defining their physiological roles has been challenging. However, recent studies using novel rodent models have provided an increased appreciation of the roles of these receptors in various metabolic and immune states [12].

Table 1. Affinity (EC50 in μM) of SCFAs at their cognate receptors.

SCFA	FFA2	FFA3
Acetate	35–431	>1000
Propionate	14–290	6–127
Butyrate	28–371	42–158
Pentanoate	>1000	42–152
Hexanoate	-	102–134

All values are for human receptors [35–37].

3.2. FFA2 and FFA3 Regulate β Cell Physiology

Both FFA2 and FFA3 are expressed in islets, predominantly in the β cells in both rodents and humans [34]. Pioneering work from the Layden Laboratory along with others has established the role of these receptors in the regulation of β cell function and mass [38–44]. Most of these effects are based upon distinct G-protein coupling preferences of FFA2 and FFA3. Upon SCFA binding, FFA2 can couple with either $G\alpha_{q/11}$ and $G\alpha_{i/o}$, thus exerting stimulatory or inhibitory effects on cellular function, respectively. FFA3, on the other hand, couples almost exclusively with $G\alpha_{i/o}$, with an inhibitory tone in its signaling [36]. Accordingly, in the islet, these receptors have opposing effects on insulin secretion: in both human and mouse islets, FFA3 inhibits insulin secretion in a $G\alpha_{i/o}$-dependent manner [38,42,45], whereas FFA2 activation may increase [39,44,46,47] or decrease insulin secretion [45], depending upon whether it couples to $G\alpha_{q/11}$ or $G\alpha_{i/o}$. Variance observed in FFA2 activity suggests that under any given condition, the effect of FFA2 activation on insulin secretion depends upon its preferred G-protein coupling [39,44]. This calls for the development of G-protein-biased ligands for FFA2. In fact, orthosteric FFA2 agonists SCA14, SCA15, and ZINC03832747 mediate the $G\alpha_{q/11}$-dependent increase in mouse islets or β cell insulin secretion in contrast to the allosteric agonists CMTB and CPTB that decrease insulin secretion via $G\alpha_{i/o}$ [39,48].

Mediation of β cell function by these receptors projects a similar profile in vivo. Whole-body deletion of FFA3 improves insulin secretion and glucose tolerance both under high fat diet induced metabolic stress [42,43,45] and a regular diet [42,43]. Correspondingly, β-cell-specific FFA3 overexpression deteriorates glucose responsiveness in mice [42]. These effects appear to be a β cell secretory phenotype, as no changes in insulin sensitivity have been observed [42,43,45]. Additionally, gene expression analysis of islets from FFA3 knockout mice [39] or β-cell-specific FFA3 overexpression mice [42] revealed complementary changes (i.e., downregulation in knockout and upregulation in overexpression model) in genes related to inflammation and immune response (such as IL1β, IL1α, CD80), besides changes in genes of calcium response and glucose utilization pathways.

Evaluation of the in vivo roles of FFA2, similar to the ex vivo data, has yielded conflicting results. Mice globally lacking FFA2 exhibited fasting hyperglycemia, reduced insulin secretion, and glucose intolerance under dietary metabolic stress [39,44]. In contrast, another study has reported a phenotype of improved glucose tolerance and enhanced insulin secretion in FFA2 knockout mice [45]. Additionally, in this same study, FFA2 and FFA3 double knockout or FFA3 knockout in combination with β-cell-specific FFA2 knockout improved glucose tolerance and insulin secretion under metabolic stress. These conflicting data may arise from differences in the G-protein coupling of activated FFA2, roles of FFA2 in other metabolically active tissues, impact of gut microbiome, and/or duration of metabolic stress, besides receptor-independent effects of SCFAs [49–51].

SCFA receptors, specifically FFA2, also modulate β cell mass [40,41,44]. FFA2 is required for the prenatal establishment of β cell mass, as FFA2 knockout mouse neonates and 21-day-old weanlings exhibit impaired β cell mass at birth and throughout adulthood [41]. Under conditions of dietary metabolic stress [44] and pregnancy [40], when β cells are compensating for insulin resistance, this deficiency in β cell mass is magnified. FFA2 activation, as a matter of fact, increases β cell proliferation [41,44], enhances the expression of genes involved in β cell differentiation [44], and reduces cytokine- and palmitate-induced β cell apoptosis [46,47]. FFA3, on the other hand, as a $G\alpha_{i/o}$-coupling receptor, may restrict β cell mass [52]. However, FFA3 knockout mice islets have been reported to be smaller with reduced proliferation and number of β cells [42], an effect not seen in a later study [43]. Similarly, β-cell-specific FFA3 overexpression in mice shows compensatory increased β cell proliferation and area [42]. Collectively, these data highlight the role of SCFA receptors FFA2 and FFA3 in modulating β cell function and mass. Importantly, defects in these two features are fundamental to the pathology of T1D.

3.3. *FFA2 and FFA3 Modulate Incretin Secretion*

In addition to their role within pancreatic islets, SCFA and their receptors are suggested to participate in the secretion of incretin hormones. In the upper intestine, SCFA concentrations range from 0.1 to 1 mM and are largely produced by oral microbiota [53]. By contrast, luminal SCFAs in the colon can reach levels of up to 100 mM due to the fermentation of dietary fibers via the gut microbiota. Within the intestine, SCFA receptors FFA2 and FFA3 are thought to act as sensors of these metabolites, and many important actions are carried out through this signaling.

Secreted by enteroendocrine cells (EECs) embedded within the intestinal epithelium, incretin hormones are peptide hormones that stimulate the release of insulin in response to nutrient intake, thereby lowering the level of circulating blood glucose [54]. Additionally, incretin hormones facilitate numerous postprandial metabolic functions, including lowering food intake, gastric emptying, and increasing cardiac output [55]. There are two primary incretin hormones: glucagon-like peptide-1 (GLP-1) and gastric inhibitory polypeptide (GIP). While both hormones carry out their various functions through the binding of their specific receptors (GLP-1R and GIPR, respectively) on the surface of various tissues, they contribute to the regulation of glucose metabolism in distinct mechanisms. Although both stimulate insulin release through $G\alpha$/cAMP at β cells, in islet α cells, GLP-1 suppresses glucagon, while GIP increases it [56]. Both hormones also protect β cell mass by inhibiting apoptosis. GLP-1 and GIP play important roles in the control of glucose levels after a meal via the physiological response known as the "incretin effect." This occurs when higher levels of glucose-stimulated insulin secretion are observed when glucose is administered orally rather than intravenously, an effect that is lost in type 2 diabetes but preserved in T1D [57].

FFA2 and FFA3 are broadly expressed within EECs throughout the gastrointestinal tract. EECs are divided into subtypes based on the peptide hormone they express and secrete [58]. Using immunohistochemical analysis and a *Ffar2*-red fluorescent protein (RFP) reporter mouse, FFA2 has been found to colocalize with peptide YY (PYY)/GLP-1 containing L cells in rodents and humans [59–62]. Using in situ hybridization and a *Ffar3*-RFP reporter mouse, FFA3 expression has been confirmed in several types of intestinal cells, including PYY-positive cells [59,63,64]. However, due to the concentrated expression of FFA3 in enteric ganglia and sympathetic ganglia, it is uncertain if its effects on EEC function arise from its expression in enterocytes or are secondary to its mediation of enteric neuronal function [59,65,66]. Further, transcriptomic analysis has found high coexpression of both receptors with *gip*, indicating their possible involvement in mediating GIP secretion [67].

Several studies have shown that stimulation of FFA2 by SCFA results in an increased secretion of GLP-1 from EECs in the intestine. Primary intestine cells harvested from global FFA2 knockout mice showed reduced GLP-1 release in vitro, and another study by the same group found that propionate was able to stimulate GLP-1 release in vivo only in the wild-type mice [68,69]. Besides GLP-1, studies have also documented the role of FFA2 in mediating GIP [44] and PYY secretion, with the latter in both humans and mice [44,70,71]. Controversy exists for this role of FFA2, however, with some studies reporting no difference in basal- and glucose-stimulated GLP-1 levels in FFA2 knockout mice compared to control mice [40,45].

For FFA3, while there is a paucity of data regarding its role in gut hormone secretion, FFA3 knockout mice have reduced GLP-1 and PYY secretion, and primary colonic cultures derived from these mice display impaired secretory response upon SCFA stimulation [63,69]. Predictably, FFA3-specific agonist enhances GLP-1 release from primary colonic cultures [59]. FFA3 is also implicated in the inhibition of GIP secretion, an effect more likely with the predominant $G\alpha_{i/o}$ coupling of the receptor [72].

More research is needed to ascertain the respective roles of FFA2 and FFA3 in the regulation of incretin hormones in the intestine. This includes the development of potent selective ligands and tissue-specific knockout mouse models. Some progress has already been made in this direction. Selective and potent human FFA2 inverse agonists have been developed and shown to stimulate GLP-1 secretion in the human EEC line, NCI-H716 [73]. Through a chemogenetic knock-in strategy, mice with designer receptors exclusively activated by the designer drugs (DREADD) variant of human *FFAR2* replacing the mouse *Ffar2* locus have been generated. DREADD activation in these mice has been shown to augment GLP-1 secretion in colonic crypt cultures and in vivo [74].

3.4. SCFAs, FFA2, and FFA3 Educate the Gut Immune Cells and Regulate Inflammation

SCFAs can regulate immune cell function via two major processes, either through their cognate GPCRs such as FFA2 and FFA3 or by modulating histone deacetylase (HDAC) activity [75]. Here, we emphasize the first role. With the highest SCFA concentration, a rich and diverse population of immune cells with the majority expressing FFA2 and few, such as dendritic cells expressing FFA3, the gut is an important site where SCFAs can impact the immune cells through these receptors.

FFA2 has been reported to affect neutrophil chemotaxis in gut inflammation models, where a deficiency in FFA2 increases neutrophil infiltration to sites of inflammation [76,77]. Accordingly, engagement of FFA2 with acetate mitigates such a response [76,77]. Similar FFA2-dependent neutrophil-driven responses are seen in pulmonary and joint inflammation models [77]. Neutrophil FFA2 engagement by SCFAs in the presence of allosteric modulators can also activate NADPH oxidase and enhance the production of reactive oxygen species, which is deemed necessary for phagocytic activity [77,78]. FFA2-derived neutrophil responses are required for the regulation of inflammatory responses. As recently shown, FFA2 activity promotes coordination between neutrophils and colonic group 3 innate lymphoid cells (ILC3). In neutrophils, inflammasome activation helps in pathogen clearance with the concomitant enhancement of IL1β production, where IL1β leads to IL22 production from ILC3, driving gut repair mechanisms [79]. FFA2 can also promote ILC3 expansion and function independent of neutrophils [80]. Immunomodulation by neutrophil FFA2, thus, appears to strike a balance between pro- and anti-inflammatory effects, potentially in a disease-centric manner.

SCFAs through their HDAC inhibitory activity are considered to be the main players in maintaining the regulatory T cell (Treg) pools [75]. These effects are, in part, mediated through FFA2. It is suggested that FFA2 exerts immune suppression by regulating the number, function, and differentiation of Tregs [81,82]. FFA2 also modulates gut homeostasis by modifying immunoglobulin A (IgA) production [83] and through direct effects on inflammasome activation in intestinal epithelial cells [84].

The role of FFA3 in immune regulation is less explored, likely due to its limited expression in immune cells. It has been suggested to be involved in the resolution of lung inflammation through effects on macrophage and dendritic cell populations [85] and in promoting thymic Treg differentiation in mouse offspring [86]. The function and expansion of $CD8^+$ T cells can also be regulated by FFA3, and this has been suggested to aid the resolution of influenza infection [87]. Both FFA3 and FFA2 have been suggested to enhance T cell memory [88], with the engagement of both receptors by butyrate appearing to mediate this effect. However, as the mouse isoform of FFA2 shows a low affinity for butyrate [31], the use of synthetic ligands is required to further delineate the role of the two receptors in this process.

3.5. SCFAs, FFA2, and FFA3, and Gut Microbiome: It Takes Three to Tango

Obliterating the GM in mice wipes off some of the physiological effects discussed above. For instance, in germ-free (GF) mice, antigen-activated T cells fail to transition to memory cells [88]. As the GM does not directly interact with the host cells except at the gut mucosal surfaces, these effects are likely indirect, being mediated via GM-derived factors such as SCFAs. The GM-derived SCFAs acting through their receptors FFA2 and FFA3 project the link, GM→SCFAs→FFA2 and FFA3.

Highlighting this relationship, whole-body FFA2 and FFA3 knockout mice have different gut microbiota profiles as compared to wild-type mice [44,63,82,84,89]. As expected, differences in fecal SCFA profiles accompany these differences in GM profiles due to adaptation to receptor deficiency. More direct evidence for roles of FFA2 and FFA3 in the GM→SCFAs→Receptor link is provided by immune function and GM studies. Both GF and FFA2 knockout mice exhibit a dysregulated immune response to induced colitis, gout, and arthritis [77,90]. While this response is mitigated by acetate supplemented drinking water in GF mice [77], FFA2 knockout mice remain refractory to acetate treatment [81].

Likewise, high-fiber diets that tend to increase GM function and SCFA levels in wild-type mice fail to promote gut homeostasis, alleviate food allergy, and prevent respiratory viral infection in FFA2 knockout mice [82,84,91]. Similar findings have been reported for FFA3. High fiber diet mediated protection against allergic airway disease and influenza virus is not observed in FFA3 knockout mice [85,87].

Metabolic studies have also highlighted the importance of the GM→SCFAs→FFA2 and FFA3 relationship. Reduced adiposity and PYY levels in FFA3 knockout mice are GM dependent, with the effect being lost in GF FFA3 knockout mice [63]. Similarly, GM-derived SCFAs mediate suppression of GIP secretion in FFA3 dependent manner, an effect lost in GF, antibiotic-treated mice (pseudo-GF), and FFA3 knockout mice [72]. In mice, a low-fiber diet or GF status during pregnancy increases the vulnerability of the offspring to obesity and insulin resistance later in life [92]. This effect could be rescued by propionate treatment or high-fiber feeding but not in absence of FFA3 or FFA2. Furthermore, FFA3 and FFA2 SCFA signaling was found to be responsible for normal embryonic development of neural, pancreatic β cell and intestine tissues. Collectively, these data suggest that the GM modulates metabolic and immune features affecting health via SCFA-FFA2 and SCFA-FFA3 axes.

4. FFA2 and FFA3 Mediating GM–Host Crosstalk in T1D

The identification of a role for FFA2 and FFA3 signaling in T1D is a budding area of research. Although there is still only sparse and indirect evidence, there is clinical interest in pursuing this area in the fight against T1D (discussed under "FFA2, FFA3, and T1D: Clinical Interests" (Section 5)). The main mechanisms linking GM to T1D include the GM-mediated influence on the development and homeostasis of the immune system and the effects of the GM upon influence on gut barrier integrity. As noted from GF and gnotobiotic mice studies, GM composition affects the development of gut-associated lymphoid tissue (GALT), the expansion and differentiation of specific T cell subsets (Foxp3$^+$ Tregs and Th17 cells), and IgA-secreting B cells [93–96]. The gut barrier guards against the entry of pathogenic microbes and their components into the host circulation and tissues. Disruption of the gut barrier has been noted in both human T1D and T1D animal models [97–99]. This is manifested in humans by increased gut permeability [100] and serum levels of the gut barrier marker, zonulin [101]. In mice, it has been reported that there is activation of islet-specific diabetogenic T cells in the gut and their translocation along with gut bacteria to pancreatic lymph nodes [99,102].

Considering (1) the role FFA2 and FFA3 play in immune homeostasis and gut epithelial integrity, (2) dysbiotic gut microbiome of T1D, and (3) altered serum and fecal SCFAs in T1D, it can be argued that these receptors are likely to be important regulators of T1D immune responses (Figure 1). Along these lines, peripheral blood monocytes in T1D subjects show high FFA2 expression [103], and FFA3 expression has been correlated with inflammation and metabolic markers [104]. The first indication of the involvement of these receptors in T1D pathogenesis came from the work of Marino et al., [15], where the feeding of acetate yielding diets to NOD mice promoted immune tolerance and delayed progression to T1D by reducing autoreactive T cells, increasing Tregs, and improving gut barrier integrity. These effects were FFA2 dependent, as the anti-T1D potency of these diets was lost in NOD mice deficient in FFA2. Feeding a butyrate yielding diet, on the other hand, could confer partial protection from T1D to NOD FFA2-deficient mice. This indirectly suggested the involvement of butyrate-favoring receptor FFA3 and/or the butyrate-activated GPR109a or receptor-independent effects of butyrate. This study thus suggests that engagement of FFA2 and/or FFA3 by dietary SCFAs plays an important role in modulating inflammatory responses in T1D. A similar protective role of FFA2 was also observed in a streptozotocin-induced mouse model of T1D [103]. In diseased mice, treatment with specific FFA2 agonists attenuated islet inflammation by inducing apoptosis of infiltrating macrophages. Furthermore, these receptors, possibly through the $G\alpha_{i/o}$-dependent pathway, were shown to promote a tolerogenic pancreatic immune environment by regulating islet production of immunoregulatory cathelicidin-related antimicrobial peptide (CRAMP) [105]. These effects, however, have not been replicated in humans. The use of FFA2 agonists in human studies is hampered due to differences in the signaling of mouse and human isoforms of the receptor. The more attractive option, oral administration of SCFAs, for example butyrate, in long-standing T1D subjects, unfortunately, has also not shown any benefits [106]. Whether or not oral SCFA administration prior to T1D onset can delay or lessen T1D pathology needs exploration. Moreover, FFA2 can also augment β cell function and protection, and both FFA2 and FFA3 can affect incretin secretion. However, it has not yet been explored whether these effects are pertinent to T1D.

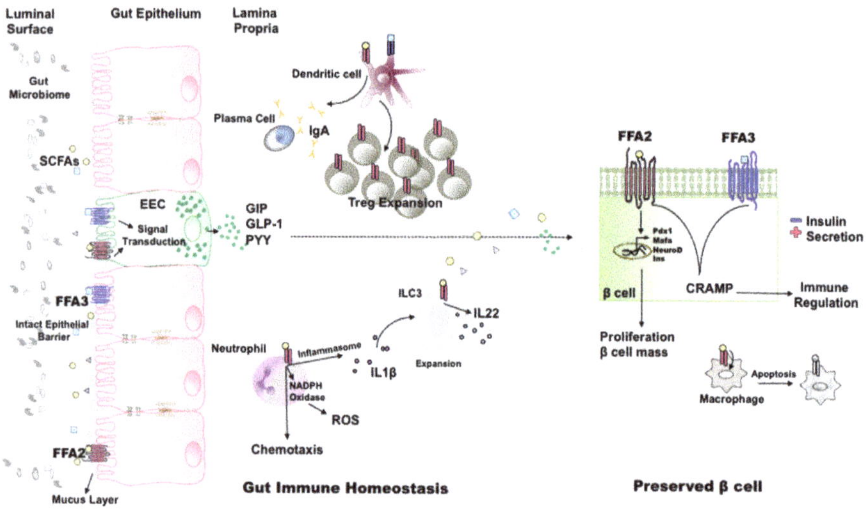

Figure 1. Role of FFA2 and FFA3 in gut immune homeostasis and β cell physiology in the context of T1D. The engagement of FFA2 and FFA3 on gut epithelial and enteroendocrine (EEC) cells by gut microbial metabolites, short-chain fatty acids (SCFAs), regulates epithelial barrier integrity and the secretion of various incretin hormones. FFA2 signaling on various gut immune cells promotes an anti-inflammatory and tolerogenic environment. Neutrophil FFA2 affects chemotaxis, the production of reactive oxygen species (ROS), and IL1β. FFA2 activation on innate lymphoid cell 3 (ILC3) directly and in conjugation with neutrophil-released IL1β promotes IL22 production. FFA2 also promotes ILC3 and regulatory T cell (Treg) expansion. Receptor activation on dendritic cells contributes to B cell (plasma cell) differentiation and IgA release. Altogether, SCFA receptor activity promotes an anti-inflammatory state that, in turn, suppresses the immune destruction of β cells. Factors produced in the gut, such as incretin hormones and SCFAs, and possibly immunosuppressive immune cells such as Tregs, travel to pancreatic β cells, influencing their physiology. In β cells, FFA2 stimulates insulin secretion and proliferation and is essential for the establishment and preservation of β cell mass. FFA3 activation reduces insulin secretion. Both receptors possibly stimulate the production of immunomodulatory cathelicidin-related antimicrobial peptide (CRAMP). FFA2 activation on infiltrating proinflammatory macrophages causes immune cell apoptosis.

5. FFA2, FFA3, and T1D: Clinical Interests

From the above discussion, a GM→SCFAs→FFA2 and FFA3→T1D link is apparent. This link has opened exciting avenues of research for identifying novel targets to treat and prevent T1D. The first question raised is if the modification of GM, which is achievable through the use of probiotics and prebiotics [107], impacts aspects of T1D disease. Probiotics are live microorganisms that, when administered in adequate amounts, confer a health benefit on the host, while prebiotics are substrates utilized by host microorganisms conferring a health benefit [108]. Several studies on spontaneous and pharmacologic rodent models of T1D have revealed that pro- and prebiotics favor a tolerogenic gut immune environment by promoting gut barrier integrity, stimulating the secretion of anti-inflammatory cytokines, and restricting the number of inflammatory T cell subsets besides increasing the abundance of beneficial gut bacteria [107]. Similar effects have been observed in human T1D trials [14,107]. Notably, early probiotic exposure in children genetically predisposed to T1D reduced the risk of the disease [109]. Interestingly, some probiotics such as *L. kefiranofaciens M* and *L. kefiri K* promoted GLP-1 secretion in a streptozotocin-induced T1D mouse model [110]. Presumably, FFA2 and FFA3 in EECs and gut immune cells are involved in some of these effects. This presumption may especially hold true for FFA2. For example, *Bifidobacterium animalis subsp. lactis* GCL2505 (GCL2505), a probiotic, increases acetate levels and engages FFA2 to exert beneficial metabolic effects in a diet-induced obesity mouse model [111]. Similarly, prebiotic fructo-oligosaccharide supplementation in rats increased the density of GLP-1 producing L cells coexpressing FFA2 [60]. Additionally, dietary supplementation with the microbially derived SCFAs, acetate and butyrate, ameliorated T1D β cell damage and immune dysfunction in an FFA2- and possibly FFA3-dependent manner [15]. Finally, FMT has appeared on the landscape of numerous anti-T1D interventions as another method to re-engineer the GM to positively impact T1D [30]. These different methods of modifying the GM for modulating the course of T1D seem potentially promising. However, we are still far from their actual clinical use. Human trials and animal studies have yielded variable outcomes ranging from no to even adverse effects [106,112]. Furthermore, the durability of these effects is unclear, while our understanding of gut microbial metabolite mediated cellular effects is incomplete.

Another new area of research in T1D therapeutics is the specific targeting of the receptors FFA2 and FFA3. Although this is seemingly straightforward compared to T1D GM modification, it comes with added complexities, as enumerated before [12]. Briefly, these are, biased G-protein coupling (FFA2) [113], species differences in GPCR signaling (FFA2) [37,114], multiple downstream effectors of the activated receptors yielding discrete physiological responses (FFA2 and FFA3) [36,37,44,66,115–117], lack of species-specific and G-protein-specific ligands (FFA2), lack of understanding of tissue-specific roles of these receptors, dependence of receptor expression on various factors such as diet [118], multiple factors such as diet, and multiple genetic polymorphisms in *FFA2* and *FFA3* without known associations with clinical/disease phenotypes [119]. Furthermore, how genetic predisposition to T1D affects SCFA receptor expression and activity in various tissues and conversely if polymorphisms in SCFA receptors confer risk to develop autoimmunity is not known. Despite these complexities, owing to the multiple ways these receptors can affect T1D, it is worthwhile to try closing these gaps in our knowledge and develop receptor-based T1D interventions. One step forward in this direction is to develop preclinical mouse models, such as NOD with double receptor knockout and the tissue-specific knockout of FFA2 and FFA3, for the precise delineation of receptor-mediated effects. Another approach is to generate humanized mouse models expressing human isoforms of the receptors globally and in tissue-specific manners through knock-in and chemogenetic approaches that will provide excellent ways to monitor effects of human receptor signaling in in vivo settings and identify unique signaling responses to new ligands. The generation of such models can be used to demonstrate the functional implications of GM changes occurring in T1D.

6. Conclusions

Mounting evidence indicates an effect of the GM upon T1D pathology mediated through the modulation of gut and pancreatic immune environments. With possible roles of the SCFA-activated GPCRs, FFA2 and FFA3, in mediating these effects, a highly relevant GM→SCFAs→FFA2/FFA3→T1D link is apparent. Although a mechanistic understanding of the interrelationships of different entities in this link is still not complete, novel therapeutic interventions against T1D based on this link are likely to emerge.

Author Contributions: Conceptualization: M.P.; funding acquisition: B.T.L.; resources: B.T.L.; supervision: M.P., B.T.L.; writing (original draft): M.P., K.L., K.X., S.G., B.W. and B.T.L.; writing (editing and revising): M.P., K.L., K.X., B.W. and B.T.L. All authors have read and agreed to the published version of the manuscript.

Funding: National Institutes of Health under Award Number R01DK104927-01A1; University of Chicago Diabetes Research and Training Center (P30DK020595); and Department of Veterans' Affairs, Veterans Health Administration, Office of Research and Development, VA merit (Grant No. 1I01BX003382-01-A1) for B.T.L.

Conflicts of Interest: The authors declare no conflict of interest.

References

1. Redondo, M.J.; Steck, A.K.; Pugliese, A. Genetics of type 1 diabetes. *Pediatric Diabetes* **2018**, *19*, 346–353. [CrossRef] [PubMed]
2. Vieira, A.T.; Fukumori, C.; Ferreira, C.M. New insights into therapeutic strategies for gut microbiota modulation in inflammatory diseases. *Clin. Transl. Immunol.* **2016**, *5*, e87. [CrossRef]
3. Livanos, A.E.; Greiner, T.U.; Vangay, P.; Pathmasiri, W.; Stewart, D.; McRitchie, S.; Li, H.; Chung, J.; Sohn, J.; Kim, S.; et al. Antibiotic-mediated gut microbiome perturbation accelerates development of type 1 diabetes in mice. *Nat. Microbiol.* **2016**, *1*, 16140. [CrossRef] [PubMed]
4. Boursi, B.; Mamtani, R.; Haynes, K.; Yang, Y.-X. The effect of past antibiotic exposure on diabetes risk. *Eur. J. Endocrinol.* **2015**, *172*, 639–648. [CrossRef]
5. Howard, S.G. Developmental Exposure to Endocrine Disrupting Chemicals and Type 1 Diabetes Mellitus. *Front. Endocrinol.* **2018**, *9*, 513. [CrossRef]
6. Aguilera, M.; Gálvez-Ontiveros, Y.; Rivas, A. Endobolome, a New Concept for Determining the Influence of Microbiota Disrupting Chemicals (MDC) in Relation to Specific Endocrine Pathogenesis. *Front. Microbiol.* **2020**, *11*, 578007. [CrossRef]
7. Siljander, H.; Honkanen, J.; Knip, M. Microbiome and type 1 diabetes. *EBioMedicine* **2019**, *46*, 512–521. [CrossRef]
8. Fan, Y.; Pedersen, O. Gut microbiota in human metabolic health and disease. *Nat. Rev. Microbiol.* **2021**, *19*, 55–71. [CrossRef]
9. Koh, A.; de Vadder, F.; Kovatcheva-Datchary, P.; Bäckhed, F. From Dietary Fiber to Host Physiology: Short-Chain Fatty Acids as Key Bacterial Metabolites. *Cell* **2016**, *165*, 1332–1345. [CrossRef]
10. Venegas, D.P.; de la Fuente, M.K.; Landskron, G.; González, M.J.; Quera, R.; Dijkstra, G.; Harmsen, H.J.M.; Faber, K.N.; Hermoso, M.A. Short Chain Fatty Acids (SCFAs)-Mediated Gut Epithelial and Immune Regulation and Its Relevance for Inflammatory Bowel Diseases. *Front. Immunol.* **2019**, *10*, 277. [CrossRef]
11. Fenneman, A.C.; Rampanelli, E.; Yin, Y.S.; Ames, J.; Blaser, M.J.; Fliers, E.; Nieuwdorp, M. Gut microbiota and metabolites in the pathogenesis of endocrine disease. *Biochem. Soc. Trans.* **2020**, *48*, 915–931. [CrossRef] [PubMed]
12. Priyadarshini, M.; Kotlo, K.U.; Dudeja, P.K.; Layden, B.T. Role of Short Chain Fatty Acid Receptors in Intestinal Physiology and Pathophysiology. *Compr. Physiol.* **2018**, *8*, 1091–1115. [CrossRef] [PubMed]
13. Priyadarshini, M.; Navarro, G.; Layden, B.T. Gut Microbiota: FFAR Reaching Effects on Islets. *Endocrinology* **2018**, *159*, 2495–2505. [CrossRef] [PubMed]
14. Yap, Y.A.; Mariño, E. Dietary SCFAs Immunotherapy: Reshaping the Gut Microbiota in Diabetes. *Adv. Exp. Med. Biol.* **2021**, *1307*, 499–519. [CrossRef] [PubMed]
15. Mariño, E.; Richards, J.L.; McLeod, K.H.; Stanley, D.; Yap, Y.A.; Knight, J.; McKenzie, C.; Kranich, J.; Oliveira, A.C.; Rossello, F.J.; et al. Gut microbial metabolites limit the frequency of autoimmune T cells and protect against type 1 diabetes. *Nat. Immunol.* **2017**, *18*, 552–562. [CrossRef]
16. Kriegel, M.A.; Sefik, E.; Hill, J.A.; Wu, H.-J.; Benoist, C.; Mathis, D. Naturally transmitted segmented filamentous bacteria segregate with diabetes protection in nonobese diabetic mice. *Proc. Natl. Acad. Sci. USA* **2011**, *108*, 11548–11553. [CrossRef] [PubMed]
17. Roesch, L.F.W.; Lorca, G.L.; Casella, G.; Giongo, A.; Naranjo, A.; Pionzio, A.M.; Li, N.; Mai, V.; Wasserfall, C.H.; Schatz, D.; et al. Culture-independent identification of gut bacteria correlated with the onset of diabetes in a rat model. *ISME J.* **2009**, *3*, 536–548. [CrossRef]
18. Paun, A.; Yau, C.; Danska, J.S. The Influence of the Microbiome on Type 1 Diabetes. *J. Immunol.* **2017**, *198*, 590–595. [CrossRef]
19. Brown, K.; Godovannyi, A.; Ma, C.; Zhang, Y.; Ahmadi-Vand, Z.; Dai, C.; Gorzelak, M.A.; Chan, Y.; Chan, J.M.; Lochner, A.; et al. Prolonged antibiotic treatment induces a diabetogenic intestinal microbiome that accelerates diabetes in NOD mice. *ISME J.* **2016**, *10*, 321–332. [CrossRef]
20. Wen, L.; Ley, R.E.; Volchkov, P.Y.; Stranges, P.B.; Avanesyan, L.; Stonebraker, A.C.; Hu, C.; Wong, F.S.; Szot, G.L.; Bluestone, J.A.; et al. Innate immunity and intestinal microbiota in the development of Type 1 diabetes. *Nature* **2008**, *455*, 1109–1113. [CrossRef]
21. Vatanen, T.; Franzosa, E.A.; Schwager, R.; Tripathi, S.; Arthur, T.D.; Vehik, K.; Lernmark, Å.; Hagopian, W.A.; Rewers, M.J.; She, J.-X.; et al. The human gut microbiome in early-onset type 1 diabetes from the TEDDY study. *Nature* **2018**, *562*, 589–594. [CrossRef]
22. Stewart, C.J.; Ajami, N.J.; O'Brien, J.L.; Hutchinson, D.S.; Smith, D.P.; Wong, M.C.; Ross, M.C.; Lloyd, R.E.; Doddapaneni, H.; Metcalf, G.A.; et al. Temporal development of the gut microbiome in early childhood from the TEDDY study. *Nature* **2018**, *562*, 583–588. [CrossRef]
23. De Goffau, M.C.; Luopajärvi, K.; Knip, M.; Ilonen, J.; Ruohtula, T.; Härkönen, T.; Orivuori, L.; Hakala, S.; Welling, G.W.; Harmsen, H.J.; et al. Fecal Microbiota Composition Differs Between Children With -Cell Autoimmunity and Those Without. *Diabetes* **2013**, *62*, 1238–1244. [CrossRef] [PubMed]

24. Dedrick, S.; Sundaresh, B.; Huang, Q.; Brady, C.; Yoo, T.; Cronin, C.; Rudnicki, C.; Flood, M.; Momeni, B.; Ludvigsson, J.; et al. The Role of Gut Microbiota and Environmental Factors in Type 1 Diabetes Pathogenesis. *Front. Endocrinol.* **2020**, *11*, 78. [CrossRef]
25. Kostic, A.D.; Gevers, D.; Siljander, H.; Vatanen, T.; Hyötyläinen, T.; Hämäläinen, A.-M.; Peet, A.; Tillmann, V.; Pöhö, P.; Mattila, I.; et al. The Dynamics of the Human Infant Gut Microbiome in Development and in Progression toward Type 1 Diabetes. *Cell Host Microbe* **2015**, *17*, 260–273. [CrossRef]
26. De Groot, P.F.; Belzer, C.; Aydin, O.; Levin, E.; Levels, J.H.; Aalvink, S.; Boot, F.; Holleman, F.; van Raalte, D.H.; Scheithauer, T.P.; et al. Distinct fecal and oral microbiota composition in human type 1 diabetes, an observational study. *PLoS ONE* **2017**, *12*, e0188475. [CrossRef]
27. Xiao, L.; Land, B.V.; Engen, P.A.; Naqib, A.; Green, S.J.; Nato, A.; Leusink-Muis, T.; Garssen, J.; Keshavarzian, A.; Stahl, B.; et al. Human milk oligosaccharides protect against the development of autoimmune diabetes in NOD-mice. *Sci. Rep.* **2018**, *8*, 3829. [CrossRef]
28. Needell, J.C.; Ir, D.; Robertson, C.E.; Kroehl, M.E.; Frank, D.N.; Zipris, D. Maternal treatment with short-chain fatty acids modulates the intestinal microbiota and immunity and ameliorates type 1 diabetes in the offspring. *PLoS ONE* **2017**, *12*, e0183786. [CrossRef] [PubMed]
29. Chhabra, P.; Spano, A.J.; Bowers, D.; Ren, T.; Moore, D.J.; Timko, M.P.; Wu, M.; Brayman, K.L. Evidence for the Role of the Cecal Microbiome in Maintenance of Immune Regulation and Homeostasis. *Ann. Surg.* **2018**, *268*, 541–549. [CrossRef]
30. De Groot, P.; Nikolic, T.; Pellegrini, S.; Sordi, V.; Imangaliyev, S.; Rampanelli, E.; Hanssen, N.; Attaye, I.; Bakker, G.; Duinkerken, G.; et al. Faecal microbiota transplantation halts progression of human new-onset type 1 diabetes in a randomised controlled trial. *Gut* **2021**, *70*, 92–105. [CrossRef]
31. Priyadarshini, M.; Wicksteed, B.; Schiltz, G.E.; Gilchrist, A.; Layden, B.T. SCFA Receptors in Pancreatic β Cells: Novel Diabetes Targets? *Trends Endocrinol. Metab.* **2016**, *27*, 653–664. [CrossRef]
32. Krautkramer, K.A.; Fan, J.; Bäckhed, F. Gut microbial metabolites as multi-kingdom intermediates. *Nat. Rev. Microbiol.* **2021**, *19*, 77–94. [CrossRef]
33. Husted, A.S.; Trauelsen, M.; Rudenko, O.; Hjorth, S.A.; Schwartz, T.W. GPCR-Mediated Signaling of Metabolites. *Cell Metab.* **2017**, *25*, 777–796. [CrossRef]
34. Tang, C.; Offermanns, S. FFA2 and FFA3 in Metabolic Regulation. *Handb. Exp. Pharmacol.* **2016**, *236*, 205–220. [CrossRef]
35. Brown, A.J.; Goldsworthy, S.M.; Barnes, A.A.; Eilert, M.M.; Tcheang, L.; Daniels, D.; Muir, A.I.; Wigglesworth, M.J.; Kinghorn, I.; Fraser, N.J.; et al. The Orphan G Protein-coupled Receptors GPR41 and GPR43 Are Activated by Propionate and Other Short Chain Carboxylic Acids. *J. Biol. Chem.* **2003**, *278*, 11312–11319. [CrossRef]
36. Le Poul, E.; Loison, C.; Struyf, S.; Springael, J.-Y.; Lannoy, V.; Decobecq, M.-E.; Brezillon, S.; Dupriez, V.; Vassart, G.; van Damme, J.; et al. Functional Characterization of Human Receptors for Short Chain Fatty Acids and Their Role in Polymorphonuclear Cell Activation. *J. Biol. Chem.* **2003**, *278*, 25481–25489. [CrossRef]
37. Hudson, B.D.; Tikhonova, I.G.; Pandey, S.K.; Ulven, T.; Milligan, G. Extracellular Ionic Locks Determine Variation in Constitutive Activity and Ligand Potency between Species Orthologs of the Free Fatty Acid Receptors FFA2 and FFA3. *J. Biol. Chem.* **2012**, *287*, 41195–41209. [CrossRef] [PubMed]
38. Priyadarshini, M.; Layden, B.T. FFAR3 modulates insulin secretion and global gene expression in mouse islets. *Islets* **2015**, *7*, e1045182. [CrossRef] [PubMed]
39. Priyadarshini, M.; Villa, S.R.; Fuller, M.; Wicksteed, B.; Mackay, C.R.; Alquier, T.; Poitout, V.; Mancebo, H.; Mirmira, R.G.; Gilchrist, A.; et al. An Acetate-Specific GPCR, FFAR2, Regulates Insulin Secretion. *Mol. Endocrinol.* **2015**, *29*, 1055–1066. [CrossRef] [PubMed]
40. Fuller, M.; Priyadarshini, M.; Gibbons, S.M.; Angueira, A.R.; Brodsky, M.; Hayes, M.G.; Kovatcheva-Datchary, P.; Bäckhed, F.; Gilbert, J.A.; Lowe, W.L., Jr.; et al. The short-chain fatty acid receptor, FFA2, contributes to gestational glucose homeostasis. *Am. J. Physiol. Endocrinol. Metab.* **2015**, *309*, E840–E851. [CrossRef]
41. Villa, S.R.; Priyadarshini, M.; Fuller, M.H.; Bhardwaj, T.; Brodsky, M.R.; Angueira, A.R.; Mosser, R.E.; Carboneau, B.A.; Tersey, S.A.; Mancebo, H.; et al. Loss of Free Fatty Acid Receptor 2 leads to impaired islet mass and beta cell survival. *Sci. Rep.* **2016**, *6*, 28159. [CrossRef]
42. Veprik, A.; Laufer, D.; Weiss, S.; Rubins, N.; Walker, M.D. GPR41 modulates insulin secretion and gene expression in pancreatic β-cells and modifies metabolic homeostasis in fed and fasting states. *FASEB J.* **2016**, *30*, 3860–3869. [CrossRef]
43. Priyadarshini, M.; Cole, C.; Oroskar, G.; Ludvik, A.E.; Wicksteed, B.; He, C.; Layden, B.T. Free fatty acid receptor 3 differentially contributes to β-cell compensation under high-fat diet and streptozotocin stress. *Am. J. Physiol. Integr. Comp. Physiol.* **2020**, *318*, R691–R700. [CrossRef]
44. McNelis, J.C.; Lee, Y.S.; Mayoral, R.; van der Kant, R.; Johnson, A.M.F.; Wollam, J.; Olefsky, J.M. GPR43 Potentiates β-Cell Function in Obesity. *Diabetes* **2015**, *64*, 3203–3217. [CrossRef] [PubMed]
45. Tang, C.; Ahmed, K.; Gille, A.; Lu, S.; Gröne, H.-J.; Tunaru, S.; Offermanns, S. Loss of FFA2 and FFA3 increases insulin secretion and improves glucose tolerance in type 2 diabetes. *Nat. Med.* **2015**, *21*, 173–177. [CrossRef]
46. Pingitore, A.; Chambers, E.S.; Hill, T.; Maldonado, I.R.; Liu, B.; Bewick, G.; Morrison, D.J.; Preston, T.; Wallis, G.A.; Tedford, C.; et al. The diet-derived short chain fatty acid propionate improves beta-cell function in humans and stimulates insulin secretion from human islets in vitro. *Diabetes Obes. Metab.* **2016**, *19*, 257–265. [CrossRef]

47. Pingitore, A.; Gonzalez-Abuin, N.; Ruz-Maldonado, I.; Huang, G.C.; Frost, G.; Persaud, S.J. Short chain fatty acids stimulate insulin secretion and reduce apoptosis in mouse and human islets in vitro: Role of free fatty acid receptor 2. *Diabetes Obes. Metab.* **2019**, *21*, 330–339. [CrossRef] [PubMed]
48. Villa, S.R.; Mishra, R.K.; Zapater, J.L.; Priyadarshini, M.; Gilchrist, A.; Mancebo, H.; E Schiltz, G.; Layden, B.T. Homology modeling of FFA2 identifies novel agonists that potentiate insulin secretion. *J. Investig. Med.* **2017**, *65*, 1116–1124. [CrossRef] [PubMed]
49. Tan, J.K.; McKenzie, C.; Mariño, E.; Macia, L.; Mackay, C.R. Metabolite-Sensing G Protein–Coupled Receptors—Facilitators of Diet-Related Immune Regulation. *Annu. Rev. Immunol.* **2017**, *35*, 371–402. [CrossRef]
50. Perry, R.J.; Peng, L.; Barry, N.A.; Cline, G.W.; Zhang, D.; Cardone, R.L.; Petersen, K.F.; Kibbey, R.G.; Goodman, N.A.B.A.L.; Shulman, R.J. Acetate mediates a microbiome–brain–β-cell axis to promote metabolic syndrome. *Nature* **2016**, *534*, 213–217. [CrossRef]
51. Kimura, I.; Ozawa, K.; Inoue, D.; Imamura, T.; Kimura, K.; Maeda, T.; Terasawa, K.; Kashihara, D.; Hirano, K.; Tani, T.; et al. The gut microbiota suppresses insulin-mediated fat accumulation via the short-chain fatty acid receptor GPR43. *Nat. Commun.* **2013**, *4*, 1829. [CrossRef] [PubMed]
52. Ahrén, B. Islet G protein-coupled receptors as potential targets for treatment of type 2 diabetes. *Nat. Rev. Drug Discov.* **2009**, *8*, 369–385. [CrossRef] [PubMed]
53. Høverstad, T.; Bjørneklett, A.; Midtvedt, T.; Fausa, O.; Bøhmer, T. Short-chain fatty acids in the proximal gastrointestinal tract of healthy subjects. *Scand. J. Gastroenterol.* **1984**, *19*, 1053–1058. [PubMed]
54. Nauck, M.A.; Meier, J.J. Incretin hormones: Their role in health and disease. *Diabetes Obes. Metab.* **2018**, *20* (Suppl. 1), 5–21. [CrossRef]
55. Seino, Y.; Fukushima, M.; Yabe, D. GIP and GLP-1, the two incretin hormones: Similarities and differences. *J. Diabetes Investig.* **2010**, *1*, 8–23. [CrossRef] [PubMed]
56. El, K.; Campbell, J.E. The role of GIP in α-cells and glucagon secretion. *Peptides* **2020**, *125*, 170213. [CrossRef]
57. Greenbaum, C.J.; Prigeon, R.L.; D'Alessio, D.A. Impaired -Cell Function, Incretin Effect, and Glucagon Suppression in Patients with Type 1 Diabetes Who Have Normal Fasting Glucose. *Diabetes* **2002**, *51*, 951–957. [CrossRef]
58. Gribble, F.M.; Reimann, F. Enteroendocrine Cells: Chemosensors in the Intestinal Epithelium. *Annu. Rev. Physiol.* **2016**, *78*, 277–299. [CrossRef]
59. Nøhr, M.K.; Pedersen, M.H.; Gille, A.; Egerod, K.L.; Engelstoft, M.S.; Husted, A.S.; Sichlau, R.M.; Grunddal, K.V.; Poulsen, S.S.; Han, S.; et al. GPR41/FFAR3 and GPR43/FFAR2 as Cosensors for Short-Chain Fatty Acids in Enteroendocrine Cells vs. FFAR3 in Enteric Neurons and FFAR2 in Enteric Leukocytes. *Endocrinology* **2013**, *154*, 3552–3564. [CrossRef]
60. Kaji, I.; Karaki, S.-I.; Tanaka, R.; Kuwahara, A. Density distribution of free fatty acid receptor 2 (FFA2)-expressing and GLP-1-producing enteroendocrine L cells in human and rat lower intestine, and increased cell numbers after ingestion of fructo-oligosaccharide. *J. Mol. Histol.* **2011**, *42*, 27–38. [CrossRef]
61. Karaki, S.-I.; Tazoe, H.; Hayashi, H.; Kashiwabara, H.; Tooyama, K.; Suzuki, Y.; Kuwahara, A. Expression of the short-chain fatty acid receptor, GPR43, in the human colon. *J. Mol. Histol.* **2008**, *39*, 135–142. [CrossRef]
62. Karaki, S.-I.; Mitsui, R.; Hayashi, H.; Kato, I.; Sugiya, H.; Iwanaga, T.; Furness, J.B.; Kuwahara, A. Short-chain fatty acid receptor, GPR43, is expressed by enteroendocrine cells and mucosal mast cells in rat intestine. *Cell Tissue Res.* **2006**, *324*, 353–360. [CrossRef] [PubMed]
63. Samuel, B.S.; Shaito, A.; Motoike, T.; Rey, F.E.; Backhed, F.; Manchester, J.K.; Hammer, R.E.; Williams, S.C.; Crowley, J.; Yanagisawa, M.; et al. Effects of the gut microbiota on host adiposity are modulated by the short-chain fatty-acid binding G protein-coupled receptor, Gpr41. *Proc. Natl. Acad. Sci. USA* **2008**, *105*, 16767–16772. [CrossRef] [PubMed]
64. Tazoe, H.; Otomo, Y.; Karaki, S.-I.; Kato, I.; Fukami, Y.; Terasaki, M.; Kuwahara, A. Expression of short-chain fatty acid receptor GPR41 in the human colon. *Biomed. Res.* **2009**, *30*, 149–156. [CrossRef]
65. Won, Y.-J.; Lu, V.B.; Puhl, H.L., 3rd; Ikeda, S.R. β-Hydroxybutyrate Modulates N-Type Calcium Channels in Rat Sympathetic Neurons by Acting as an Agonist for the G-Protein-Coupled Receptor FFA3. *J. Neurosci.* **2013**, *33*, 19314–19325. [CrossRef] [PubMed]
66. Kimura, I.; Inoue, D.; Maeda, T.; Hara, T.; Ichimura, A.; Miyauchi, S.; Kobayashi, M.; Hirasawa, A.; Tsujimoto, G. Short-chain fatty acids and ketones directly regulate sympathetic nervous system via G protein-coupled receptor 41 (GPR41). *Proc. Natl. Acad. Sci. USA* **2011**, *108*, 8030–8035. [CrossRef]
67. Iwasaki, K.; Harada, N.; Sasaki, K.; Yamane, S.; Iida, K.; Suzuki, K.; Hamasaki, A.; Nasteska, D.; Shibue, K.; Joo, E.; et al. Free Fatty Acid Receptor GPR120 Is Highly Expressed in Enteroendocrine K Cells of the Upper Small Intestine and Has a Critical Role in GIP Secretion After Fat Ingestion. *Endocrinology* **2015**, *156*, 837–846. [CrossRef]
68. Psichas, A.; Sleeth, M.L.; Murphy, K.G.; Brooks, L.; Bewick, G.A.; Hanyaloglu, A.C.; Ghatei, M.A.; Bloom, S.R.; Frost, G. The short chain fatty acid propionate stimulates GLP-1 and PYY secretion via free fatty acid receptor 2 in rodents. *Int. J. Obes.* **2015**, *39*, 424–429. [CrossRef]
69. Tolhurst, G.; Heffron, H.; Lam, Y.S.; Parker, H.E.; Habib, A.M.; Diakogiannaki, E.; Cameron, J.; Grosse, J.; Reimann, F.; Gribble, F.M. Short-Chain Fatty Acids Stimulate Glucagon-Like Peptide-1 Secretion via the G-Protein-Coupled Receptor FFAR2. *Diabetes* **2011**, *61*, 364–371. [CrossRef]

0. Forbes, S.; Stafford, S.; Coope, G.; Heffron, H.; Real, K.; Newman, R.; Davenport, R.; Barnes, M.; Grosse, J.; Cox, H. Selective FFA2 Agonism Appears to Act via Intestinal PYY to Reduce Transit and Food Intake but Does Not Improve Glucose Tolerance in Mouse Models. *Diabetes* **2015**, *64*, 3763–3771. [CrossRef]
71. Larraufie, P.; Martin-Gallausiaux, C.; Lapaque, N.; Dore, J.; Gribble, F.M.; Reimann, F.; Blottiere, H.M. SCFAs strongly stimulate PYY production in human enteroendocrine cells. *Sci. Rep.* **2018**, *8*, 74. [CrossRef] [PubMed]
72. Lee, E.-Y.; Zhang, X.; Miyamoto, J.; Kimura, I.; Taknaka, T.; Furusawa, K.; Jomori, T.; Fujimoto, K.; Uematsu, S.; Miki, T. Gut carbohydrate inhibits GIP secretion via a microbiota/SCFA/FFAR3 pathway. *J. Endocrinol.* **2018**, *239*, 267–276. [CrossRef] [PubMed]
73. Park, B.-O.; Kim, S.H.; Kong, G.Y.; Kim, D.H.; Kwon, M.S.; Lee, S.U.; Kim, M.-O.; Cho, S.; Lee, S.; Lee, H.-J.; et al. Selective novel inverse agonists for human GPR43 augment GLP-1 secretion. *Eur. J. Pharmacol.* **2016**, *771*, 1–9. [CrossRef] [PubMed]
74. Bolognini, D.; Barki, N.; Butcher, A.J.; Hudson, B.D.; Sergeev, E.; Molloy, C.; Moss, C.E.; Bradley, S.J.; le Gouill, C.; Bouvier, M.; et al. Chemogenetics defines receptor-mediated functions of short chain free fatty acids. *Nat. Chem. Biol.* **2019**, *15*, 489–498. [CrossRef] [PubMed]
75. Kim, C.H. Immune regulation by microbiome metabolites. *Immunology* **2018**, *154*, 220–229. [CrossRef]
76. Kamp, M.E.; Shim, R.; Nicholls, A.J.; Oliveira, A.C.; Mason, L.J.; Binge, L.; Mackay, C.R.; Wong, C.H.Y. G Protein-Coupled Receptor 43 Modulates Neutrophil Recruitment during Acute Inflammation. *PLoS ONE* **2016**, *11*, e0163750. [CrossRef] [PubMed]
77. Maslowski, K.M.; Vieira, A.T.; Ng, A.; Kranich, J.; Sierro, F.; Yu, D.; Schilter, H.C.; Rolph, M.S.; Mackay, F.; Artis, D.; et al. Regulation of inflammatory responses by gut microbiota and chemoattractant receptor GPR43. *Nature* **2009**, *461*, 1282–1286. [CrossRef] [PubMed]
78. Dahlgren, C.; Holdfeldt, A.; Lind, S.; Mårtensson, J.; Gabl, M.; Björkman, L.; Sundqvist, M.; Forsman, H. Neutrophil Signaling That Challenges Dogmata of G Protein-Coupled Receptor Regulated Functions. *ACS Pharmacol. Transl. Sci.* **2020**, *3*, 203–220. [CrossRef]
79. Fachi, J.L.; Sécca, C.; Rodrigues, P.B.; de Mato, F.C.P.; di Luccia, B.; Felipe, J.D.S.; Pral, L.P.; Rungue, M.; Rocha, V.D.M.; Sato, F.T.; et al. Acetate coordinates neutrophil and ILC3 responses against C. difficile through FFAR2. *J. Exp. Med.* **2020**, *217*. [CrossRef]
80. Chun, E.; Lavoie, S.; Fonseca-Pereira, D.; Bae, S.; Michaud, M.; Hoveyda, H.R.; Fraser, G.L.; Comeau, C.A.G.; Glickman, J.N.; Fuller, M.H.; et al. Metabolite-Sensing Receptor Ffar2 Regulates Colonic Group 3 Innate Lymphoid Cells and Gut Immunity. *Immunity* **2019**, *51*, 871–884.e6. [CrossRef]
81. Smith, P.M.; Howitt, M.R.; Panikov, N.; Michaud, M.; Gallini, C.A.; Bohlooly-Y, M.; Glickman, J.N.; Garrett, W.S. The Microbial Metabolites, Short-Chain Fatty Acids, Regulate Colonic Treg Cell Homeostasis. *Science* **2013**, *341*, 569–573. [CrossRef] [PubMed]
82. Tan, J.; McKenzie, C.; Vuillermin, P.J.; Goverse, G.; Vinuesa, C.G.; Mebius, R.E.; Macia, L.; Mackay, C.R. Dietary Fiber and Bacterial SCFA Enhance Oral Tolerance and Protect against Food Allergy through Diverse Cellular Pathways. *Cell Rep.* **2016**, *15*, 2809–2824. [CrossRef]
83. Wu, W.; Sun, M.; Chen, F.; Cao, A.T.; Liu, H.; Zhao, Y.; Huang, X.; Xiao, Y.; Yao, S.; Zhao, Q.; et al. Microbiota metabolite short-chain fatty acid acetate promotes intestinal IgA response to microbiota which is mediated by GPR43. *Mucosal Immunol.* **2017**, *10*, 946–956. [CrossRef] [PubMed]
84. Macia, L.; Tan, J.; Vieira, A.T.; Leach, K.; Stanley, D.; Luong, S.; Maruya, M.; McKenzie, C.I.; Hijikata, A.; Wong, C.; et al. Metabolite-sensing receptors GPR43 and GPR109A facilitate dietary fibre-induced gut homeostasis through regulation of the inflammasome. *Nat. Commun.* **2015**, *6*, 6734. [CrossRef] [PubMed]
85. Trompette, A.; Gollwitzer, E.S.; Yadava, K.; Sichelstiel, A.K.; Sprenger, N.; Ngom-Bru, C.; Blanchard, C.; Junt, T.; Nicod, L.P.; Harris, N.L.; et al. Gut microbiota metabolism of dietary fiber influences allergic airway disease and hematopoiesis. *Nat. Med.* **2014**, *20*, 159–166. [CrossRef]
86. Nakajima, A.; Kaga, N.; Nakanishi, Y.; Ohno, H.; Miyamoto, J.; Kimura, I.; Hori, S.; Sasaki, T.; Hiramatsu, K.; Okumura, K.; et al. Maternal High Fiber Diet during Pregnancy and Lactation Influences Regulatory T Cell Differentiation in Offspring in Mice. *J. Immunol.* **2017**, *199*, 3516–3524. [CrossRef] [PubMed]
87. Trompette, A.; Gollwitzer, E.S.; Pattaroni, C.; Lopez-Mejia, I.C.; Riva, E.; Pernot, J.; Ubags, N.; Fajas, L.; Nicod, L.P.; Marsland, B.J. Dietary Fiber Confers Protection against Flu by Shaping Ly6c− Patrolling Monocyte Hematopoiesis and CD8+ T Cell Metabolism. *Immunity* **2018**, *48*, 992–1005.e8. [CrossRef] [PubMed]
88. Bachem, A.; Makhlouf, C.; Binger, K.J.; de Souza, D.P.; Tull, D.; Hochheiser, K.; Whitney, P.G.; Fernandez-Ruiz, D.; Dähling, S.; Kastenmüller, W.; et al. Microbiota-Derived Short-Chain Fatty Acids Promote the Memory Potential of Antigen-Activated CD8+ T Cells. *Immunity* **2019**, *51*, 285–297.e5. [CrossRef]
89. Sivaprakasam, S.; Gurav, A.; Paschall, A.V.; Coe, G.L.; Chaudhary, K.; Cai, Y.; Kolhe, R.; Martin, P.; Browning, D.; Huang, L.; et al. An essential role of Ffar2 (Gpr43) in dietary fibre-mediated promotion of healthy composition of gut microbiota and suppression of intestinal carcinogenesis. *Oncogenesis* **2016**, *5*, e238. [CrossRef]
90. Vieira, A.T.; Macia, L.; Galvão, I.; Martins, F.S.; Canesso, M.C.C.; Amaral, F.A.; Garcia, C.C.; Maslowski, K.M.; de Leon, E.; Shim, D.; et al. A Role for Gut Microbiota and the Metabolite-Sensing Receptor GPR43 in a Murine Model of Gout. *Arthritis Rheumatol.* **2015**, *67*, 1646–1656. [CrossRef] [PubMed]
91. Antunes, K.H.; Fachi, J.L.; de Paula, R.; da Silva, E.F.; Pral, L.P.; dos Santos, A.Á.; Dias, G.B.M.; Vargas, J.E.; Puga, R.; Mayer, F.Q.; et al. Microbiota-derived acetate protects against respiratory syncytial virus infection through a GPR43-type 1 interferon response. *Nat. Commun.* **2019**, *10*, 3273. [CrossRef]

92. Kimura, I.; Miyamoto, J.; Ohue-Kitano, R.; Watanabe, K.; Yamada, T.; Onuki, M.; Aoki, R.; Isobe, Y.; Kashihara, D.; Inoue, D.; et al. Maternal gut microbiota in pregnancy influences offspring metabolic phenotype in mice. *Science* **2020**, *367*, eaaw8429. [CrossRef]
93. Falk, P.G.; Hooper, L.V.; Midtvedt, T.; Gordon, J.I. Creating and Maintaining the Gastrointestinal Ecosystem: What We Know and Need to Know from Gnotobiology. *Microbiol. Mol. Biol. Rev.* **1998**, *62*, 1157–1170. [CrossRef] [PubMed]
94. Hapfelmeier, S.; Lawson, M.A.E.; Slack, E.; Kirundi, J.K.; Stoel, M.; Heikenwalder, M.; Cahenzli, J.; Velykoredko, Y.; Balmer, M.L.; Endt, K.; et al. Reversible Microbial Colonization of Germ-Free Mice Reveals the Dynamics of IgA Immune Responses. *Science* **2010**, *328*, 1705–1709. [CrossRef]
95. Round, J.L.; Mazmanian, S.K. The gut microbiota shapes intestinal immune responses during health and disease. *Nat. Rev. Immunol.* **2009**, *9*, 313–323. [CrossRef] [PubMed]
96. Ivanov, I.I.; Atarashi, K.; Manel, N.; Brodie, E.L.; Shima, T.; Karaoz, U.; Wei, D.; Goldfarb, K.C.; Santee, C.A.; Lynch, S.V.; et al. Induction of Intestinal Th17 Cells by Segmented Filamentous Bacteria. *Cell* **2009**, *139*, 485–498. [CrossRef] [PubMed]
97. Li, X.; Atkinson, M.A. The role for gut permeability in the pathogenesis of type 1 diabetes—A solid or leaky concept? *Pediatr. Diabetes* **2015**, *16*, 485–492. [CrossRef]
98. Gavin, P.G.; Mullaney, J.A.; Loo, D.; Cao, K.-A.L.; Gottlieb, P.A.; Hill, M.M.; Zipris, D.; Hamilton-Williams, E.E. Intestinal Metaproteomics Reveals Host-Microbiota Interactions in Subjects at Risk for Type 1 Diabetes. *Diabetes Care* **2018**, *41*, 2178–2186. [CrossRef] [PubMed]
99. Sorini, C.; Cosorich, I.; Conte, M.L.; de Giorgi, L.; Facciotti, F.; Lucianò, R.; Rocchi, M.; Ferrarese, R.; Sanvito, F.; Canducci, F.; et al. Loss of gut barrier integrity triggers activation of islet-reactive T cells and autoimmune diabetes. *Proc. Natl. Acad. Sci. USA* **2019**, *116*, 15140–15149. [CrossRef]
100. Bosi, E.; Molteni, L.; Radaelli, M.G.; Folini, L.; Fermo, I.; Bazzigaluppi, E.; Piemonti, L.; Pastore, M.R.; Paroni, R. Increased intestinal permeability precedes clinical onset of type 1 diabetes. *Diabetologia* **2006**, *49*, 2824–2827. [CrossRef]
101. Sapone, A.; de Magistris, L.; Pietzak, M.; Clemente, M.G.; Tripathi, A.; Cucca, F.; Lampis, R.; Kryszak, D.; Cartenì, M.; Generoso, M.; et al. Zonulin Upregulation Is Associated with Increased Gut Permeability in Subjects with Type 1 Diabetes and Their Relatives. *Diabetes* **2006**, *55*, 1443–1449. [CrossRef]
102. Costa, F.R.; Françozo, M.C.; de Oliveira, G.G.; Ignacio, A.; Castoldi, A.; Zamboni, D.S.; Ramos, S.G.; Câmara, N.O.; de Zoete, M.R.; Palm, N.W.; et al. Gut microbiota translocation to the pancreatic lymph nodes triggers NOD2 activation and contributes to T1D onset. *J. Exp. Med.* **2016**, *213*, 1223–1239. [CrossRef] [PubMed]
103. Shi, G.; Sun, C.; Gu, W.; Yang, M.; Zhang, X.; Zhai, N.; Lu, Y.; Shou, P.; Zhang, Z.; Ning, G. Free fatty acid receptor 2, a candidate target for type 1 diabetes, induces cell apoptosis through ERK signaling. *J. Mol. Endocrinol.* **2014**, *53*, 367–380. [CrossRef] [PubMed]
104. Pivovarova, O.; Hornemann, S.; Weimer, S.; Lu, Y.; Murahovschi, V.; Zhuk, S.; Seltmann, A.-C.; Malashicheva, A.; Kostareva, A.; Kruse, M.; et al. Regulation of nutrition-associated receptors in blood monocytes of normal weight and obese humans. *Peptides* **2015**, *65*, 12–19. [CrossRef]
105. Sun, J.; Furio, L.; Mecheri, R.; van der Does, A.M.; Lundeberg, E.; Saveanu, L.; Chen, Y.; van Endert, P.; Agerberth, B.; Diana, J. Pancreatic β-Cells Limit Autoimmune Diabetes via an Immunoregulatory Antimicrobial Peptide Expressed under the Influence of the Gut Microbiota. *Immunity* **2015**, *43*, 304–317. [CrossRef] [PubMed]
106. De Groot, P.F.; Nikolic, T.; Imangaliyev, S.; Bekkering, S.; Duinkerken, G.; Keij, F.M.; Herrema, H.; Winkelmeijer, M.; Kroon, J.; Levin, E.; et al. Oral butyrate does not affect innate immunity and islet autoimmunity in individuals with longstanding type 1 diabetes: A randomised controlled trial. *Diabetologia* **2020**, *63*, 597–610. [CrossRef] [PubMed]
107. Mishra, S.P.; Wang, S.; Nagpal, R.; Miller, B.; Singh, R.; Taraphder, S.; Yadav, H. Probiotics and Prebiotics for the Amelioration of Type 1 Diabetes: Present and Future Perspectives. *Microorganisms* **2019**, *7*, 67. [CrossRef] [PubMed]
108. Sanders, M.E.; Merenstein, D.J.; Reid, G.; Gibson, G.R.; Rastall, R.A. Probiotics and prebiotics in intestinal health and disease: From biology to the clinic. *Nat. Rev. Gastroenterol. Hepatol.* **2019**, *16*, 605–616. [CrossRef]
109. Uusitalo, U.; Liu, X.; Yang, J.; Aronsson, C.A.; Hummel, S.; Butterworth, M.; Lernmark, Å.; Rewers, M.; Hagopian, W.; She, J.-X.; et al. Association of Early Exposure of Probiotics and Islet Autoimmunity in the TEDDY Study. *JAMA Pediatrics* **2016**, *170*, 20–28. [CrossRef]
110. Wei, S.-H.; Chen, Y.-P.; Chen, M.-J. Selecting probiotics with the abilities of enhancing GLP-1 to mitigate the progression of type 1 diabetes in vitro and in vivo. *J. Funct. Foods* **2015**, *18*, 473–486. [CrossRef]
111. Horiuchi, H.; Kamikado, K.; Aoki, R.; Suganuma, N.; Nishijima, T.; Nakatani, A.; Kimura, I. Bifidobacterium animalis subsp. lactis GCL2505 modulates host energy metabolism via the short-chain fatty acid receptor GPR43. *Sci. Rep.* **2020**, *10*, 1–8. [CrossRef]
112. Yang, J.; Tamura, R.N.; Uusitalo, U.M.; Aronsson, C.A.; Silvis, K.; Riikonen, A.; Frank, N.; Joslowski, G.; Winkler, C. Vitamin D and probiotics supplement use in young children with genetic risk for type 1 diabetes. *Eur. J. Clin. Nutr.* **2017**, *71*, 1449–1454. [CrossRef] [PubMed]
113. Bolognini, D.; Tobin, A.B.; Milligan, G.; Moss, C.E. The Pharmacology and Function of Receptors for Short-Chain Fatty Acids. *Mol. Pharmacol.* **2016**, *89*, 388–398. [CrossRef]
114. Suckow, A.T.; Briscoe, C.P. Key Questions for Translation of FFA Receptors: From Pharmacology to Medicines. *Handb. Exp. Pharmacol.* **2016**, *236*, 101–131. [CrossRef]

115. Ang, Z.; Er, J.Z.; Tan, N.S.; Lu, J.; Liou, Y.-C.; Grosse, J.; Ding, J.L. Human and mouse monocytes display distinct signalling and cytokine profiles upon stimulation with FFAR2/FFAR3 short-chain fatty acid receptor agonists. *Sci. Rep.* **2016**, *6*, srep34145. [CrossRef] [PubMed]
116. Ang, Z.; Xiong, D.; Wu, M.; Ding, J.L. FFAR2-FFAR3 receptor heteromerization modulates short-chain fatty acid sensing. *FASEB J.* **2018**, *32*, 289–303. [CrossRef] [PubMed]
117. Lee, S.U.; In, H.J.; Kwon, M.S.; Park, B.-O.; Jo, M.; Kim, M.-O.; Cho, S.; Lee, S.; Lee, H.-J.; Kwak, Y.S.; et al. β-Arrestin 2 Mediates G Protein-Coupled Receptor 43 Signals to Nuclear Factor-κB. *Biol. Pharm. Bull.* **2013**, *36*, 1754–1759. [CrossRef]
118. Lu, Y.; Fan, C.; Li, P.; Lu, Y.; Chang, X.; Qi, K. Short Chain Fatty Acids Prevent High-fat-diet-induced Obesity in Mice by Regulating G Protein-coupled Receptors and Gut Microbiota. *Sci. Rep.* **2016**, *6*, 37589. [CrossRef]
119. Hudson, B.D.; Murdoch, H.; Milligan, G. Minireview: The effects of species ortholog and SNP variation on receptors for free fatty acids. *Mol. Endocrinol.* **2013**, *27*, 1177–1187. [CrossRef] [PubMed]

metabolites

Article

The Capacity to Secrete Insulin Is Dose-Dependent to Extremely High Glucose Concentrations: A Key Role for Adenylyl Cyclase

Katherine M. Gerber [1,†], Nicholas B. Whitticar [2,3,†], Daniel R. Rochester [2], Kathryn L. Corbin [2], William J. Koch [2,3] and Craig S. Nunemaker [2,4,*]

1 Translational Health, Honors Tutorial College, Ohio University, Athens, OH 45701, USA; kg010716@ohio.edu
2 Biomedical Sciences, Heritage College of Osteopathic Medicine, Ohio University, Athens, OH 45701, USA; nw575612@ohio.edu (N.B.W.); dr935216@ohio.edu (D.R.R.); corbinkl@ohio.edu (K.L.C.); wk922015@ohio.edu (W.J.K.)
3 Translational Biomedical Sciences Program, Heritage College of Osteopathic Medicine, Ohio University, Athens, OH 45701, USA
4 Diabetes Institute, Heritage College of Osteopathic Medicine, Ohio University, Athens, OH 45701, USA
* Correspondence: nunemake@ohio.edu; Tel.: +740-593-2387
† These authors contributed equally to this work.

Abstract: Insulin secretion is widely thought to be maximally stimulated in glucose concentrations of 16.7-to-30 mM (300-to-540 mg/dL). However, insulin secretion is seldom tested in hyperglycemia exceeding these levels despite the Guinness World Record being 147.6 mM (2656 mg/dL). We investigated how islets respond to 1-h exposure to glucose approaching this record. Insulin secretion from human islets at 12 mM glucose intervals dose-dependently increased until at least 72 mM glucose. Murine islets in 84 mM glucose secreted nearly double the insulin as in 24 mM ($p < 0.001$). Intracellular calcium was maximally stimulated in 24 mM glucose despite a further doubling of insulin secretion in higher glucose, implying that insulin secretion above 24 mM occurs through amplifying pathway(s). Increased osmolarity of 425-mOsm had no effect on insulin secretion (1-h exposure) or viability (48-h exposure) in murine islets. Murine islets in 24 mM glucose treated with a glucokinase activator secreted as much insulin as islets in 84 mM glucose, indicating that glycolytic capacity exists above 24 mM. Using an incretin mimetic and an adenylyl cyclase activator in 24 mM glucose enhanced insulin secretion above that observed in 84 mM glucose while adenylyl cyclase inhibitor reduced stimulatory effects. These results highlight the underestimated ability of islets to secrete insulin proportionally to extreme hyperglycemia through adenylyl cyclase activity.

Keywords: amplifying pathway; hyperglycemia; adenylyl cyclase; incretins; glucokinase; forskolin; cAMP; exenatide; diabetes; insulin; islets

Citation: Gerber, K.M.; Whitticar, N.B.; Rochester, D.R.; Corbin, K.L.; Koch, W.J.; Nunemaker, C.S. The Capacity to Secrete Insulin Is Dose-Dependent to Extremely High Glucose Concentrations: A Key Role for Adenylyl Cyclase. *Metabolites* **2021**, *11*, 401. https://doi.org/10.3390/metabo11060401

Academic Editors: Belinda Yau and Melkam Kebede

Received: 31 May 2021
Accepted: 18 June 2021
Published: 19 June 2021

1. Introduction

Pancreatic beta cells secrete insulin in response to glucose stimulation to maintain blood glucose levels within a relatively narrow range [1]. Insulin is required to transport glucose from the bloodstream to target tissues. High blood glucose levels in the body are caused by problems with insulin secretion, insulin action, or both. Extremely high levels of glucose lead to the presentation of ketoacidosis or hyperosmolar hyperglycemic nonketoic syndrome, which are key indicators of the metabolic disease diabetes [2]. Poor control of glucose regulation in this disease can bring potential stupor, coma, or death [3].

In humans, insulin secretion is typically stimulated by glucose concentrations ranging from 4.4 to 6.6 mM (80–120 mg/dL) [2,4]. Moreover, it is generally accepted that the effective concentration of glucose for half of the maximal insulin secretion (EC_{50}) is approximately 5 mM. These EC_{50} estimations are based on dose-response curves with the highest stimulation typically being described at 16.7 (300 mg/dL) to 30 mM (540 mg/dL). However, close inspection of published glucose dose-response curves suggests that even though a classic sigmoid dose-response curve should flatten out, the curves typically show an increasing trend of higher insulin secretion near the maximal glucose level tested. This provokes the hypothesis that insulin secretion may be sensitive to a much wider range of glucose concentrations than commonly thought [5–9].

Interestingly, there are multiple case reports of individuals who have had blood glucose levels greater than 100 mM (1800 mg/dL) and survived [2,4,10,11]. This includes the world record blood glucose level of 147.6 mM (2656 mg/dL) which was set by a young boy when admitted to the hospital [11]. The pathways involving insulin secretion at extremely high glucose concentrations have not been examined to our knowledge.

Glucose is the primary stimulator of insulin release from pancreatic beta cells. The ability of glucose to elicit an increase in intracellular calcium leading to insulin secretion is known as the triggering pathway [12–14]. This pathway begins when glucose enters the beta cell through both the GLUT 1 and GLUT 3 glucose transporters in human islets and the GLUT 2 transporter in mouse islets [13,14]. Glucose is then phosphorylated by glucokinase and yields glucose-6-phosphate which travels through glycolysis to yield pyruvate and ATP. Glycolysis and downstream mitochondrial metabolism drive the ratio of ATP to ADP to rise, leading to the closure of K_{ATP} channels. This closure activates the voltage-dependent calcium channels to allow the influx of calcium that constitutes the triggering pathway for insulin secretion.

In addition to the triggering pathway, many different intermediate metabolites of glucose and other cellular components are thought to participate in a series of events known as the amplification pathway(s) in which K_{ATP} channel closure is not the source of the increased insulin secretion. A review on this topic contained within this special issue describes this pathway as "the sequence of events that enables the secretory response to a nutrient secretagogue to exceed the secretory response of a purely depolarizing stimulus" [15]. Cyclic adenosine monophosphate (cAMP) is a powerful player in the amplification pathway-related insulin secretion. Glucose is a known factor leading to the upregulation of cAMP which is formed from ATP via adenylyl cyclases [16], although the exact mechanism of metabolism stimulated cAMP is unknown. Additionally, there is a glucose-linked amplification pathway that augments insulin secretion through adenylyl cyclase (AC) activation caused by incretin stimulation. Incretins are hormones secreted by endocrine cells in the small intestine after meal ingestion that lead to insulin secretion [12]. Gastric inhibitory polypeptide (GIP) and glucagon-like peptide-1 (GLP-1) are specific incretins that act on the beta cell. Incretins interact with G-protein-coupled receptors at the cell membrane and lead to an upregulation of cAMP [17,18]. The common theme is that cAMP is involved in the early steps leading to insulin secretion in these amplifying pathway(s).

In this study, we aim to determine the relationship between glucose stimulation and insulin secretion in a much higher glucose range than typically examined. We show that insulin secretion is maintained in response to extreme glucose, osmolarity does not affect insulin secretion or islet viability at extreme levels, and that intracellular calcium is maximized at 24 mM glucose. We also sought to identity possible pathway(s) used in extreme glucose conditions and found that adenylyl cyclase plays an important role.

2. Results

2.1. Insulin Secretion Is Maintained in Response to Extreme Glucose

Identifying the maximal physiological response of biological systems often gives insight into function. We wanted to determine what happens to insulin secretion above the commonly accepted maximum concentration of glucose for stimulation. We thus examined insulin secretion from human and murine islets in conditions of extremely high glucose. Islets from human donors were placed in mannitol-balanced Krebs-Ring buffer (KRB) solutions containing glucose ranging from 0 mM to 144 mM for 1 h. As shown in Figure 1A, insulin secretion from each of the four donors was normalized to their respective maximal value and averaged amongst donors (see Table 1 for donor information). From 0 to 24 mM, there was a significant upward trend (rho = 0.93, $p < 0.001$) and from 24–84 mM, there was still a positive correlation, though not as strong (rho = 0.73, $p < 0.001$). Insulin secretion from murine islets was also measured (Figure 1B). From 0–24 mM, there was a strong upward trend (rho = 0.75, $p < 0.001$), and from 24–84 mM, there was still a significant positive trend (rho = 0.63, $p < 0.001$). Overall, there was a positive correlation between insulin and glucose for 0–24 mM, as expected, but a similar relationship exists from 24 mM glucose and beyond.

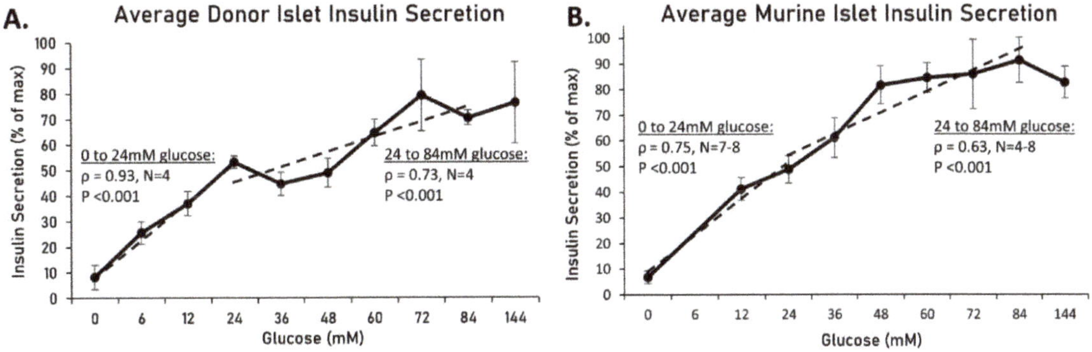

Figure 1. Human and murine islet insulin secretion in conditions of extremely high glucose. Islets from human donors (**A**) and mice (**B**) were placed in mannitol balanced KRB solutions containing glucose ranging from 0 mM to 144 mM for 1 h. Insulin secretion from each donor or mouse replicate was normalized to their respective maximal value and averaged. Dotted linear trendlines are drawn for 0–24 mM and 24–84 mM glucose to indicate slope, Spearman's rho, and the respective *p*-value.

Table 1. Human islet donor information.

Source	Isolation ID	Purity	Donor Age	BMI	Height (m)	Weight (kg)	HbA1c	Sex	Diabetes	Cold Ischemia Time (h)	Donation Type
IsletCore, University of Alberta	R286	95%	41	20.4	1.80	66	5.2	M	No	13.5	Neurological
IsletCore, University of Alberta	R318	90%	54	20.5	1.85	70	5.0	M	No	16	Neurological
IsletCore, University of Alberta	R322	90%	44	23.2	1.58	58	4.9	F	No	11.5	Neurological

2.2. *Increased Osmolarity Does Not Alter Insulin Secretion or Islet Viability*

To control for osmolarity as a variable, solutions were balanced to ~425 mOsm/L using the sugar alcohol mannitol. Mannitol was used to account for the difference in osmolarity between low and high glucose conditions, and here the effect of mannitol on its own was explored. Mannitol is not metabolized and therefore should not increase intracellular calcium or insulin secretion, although indirect factors such as membrane potential or ion flux might alter insulin secretion [19]. To investigate this possibility, islets were treated for 1 h in modified KRB with 0 mM glucose or 84 mM glucose with and without mannitol balanced to 144 mM. While it is known that various osmotic receptors on beta-cells have been shown to alter insulin secretion [20], as shown in Figure 2A, large increases in osmolarity caused no significant change in insulin secretion, and mannitol alone did not have a stimulatory effect. As expected, islets in each 84 mM glucose group secreted significantly more insulin (p <0.01) than islets in each glucose-free group, but osmolarity changes due to mannitol did not significantly affect insulin secretion.

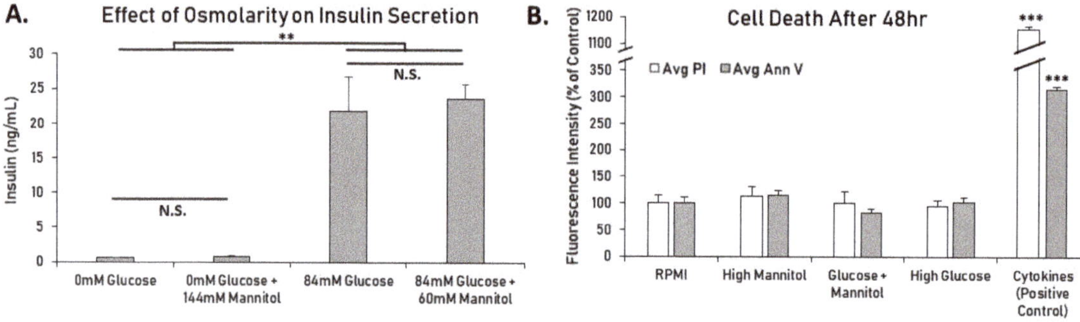

Figure 2. Effects of high glucose and mannitol on mouse islet insulin secretion and cell death. (**A**) Mouse islets were placed in wells containing modified KRB with 0 mM or 84 mM glucose ± mannitol balanced to 144 mM total for 1h. Increasing osmolarity using mannitol did not significantly alter insulin secretion in glucose-free or high-glucose solutions. Twenty islets per condition in duplicate ran in 2 separate trials (N = 4). (**B**) Islets were placed in either standard RPMI, high glucose (144 mM), high mannitol (144 mM), or both glucose (60mM) and mannitol (84 mM). Islets treated overnight with 5 ng/mL IL-1beta and 10 ng/mL TNF-alpha were used as a positive control to show typical fluorescence levels of induced cell death. Islets in each condition ranged from N=31–37 after combining two separate trials. All data are presented as mean ± SEM. ** $p < 0.01$, *** $p < 0.001$, N.S. = Not Significant.

Next, we measured cell death using fluorescent microscopy to check for toxicity. Isolated mouse islets treated in standard RPMI, high glucose (144 mM), high osmolarity (144 mM), or a combination of both (84 mM glucose + 60 mM mannitol) for 48 h displayed no significant differences in cell death (propidium iodide) or apoptosis (annexin V) (Figure 2B). Proinflammatory cytokines were used as a positive control to induce beta-cell death. Collectively, these results indicate that substantial increases in osmolarity have no effect on cellular function or viability in mouse islets in these conditions.

2.3. Intracellular Calcium Is Maximally Stimulated in 24 mM Glucose

To determine whether these responses were calcium dependent, mouse islets were loaded with the calcium probe fura-2AM and exposed to increasing stimulation as indicated by the black bars in Figure 3. As expected, a large increase was observed in response to 24 mM glucose stimulation. When the stimulus was increased from 24 to 84 mM glucose, average calcium levels of all islets were not significantly different (Figure 3A). Large increases in glucose above 24 mM do not appear to impact calcium handling. Furthermore, there was no significant difference between calcium levels in the 3 mM and the 3 mM wash ($p > 0.05$) showing that the extreme glucose did not alter the islet's ability to control the secretion process.

Tolbutamide, a sulfonylurea known to depolarize beta cells via reduction of potassium permeability, leads to the opening of voltage-dependent calcium channels to subsequently trigger insulin release [21]. As anticipated, there was no significant difference ($p > 0.05$) between 24 mM and 24 mM combined with 250 μM tolbutamide in terms of intracellular calcium influx (Figure 3B). This supports the observation that intracellular calcium is maximally stimulated in 24 mM glucose.

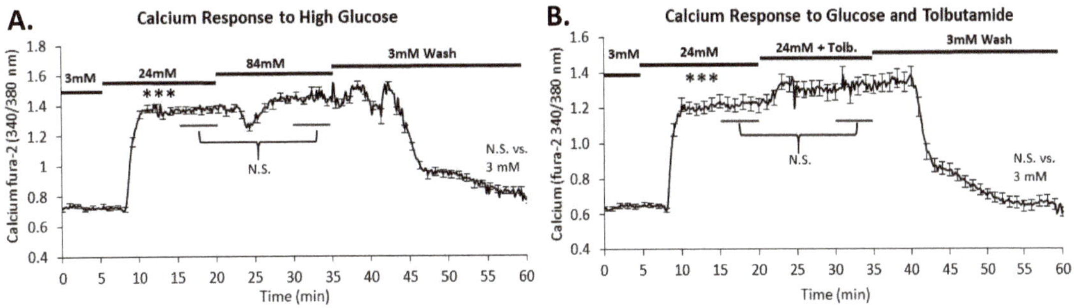

Figure 3. Effects of high glucose on intracellular calcium. (**A**) Intracellular calcium was measured in islets exposed to the increasing glucose conditions noted by the horizontal bars. Calcium levels increased sharply from 3 mM to 24 mM glucose but did not increase further when exposed to 84 mM glucose. Average calcium levels from 23 islets were calculated from the 15–20- and 30–35 min time points and showed no significant difference. (**B**) Intracellular calcium was measured in islets exposed to the increasing glucose conditions and tolbutamide noted by the horizontal bars. Calcium levels increased sharply from 3 mM to 24 mM but did not increase further when exposed to 24 mM glucose combined with 250 µM tolbutamide. Average calcium levels from 23 islets were calculated from the 15–20 and 30–35 min time points and showed no significant difference. *** $p < 0.001$ difference between 3 and 24 mM glucose.

2.4. Glycolytic Capacity to Secrete Insulin Dose-Dependently Extends above 24 mM Glucose

We next conducted a series of trials to determine what drives increased insulin secretion in high glucose. The following studies were all conducted in 24 mM glucose (regarded as maximal) and 84 mM glucose (regarded as extreme). As shown in Figure 4A, insulin secretion was significantly higher in 84 mM glucose compared to 24 mM glucose after combining all trials ($p < 0.001$). The averages are reproduced for comparisons made in Figures 4B and 5.

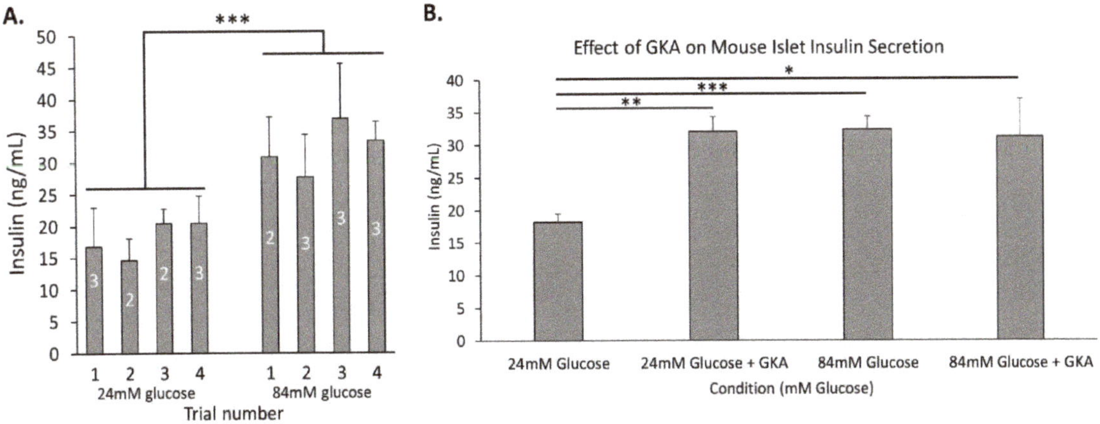

Figure 4. Insulin secretion in extremely high glucose is associated with increased glycolysis. (**A**) Insulin secretion was measured after 1h incubation in 24 mM glucose (regarded as maximal) and 84 mM glucose (regarded as extreme) shown for four separate trials with N = 2–3 replicates to demonstrate consistency of the difference in insulin secretion from trial to trial. Combining these trials, the data were significantly different with a p-value of <0.001 by two-tailed T-test between 24 and 84 mM glucose. (**B**) Mouse islets were placed in wells containing modified KRB with 24 mM or 84 mM glucose ± mannitol balanced to 84 mM total for 1 h with or without 500 nM GKA. The GKA increased insulin secretion significantly in 24 mM glucose ($p < 0.01$) but failed to do so in 84 mM glucose. All data are presented as ± SEM. N = 3–11 replicates. * $p < 0.05$, ** $p < 0.01$, *** $p < 0.001$.

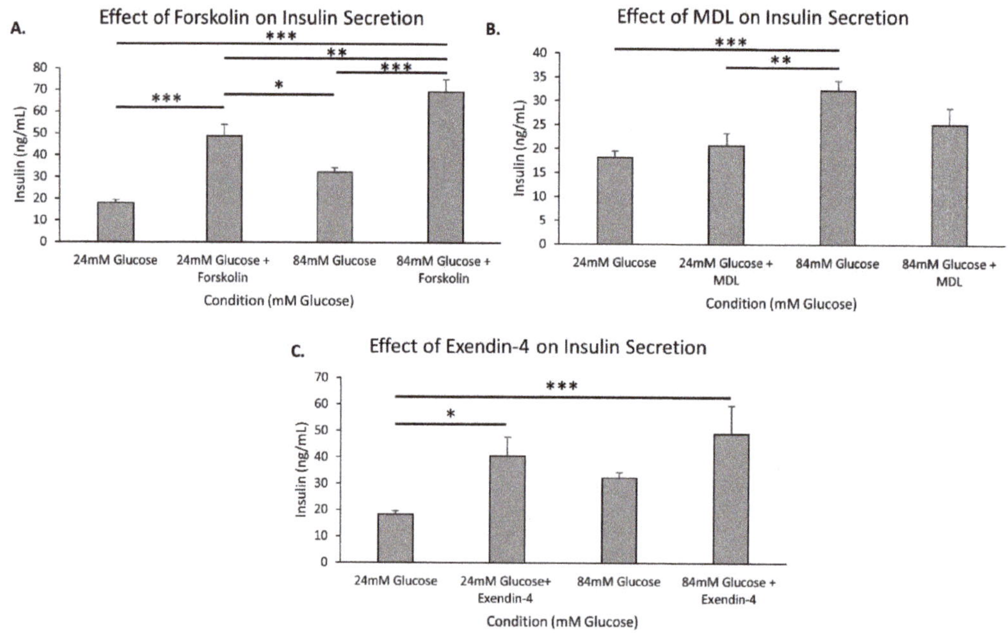

Figure 5. cAMP-related drug effect on insulin secretion in high glucose. (**A–C**) Murine islets were placed in wells containing modified KRB with 24 mM or 84 mM glucose ± mannitol balanced to 84 mM total for 1 h and then various drugs were added to each condition for 1 h of incubation before insulin was measured: 10 µM forskolin (**A**), 10 µM MDL 12,330A (**B**), and 10 nM exendin-4 (**C**) All data are presented as ± SEM. N = 5–11 replicates. * $p < 0.05$, ** $p < 0.01$, *** $p < 0.001$.

To examine the contribution of glycolytic activity leading to insulin secretion, the glucokinase activator R0-28-1675 (hereafter called GKA) was utilized, and insulin measurements were taken. As shown in Figure 4B, islets treated with GKA in 24 mM glucose displayed a significant increase in insulin secretion compared to islets in 24 mM alone ($p < 0.01$). Insulin secretion in 84 mM glucose was on par with islets given GKA in 24 mM glucose, but GKA did not provide any additional stimulation in 84 mM glucose. Together, these data indicate that glycolytically driven insulin secretion continues to dose-dependently increase, nearly doubling insulin secretion between 24 and 84 mM glucose.

2.5. The cAMP Pathway Provides Additional Capacity for Insulin Secretion in Extremely High Glucose

cAMP, a known amplifier of insulin secretion, was examined by the use of forskolin, an adenylyl cyclase activator that increases cAMP levels. As shown in Figure 5A, in 24 mM glucose for 1 h, forskolin increased insulin secretion dramatically compared to the control 24 mM glucose ($p < 0.001$). There was also a significant difference found between 84 mM glucose and 84 mM glucose combined with forskolin ($p < 0.001$). In addition, the amount of insulin secreted at 84 mM glucose with forskolin was significantly higher than the amount of insulin secreted at 24 mM glucose with forskolin ($p < 0.01$). Overall, the cAMP pathway has the capacity to dramatically increase insulin secretion, even at extreme glucose concentrations.

MDL-12,330A (MDL), an adenylyl cyclase inhibitor, was used to determine if reducing cAMP by inhibiting adenylyl cyclase could block the additional insulin secretion in extreme hyperglycemic conditions. MDL appeared to partially inhibit insulin secretion in 84 mM glucose. The significantly higher levels of insulin secretion in 84 mM glucose compared to 24 mM glucose were not significant with MDL added to 84 mM glucose (Figure 5B). Importantly, MDL did not reduce insulin secretion in 24 mM glucose compared to 24 mM glucose alone, indicating that the inhibitory effects of MDL on adenylyl cyclase occur only in extremely high glucose conditions when adenylyl cyclase is likely contributing more to overall insulin secretion.

Pancreatic islets were next stimulated with exendin-4, a GLP-1 agonist, to examine the possible contributions of incretin pathways on insulin secretion in extreme glucose conditions. There was a significant increase in insulin secretion with exendin added to 24 mM glucose compared to 24 mM glucose alone ($p < 0.05$). However, when exendin was tested with 84 mM glucose, there was no significant difference between 84 mM glucose and 84 mM glucose with exendin (Figure 5C). This indicates that insulin secretion, due to incretin effects, continues to increase between 24 mM and 84 mM glucose, revealing possible incretin involvement in the amplification pathway at extreme glucose concentrations.

3. Discussion

3.1. Maximum Glucose Concentrations for Glucose-Stimulated Insulin Secretion Are Much Higher Than Previously Reported

Our study demonstrates a surprisingly high capacity of islets to maintain stimulus-secretion coupling in glucose concentrations far exceeding what is considered normal for both mice and humans. We showed that insulin secretion doubled from 24 mM to 84 mM glucose, while calcium levels were unchanged over the same range. This suggests that insulin secretion occurs through the amplifying pathway, which can contribute 50% or more to insulin secretion when calcium is saturated [22]. In addition, our findings suggest that there is still much to be learned about the mechanisms and limits of the amplifying pathway of glucose-stimulated insulin secretion.

3.2. High Glucose Leads to Increased cAMP via Different Potential Mechanisms

Insulin secretion stimulated by cAMP can be thought of in two independent routes, the secretion that is dependent on calcium changes and that which is stimulated from cellular glucose metabolism [16]. Once cAMP accumulates, intracellular calcium rises through increased L-type-calcium-channel activity [23,24] and through release from intracellular stores [25]. Although cAMP thus has the capacity to increase intracellular calcium, our results reveal that intracellular calcium is already saturated in 24 mM glucose, so any additional stimulation of insulin secretion would likely not involve changes in calcium-dependent pathways. This study demonstrates that extremely high glucose levels can increase insulin secretion independently of the triggering pathway.

It is understood that increased glucose leads to an increase in cAMP [26]. Our data show that glucose continues to dose-dependently increase rates of glycolysis, as shown by a near doubling of insulin secretion between 24 and 84 mM glucose. As shown in Figure 6, the effects of extreme glucose can be reproduced with GKA in 24 mM glucose. This indicates that glycolytically driven metabolite formation is responsible for a large portion of insulin secreted above 24 mM glucose. Additionally, it should be noted that in Figure 6, forskolin and MDL act directly on adenylyl cyclase to alter cAMP levels. Forskolin, which acts to increase adenylyl cyclase, showed a huge potential for stimulating insulin secretion at high glucose concentrations. With that, MDL was able to partially block the insulin released in extremely high glucose conditions. MDL may not have fully reduced the insulin secreted in 84 mM glucose back down to 24 mM levels as expected, since other pathways independent of cAMP are likely working synergistically to amplify insulin secretion. Regardless, there are several ways in which increased glucose metabolism can augment insulin secretion through cAMP that do not require changes in intracellular calcium.

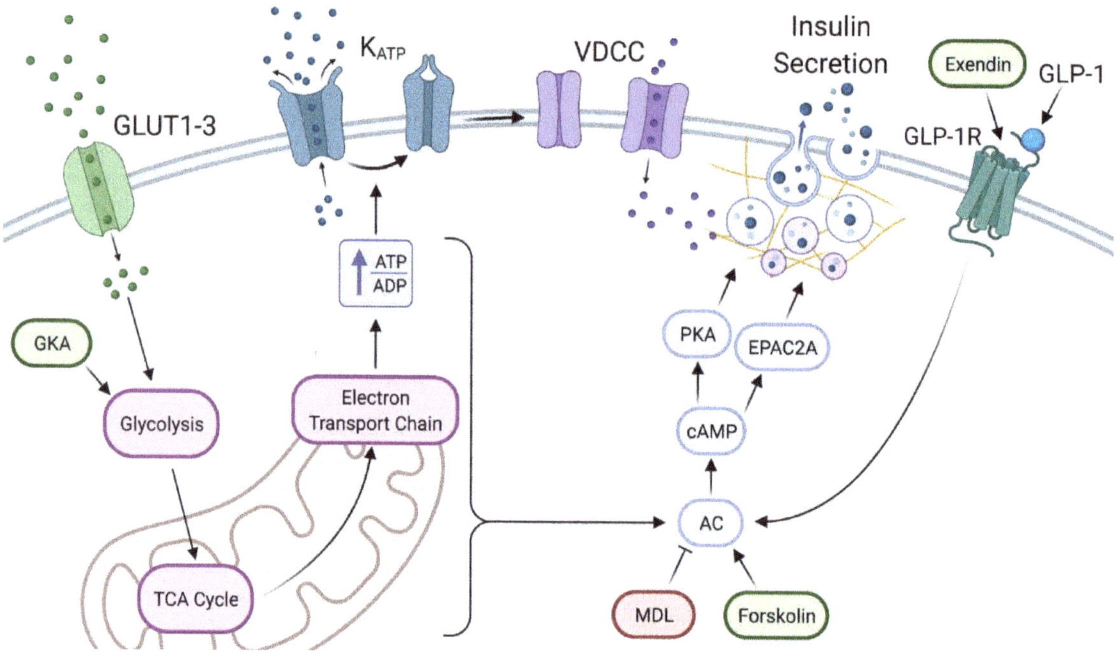

Figure 6. The triggering and amplifying pathways of insulin secretion in extreme glucose. (Left) Glucose enters the beta cell and is metabolized, causing an increase in the ATP to ADP ratio. ATP-sensitive potassium channels (KATP) close, causing membrane depolarization, the opening of voltage-dependent calcium channels (VDCC), and insulin secretion. (Right) GLP-1 from alpha cells or exendin stimulate beta-cell GLP-1 receptors which stimulate AC. Additionally, various metabolites produced from glucose metabolism can stimulate AC. Increased production of cAMP stimulates PKA and EPAC2A, causing increased insulin secretion by various mechanisms. Ovals indicate points of stimulation (green) or inhibition (red) for pharmacological agents used in these studies.

Protein kinase A (PKA) has a permissive role in increasing insulin secretion that has been shown to be glucose dependent [26]. This should be no surprise, since cAMP/PKA pathways have been long explored as targets for potential diabetes therapy [27]. As shown in Figure 6, cAMP activates downstream effectors PKA and guanine-nucleotide exchange protein (EPAC), which both lead to cAMP-mediated insulin secretion [28]. Focusing on the PKA pathway, glutamate derived from glucose through the malate aspartate shuttle is the specific signal underlying insulin secretion after being stimulated via cAMP [12]. It has been shown that the cAMP/PKA pathway potentiates the release of insulin via increased effectiveness of K_{ATP}-channel-independent actions of glucose [29], which is consistent with our observations. Intracellular calcium is unaffected by PKA activation, and PKA effects on insulin secretion are mediated by the phosphorylation of various downstream proteins [30]. The mechanisms leading to the rise in insulin secretion are studied with electrophysiological and optical methods which monitor the movements and exocytosis of individual insulin granules [31,32]. Changes in the size of distinct granular pools, facilitation of granule recruitment from the pools to the plasma membrane, and the acceleration of the priming process that confers granules with release competence may all play a role as well as the involvement of SNARE complexes. Relating this back to our study, the PKA pathway stemming from the increase in cAMP ties to the amplification pathway via glucose metabolism and is most likely a key player in insulin release at such high glucose concentrations.

In addition to what we have shown, there are other glucose-associated shunts that lead to insulin secretion through the amplification pathway. Our data suggest that glycolysis itself maintains glucose dependence well above 24 mM glucose. Although, other than direct intermediates of glucose metabolism, there are potential excess-fuel detoxification pathways dealing with glycerol and free fatty acid formation and their extracellular release [32]. Free fatty acids play a role in insulin secretion by stimulating monoacylglycerol formation, whereas the inhibition of monoacylglycerol lipase activity decreases insulin secretion [33]. The aforementioned pathways may include the diversion of glucose carbons to triglycerides and cholesterol esters. Aspects relating to mitochondrial energy metabolism independent signals, including 1-monoglycerol, diacylglycerol, and malonyl-CoA are pieces of the amplification pathway to explore at high glucose levels. [33]. Future studies should investigate glycerol release and free fatty acids in extreme glucose conditions to further understand these intricate pathways.

3.3. Paracrine Effects of Extreme Glucose via Alpha Cells

Increasing glucose levels have been shown to lead to the induction of cAMP oscillations in both alpha and beta cells [34]. The alpha cells in pancreatic islets secrete GLP-1, which generally suppresses glucagon secretion [35]. Glucose-related glucagon secretion is observed in islets and reflects direct effects on alpha cells [29,30]. L-arginine can potently stimulate GLP-1 release in islets, and there is evidence that glucose may potentiate L-arginine-stimulated insulin secretion via PKA [36]. The relationship between glucose and GLP-1R is pivotal to understanding how glucose leads to an increase in cAMP.

It is known that glucose inhibits glucagon secretion by lowering cytoplasmic calcium in the alpha-cell; however, stimulation of glucagon at high glucose concentrations does not require an increase in intracellular calcium, and at higher glucose, glucagon secretion is actually stimulated. This paradoxical stimulation of glucagon release occurs around at least 25–30 mM [30,31]. In fact, high glucose has been shown to have a stimulatory effect on glucagon secretion possibly exceeding that of the inhibitory influence [37–39]. Looking at Figure 6, our studies demonstrate the connection of increasing glucose concentrations to increased GLP-1 and increased glucagon secreted by alpha cells. Our study suggests that the increase in incretins and glucagon from the alpha cells at high glucose concentrations could act on the beta cell to augment insulin secretion through the GLP-1 receptor pathway at these extreme levels.

3.4. Clinical Relevance

Diabetic patients have survived extreme glucose levels of over 100 mM [2,4,10]. Our study shows that insulin secretion occurs at extreme glucose levels, but as the concentration of glucose in the blood increases, insulin secretion in the higher glucose range does not keep to the same rate as in lower glucose concentrations, which is evidenced by the decrease in slope (Figure 1). However, this ability of beta cells to secrete insulin in these extremes of hyperglycemia is what distinguishes hyperosmolar hyperglycemic nonketoic syndrome from diabetic ketoacidosis. There is enough insulin present to prevent ketosis but not sufficient insulin to stimulate glucose utilization in target tissues (~10× as much insulin needed) [40]. We also observed that the extreme osmolarity increase associated with extreme hyperglycemia does not appear to negatively impact insulin secretion, at least in our in vitro studies in mouse islets. Thus, although individuals have survived, their bodies endured extreme stress during these instances. It should be noted that glucotoxicity has an effect on limiting the body's ability to secrete insulin in extreme conditions, but this is considered a more chronic state than what we report.

Overall, we showed that insulin secretion from islets of both mice and human donors continues to increase in a dose-dependent manner to much higher glucose levels than previously thought. It is possible that novel pathways to insulin secretion could be identified only by stimulation in extremely high glucose. Once identified, it may be possible to develop novel therapeutics that could stimulate this secretory activity without requiring extremely high glucose.

3.5. Strengths and Limitations

An important strength of this study was the consistent observation in both murine and human islets in multiple trials which showed that the dose-dependent range of glucose-stimulated insulin secretion extends far higher than commonly thought. We further show that these increases rely on increases in glycolytic activity and cAMP, but not on changes in intracellular calcium. In addition, these studies show that osmolarity does not impact insulin secretion in vitro, which eliminates a potential confounding variable. Limitations in this study include the fact that isolated islets in vitro lack normal neural and humoral inputs found in vivo that can modulate function. Islets also lack the vasculature of their in vivo environment, which can impact how nutrients like glucose reach the islet. These are issues common to any in vitro study of pancreatic islets. Lastly, although our study shows an important role for cAMP in the insulin response to extreme glucose, many other factors of the amplification pathway could also be involved. Examining additional mechanisms will be the focus of future work.

4. Materials and Methods

4.1. Islet Sources and Isolation

Mouse islets were isolated from male CD-1 (Envigo, Indianapolis, IN, USA) mice ages 8–12 weeks, as previously described [41]. Briefly, pancreatic islets were isolated using collagenase-*p* digestion (Roche Diagnostics, Indianapolis, IN, USA) followed by centrifugation using Histopaque 1100 (Sigma-Aldrich, St. Louis, MO, USA). Islets were allowed to recover overnight in RPMI 1640 (Invitrogen, Carlsbad, CA, USA), supplemented with 10% fetal bovine serum and 1% penicillin/streptomycin before being used for experiments. All animal procedures were approved by the Ohio University Institutional Animal Care and Use Committee. Human islets from deidentified donors were obtained from the University of Alberta IsletCore and the University of Alberta/Alberta Health Services Clinical Islet Laboratory.

4.2. Calcium Imaging

Fura-2 AM fluorescence imaging was utilized to measure intracellular calcium levels. Perifused solutions first passed through an inline heater to a temperature of 35+/−3 degrees Celsius into an open diamond bath imaging chamber (Warner Instruments, Cat: 64-0288) which was mounted using a stage adapter (Warner Instruments, Cat: 64-0298). Observation of islets was performed using a Hamamatsu ORCA-Flash4.0 digital camera (Hamamatsu Photonics K.K., Hamamatsu City, Japan, Model C11440-22CU) mounted on a BX51WIF fluorescence microscope with a 10X objective (Olympus, Tokyo, Japan). Excitation light was provided by a xenon burner supplied to the image field through a light pipe and filter wheel (Sutter Instrument Co., Novato, CA, USA, Model LB-LS/30) with a Lambda 10-3 Optical Controller (Sutter Instrument Co., Novato, CA, USA, Model LB10-3-1572). Images were taken sequentially with 340 nm and 380 nm excitation to produce each ratio from emitted light at 510 nm. Data were analyzed using cellSens Dimension 1.13 imaging software (Olympus, Tokyo, Japan) [42].

Islets were exposed to KRB containing 1 μM fura-2 AM and incubated for 30 min. They then were transferred onto the calcium scope. The fura-2 signal was recorded for two different experimental protocols. Protocol 1: 3 mM glucose (G) for 5 min, 24 G for 15 min, and 84 G for 15 min. Protocol 2: 3 mM glucose (G) for 5 min, 24 G for 15 min, 24 G containing 250 μM tolbutamide for 15 min, and back to 3 G for 25 min.

4.3. Insulin Secretion

To study insulin secretion, glucose-stimulated insulin secretion (GSIS) assays were performed. All GSIS used 12-well plates with 20 islets per well in 1 mL of KRB solution. Islets were size matched to aid in normalization as discussed in [43]. Briefly, islets were placed in 0 mM glucose for one hour, then transferred to the experimental conditions for an additional hour. Supernatants were collected from experimental conditions of 0,

12, 24, 36, 48, 60, 72, 84, and 144 mM. Using the same approach, we studied additional conditions using osmolarity-matched solutions at ~425 mOsm. Osmolarity was measured with Wescor Vapor Pressure Osmometer (Model 5520). Specific stimulators or inhibitors were placed into 24 G and 84 G to examine effects of exendin (10 nM, Sigma Aldrich, St. Louis, MO, USA, [18]), forskolin (10 μM, Sigma Aldrich, St. Louis, MO, USA, [44]), MDL-12330A (10 μM, Sigma Aldrich, St. Louis, MO, USA, [45]), and Ro-28-1675 (500 nM, Axon Medchem (Reston, VA, USA)). Insulin secretion was measured using mouse (Cat#80-INSMSU-E10) and human (Cat#80-INSHU-E01.1) ELISA following the manufacturer's directions (ALPCO, Salem, NH, USA). Intra-assay variability was kept to below 15% for all studies. D-mannitol (Sigma-Aldrich, St. Louis, MO, USA) was used to balance the osmolarity of Modified KRB solutions to ~425 mOsm.

4.4. Cell Death Quantification

Islets were incubated in a glucose solution for 48 h in standard RPMI media supplemented 10% fetal bovine serum and 1% penicillin/streptomycin. Twenty islets were placed in each well per treatment in a 12-well plate. The 12-well plate contained the following treatments. Standard RPMI media, RPMI media+144 mM mannitol, RPMI media+60 mM glucose+84 mM mannitol, and finally RPMI+144 mM glucose. Cell death was measured with propidium iodide (Sigma-Aldrich, St. Louis, MO, USA) and annexin V (Invitrogen, Carlsbad, CA, USA) staining. Apoptosis was measured using annexin V (488 nm excitation/525 nm emission), which detects phosphatidylserine when it is exposed to the outer leaflet of the plasma membrane during apoptosis [46]. Propidium iodide (535 nm excitation/620 nm emission), which is a cell exclusion dye, was used to detect generalized cell death. Regions of interest were drawn around islets to measure fluorescence intensity per islet for each individual islet normalized to surface area. These techniques have been used in previous publications [47,48] including in comparison to other methods to measure cell death [49].

4.5. Statistical Analysis

Statistical analysis was performed using R Statistical Computing Software. Data are expressed as the mean ± standard error of the mean. Data were tested for normality using Shapiro-Wilk test and for equal variance using Levene's test. Henze-Zirkler's multivariate normality test was used for the insulin correlation data. All comparisons were analyzed using two-tailed *t*-test for comparisons of two groups or one-way ANOVA with Tukey's post hoc test for more than two groups. Differences between groups were considered significant at $p < 0.05$. Spearman's rank correlation coefficient was used to analyze insulin secretion patterns in Figure 1.

4.6. Ethical Approval

All animal procedures were approved by the Ohio University Institutional Animal Care and Use Committee. Human islet isolation was approved by the Human Research Ethics Board at the University of Alberta (Pro00013094). All donors' families gave informed consent for the use of pancreatic tissue in research. Donor information was deidentified prior to our acquisition.

Author Contributions: Conceptualization, C.S.N., D.R.R., K.L.C., K.M.G.; methodology, all authors; software, N.B.W., C.S.N.; validation, all authors; formal analysis, K.M.G., C.S.N., N.B.W.; investigation, K.M.G., C.S.N., N.B.W., W.J.K.; resources, K.M.G., C.S.N.; data curation, K.M.G., C.S.N., N.B.W.; writing—original draft preparation, K.M.G., N.B.W.; writing—review and editing, C.S.N., K.M.G., N.B.W.; visualization, K.M.G., N.B.W., C.S.N.; supervision, C.S.N.; project administration, C.S.N.; funding acquisition, C.S.N. All authors have read and agreed to the published version of the manuscript.

Funding: This research was funded by NIDDK R15 DK121247, the Diabetes Institute, and the Ohio University Heritage College of Osteopathic Medicine.

Institutional Review Board Statement: All animal procedures were approved by the Ohio University Institutional Animal Care and Use Committee. Human islet isolation was approved by the Human Research Ethics Board at the University of Alberta (Pro00013094 and Pro00001620). All donors' families gave informed consent for the use of pancreatic tissue in research. Donor information was deidentified prior to our acquisition.

Informed Consent Statement: All donors' families gave informed consent for the use of pancreatic tissue in research.

Data Availability Statement: All data used to support the findings of this study are available from the corresponding author upon request.

Acknowledgments: Human islets were provided by the Alberta Islet Distribution Program and by the Alberta Diabetes Institute IsletCore, University of Alberta in Edmonton. Islet isolation was approved by the Human Research Ethics Board at the University of Alberta (Pro00013094). The graphical abstract and Figure 6 were made using BioRender.

Conflicts of Interest: The authors declare no conflict of interest.

References

1. Komatsu, M.; Takei, M.; Ishii, H.; Sato, Y. Glucose-stimulated insulin secretion: A newer perspective. *J. Diabetes Investig.* **2013**, *4*, 511–516. [CrossRef] [PubMed]
2. Gupta, A.; Rohrscheib, M.; Tzamaloukas, A.H. Extreme hyperglycemia with ketoacidosis and hyperkalemia in a patient on chronic hemodialysis. *Hemodial. Int.* **2008**, *12*, S43–S47. [CrossRef]
3. Kharroubi, A.T.; Darwish, H.M. Diabetes mellitus: The epidemic of the century. *World J. Diabetes* **2015**, *6*, 850–867. [CrossRef] [PubMed]
4. Ahlsson, F.; Gedeborg, R.; Hesselager, G.; Tuvemo, T.; Enblad, P. Treatment of extreme hyperglycemia monitored with intracerebral microdialysis. *Pediatr. Crit. Care Med.* **2004**, *5*, 89–92. [CrossRef] [PubMed]
5. Alcazar, O.; Buchwald, P. Concentration-Dependency and Time Profile of Insulin Secretion: Dynamic Perifusion Studies with Human and Murine Islets. *Front. Endocrinol* **2019**, *10*, 680. [CrossRef]
6. Druet, C.; Tubiana-Rufi, N.; Chevenne, D.; Rigal, O.; Polak, M.; Levy-Marchal, C. Characterization of Insulin Secretion and Resistance in Type 2 Diabetes of Adolescents. *J. Clin. Endocrinol. Metab.* **2006**, *91*, 401–404. [CrossRef]
7. Johnson, D.; Shepherd, R.M.; Gill, D.; Gorman, T.; Smith, D.M.; Dunne, M.J. Glucose-dependent modulation of insulin secretion and intracellular calcium ions by GKA50, a glucokinase activator. *Diabetes* **2007**, *56*, 1694–1702. [CrossRef] [PubMed]
8. Ramachandran, K.; Peng, X.; Bokvist, K.; Stehno-Bittel, L. Assessment of re-aggregated human pancreatic islets for secondary drug screening. *Br. J. Pharmacol.* **2014**, *171*, 3010–3022. [CrossRef] [PubMed]
9. Lewandowski, S.L.; Cardone, R.L.; Foster, H.R.; Ho, T.; Potapenko, E.; Poudel, C.; VanDeusen, H.R.; Sdao, S.M.; Alves, T.C.; Zhao, X.; et al. Pyruvate Kinase Controls Signal Strength in the Insulin Secretory Pathway. *Cell Metab.* **2020**, *32*, 736–750. [CrossRef]
10. Gopalakrishnan, M.; Manappallil, R.G.; Ramdas, D.; Jayaraj, J. The survival story of a diabetic ketoacidosis patient with blood sugar levels of 1985 mg/dL. *Asian J. Med Sci.* **2017**, *8*, 60. [CrossRef]
11. Highest Blood Sugar Level. Guinness World Records. Available online: https://www.guinnessworldrecords.com/world-records/highest-blood-sugar-level/ (accessed on 27 March 2020).
12. Gheni, G.; Ogura, M.; Iwasaki, M.; Yokoi, N.; Minami, K.; Nakayama, Y.; Harada, K.; Hastoy, B.; Wu, X.; Takahashi, H.; et al. Glutamate Acts as a Key Signal Linking Glucose Metabolism to Incretin/cAMP Action to Amplify Insulin Secretion. *Cell Rep.* **2014**, *9*, 661–673. [CrossRef]
13. Henquin, J.C. Triggering and amplifying pathways of regulation of insulin secretion by glucose. *Diabetes* **2000**, *49*, 1751–1760. [CrossRef] [PubMed]
14. Kalwat, M.A.; Cobb, M.H. Mechanisms of the amplifying pathway of insulin secretion in the β cell. *Pharmacol. Ther.* **2017**, *179*, 17–30. [CrossRef]
15. Rustenbeck, I.; Schulze, T.; Morsi, M.; Alshafei, M.; Panten, U. What Is the Metabolic Amplification of Insulin Secretion and Is It (Still) Relevant? *Metabolites* **2021**, *11*, 355. [CrossRef]
16. Tengholm, A. Cyclic AMP dynamics in the pancreatic β-cell. *Upsala J. Med. Sci.* **2012**, *117*, 355–369. [CrossRef]
17. Evans-Molina, C.; Mirmira, R.G. Achieving 'PeaK-A' Insulin Secretion. *Diabetes* **2013**, *62*, 1389–1390. [CrossRef]
18. Peyot, M.-L.; Gray, J.P.; Lamontagne, J.; Smith, P.J.; Holz, G.G.; Madiraju, S.M.; Prentki, M.; Heart, E. Glucagon-like peptide-1 induced signaling and insulin secretion do not drive fuel and energy metabolism in primary rodent pancreatic β-cells. *PLoS ONE* **2009**, *4*, e6221. [CrossRef]
19. Tenny, S.; Patel, R.; Thorell, W. Mannitolin. In *StatPearls*; StatPearls Publishing: Treasure Island, FL, USA, 2021. Available online: http://www.ncbi.nlm.nih.gov/books/NBK470392/ (accessed on 11 May 2021).
20. Lee, B.; Jonas, J.C.; Weir, G.C.; Laychock, S.G. Glucose regulates expression of inositol 1,4,5-trisphosphate receptor isoforms in isolated rat pancreatic islets. *Endocrinology* **1999**, *140*, 2173–2182. [CrossRef] [PubMed]

21. Jonkers, F.C.; Guiot, Y.; Rahier, J.; Henquin, J.-C. Tolbutamide stimulation of pancreatic β-cells involves both cell recruitment and increase in the individual Ca^{2+} response. *Br. J. Pharmacol.* **2001**, *133*, 575–585. [CrossRef] [PubMed]
22. Henquin, J.C. Regulation of insulin secretion: A matter of phase control and amplitude modulation. *Diabetologia* **2009**, *52*, 739. [CrossRef] [PubMed]
23. Henquin, J.C.; Meissner, H.P. The ionic, electrical, and secretory effects of endogenous cyclic adenosine monophosphate in mouse pancreatic B cells: Studies with forskolin. *Endocrinology* **1984**, *115*, 1125–1134. [CrossRef] [PubMed]
24. Yada, T.; Itoh, K.; Nakata, M. Glucagon-like peptide-1-(7-36)amide and a rise in cyclic adenosine 3′,5′-monophosphate increase cytosolic free Ca^{2+} in rat pancreatic beta-cells by enhancing Ca^{2+} channel activity. *Endocrinology* **1993**, *133*, 1685–1692. [CrossRef]
25. Liu, Y.J.; Grapengiesser, E.; Gylfe, E.; Hellman, B. Crosstalk between the cAMP and inositol trisphosphate-signalling pathways in pancreatic beta-cells. *Arch. Biochem. Biophys.* **1996**, *334*, 295–302. [CrossRef]
26. Chepurny, O.G.; Kelley, G.G.; Dzhura, I.; Leech, C.A.; Roe, M.W.; Dzhura, E.; Li, X.; Schwede, F.; Genieser, H.G.; Holz, G.G. PKA-dependent potentiation of glucose-stimulated insulin secretion by Epac activator 8-pCPT-2′-O-Me-cAMP-AM in human islets of Langerhans. *Am. J. Physiol. Endocrinol. Metab.* **2009**, *298*, E622–E633. [CrossRef]
27. Yang, H.; Yang, L. Targeting cAMP/PKA pathway for glycemic control and type 2 diabetes therapy. *J. Mol. Endocrinol.* **2016**, *57*, R93–R108. [CrossRef]
28. Seino, S.; Shibasaki, T. PKA-dependent and PKA-independent pathways for cAMP-regulated exocytosis. *Physiol. Rev.* **2005**, *85*, 1303–1342. [CrossRef]
29. Yajima, H.; Komatsu, M.; Schermerhorn, T.; Aizawa, T.; Kaneko, T.; Nagai, M.; Sharp, G.W.; Hashizume, K. cAMP enhances insulin secretion by an action on the ATP-sensitive K+ channel-independent pathway of glucose signaling in rat pancreatic islets. *Diabetes* **1999**, *48*, 1006–1012. [CrossRef] [PubMed]
30. Kaihara, K.A.; Dickson, L.M.; Jacobson, D.A.; Tamarina, N.; Roe, M.W.; Philipson, L.H.; Wicksteed, B. β-Cell–Specific Protein Kinase A Activation Enhances the Efficiency of Glucose Control by Increasing Acute-Phase Insulin Secretion. *Diabetes* **2013**, *62*, 1527–1536. [CrossRef] [PubMed]
31. Gaisano, H.Y. Here come the newcomer granules, better late than never. *Trends Endocrinol. Metab.* **2014**, *25*, 381–388. [CrossRef] [PubMed]
32. Henquin, J.-C.; Nenquin, M. Activators of PKA and Epac Distinctly Influence Insulin Secretion and Cytosolic Ca^{2+} in Female Mouse Islets Stimulated by Glucose and Tolbutamide. *Endocrinology* **2014**, *155*, 3274–3287. [CrossRef]
33. Lamontagne, J.; Al-Mass, A.; Nolan, C.J.; Corkey, B.E.; Madiraju, S.M.; Joly, E.; Prentki, M. Identification of the signals for glucose-induced insulin secretion in INS1 (832/13) β-cells using metformin-induced metabolic deceleration as a model. *J. Biol. Chem.* **2017**, *292*, 19458–19468. [CrossRef] [PubMed]
34. Tian, G.; Sandler, S.; Gylfe, E.; Tengholm, A. Glucose- and hormone-induced cAMP oscillations in α- and β-cells within intact pancreatic islets. *Diabetes* **2011**, *60*, 1535–1543. [CrossRef] [PubMed]
35. Liu, P.; Song, J.; Liu, H.; Yan, F.; He, T.; Wang, L.; Shen, H.; Hou, X.; Chen, L. Insulin regulates glucagon-like peptide-1 secretion by pancreatic alpha cells. *Endocrine* **2018**, *62*, 394–403. [CrossRef]
36. Thams, P.; Capito, K. L-arginine stimulation of glucose-induced insulin secretion through membrane depolarization and independent of nitric oxide. *Eur. J. Endocrinol.* **1999**, *140*, 87–93. [CrossRef]
37. Salehi, A.; Vieira, E.; Gylfe, E. Paradoxical stimulation of glucagon secretion by high glucose concentrations. *Diabetes* **2006**, *55*, 2318–2323. [CrossRef] [PubMed]
38. Yu, Q.; Shuai, H.; Ahooghalandari, P.; Gylfe, E.; Tengholm, A. Glucose controls glucagon secretion by directly modulating cAMP in alpha cells. *Diabetologia* **2019**, *62*, 1212–1224. [CrossRef] [PubMed]
39. Gylfe, E.; Gilon, P. Glucose regulation of glucagon secretion. *Diabetes Res. Clin. Pract.* **2014**, *103*, 1–10. [CrossRef] [PubMed]
40. Gosmanov, A.R.; Gosmanova, E.O.; Kitabchi, A.E. Hyperglycemic Crises: Diabetic Ketoacidosis (DKA), and Hyperglycemic Hyperosmolar State (HHS). In *Endotext*; Feingold, K.R., Anawalt, B., Boyce, A., Chrousos, G., de Herder, W.W., Dhatariya, K., Dungan, K., Grossman, A., Hershman, J.M., Hofland, J., et al., Eds.; MDText.com, Inc.: South Dartmouth, MA, USA, 2000. Available online: http://www.ncbi.nlm.nih.gov/books/NBK279052/ (accessed on 25 May 2021).
41. Corbin, K.L.; West, H.L.; Brodsky, S.; Whitticar, N.B.; Koch, W.J.; Nunemaker, C.S. A Practical Guide to Rodent Islet Isolation and Assessment Revisited. *Biol. Proced. Online* **2021**, *23*, 7. [CrossRef]
42. Whitticar, N.B.; Strahler, E.W.; Rajan, P.; Kaya, S.; Nunemaker, C.S. An Automated Perifusion System for Modifying Cell Culture Conditions over Time. *Biol. Proced. Online* **2016**, *18*, 19. [CrossRef]
43. Slepchenko, K.G.; Corbin, K.L.; Nunemaker, C.S. Comparing methods to normalize insulin secretion shows the process may not be needed. *J. Endocrinol.* **2019**, *241*, 149–159. [CrossRef] [PubMed]
44. Ammon, H.P.T.; Muller, A.B. Effect of forskolin on islet cyclic AMP, insulin secretion, blood glucose and intravenous glucose tolerance in rats. *Naunyn-Schmiedeberg's Arch. Pharmacol.* **1984**, *326*, 364–367. [CrossRef]
45. Li, X.; Guo, Q.; Gao, J.; Yang, J.; Zhang, W.; Liang, Y.; Wu, D.; Liu, Y.; Weng, J.; Li, Q.; et al. The Adenylyl Cyclase Inhibitor MDL-12,330A Potentiates Insulin Secretion via Blockade of Voltage-Dependent K+ Channels in Pancreatic Beta Cells. *PLoS ONE* **2013**, *8*, e77934. [CrossRef] [PubMed]
46. Baskić, D.; Popović, S.; Ristić, P.; Arsenijević, N.N. Analysis of cycloheximide-induced apoptosis in human leukocytes: Fluorescence microscopy using annexin V/propidium iodide versus acridin orange/ethidium bromide. *Cell Biol. Int.* **2006**, *30*, 924–932. [CrossRef]

47. Dula, S.B.; Jecmenica, M.; Wu, R.; Jahanshahi, P.; Verrilli, G.M.; Carter, J.D.; Brayman, K.L.; Nunemaker, C.S. Evidence that low-grade systemic inflammation can induce islet dysfunction as measured by impaired calcium handling. *Cell Calcium* **2010**, *48*, 133–142. [CrossRef] [PubMed]
48. Gelin, L.; Li, J.; Corbin, K.L.; Jahan, I.; Nunemaker, C.S. Metformin Inhibits Mouse Islet Insulin Secretion and Alters Intracellular Calcium in a Concentration-Dependent and Duration-Dependent Manner near the Circulating Range. *J. Diabetes Res.* **2018**, *2018*, 9163052. [CrossRef] [PubMed]
49. O'Neill, C.M.; Lu, C.; Corbin, K.L.; Sharma, P.R.; Dula, S.B.; Carter, J.D.; Ramadan, J.W.; Xin, W.; Lee, J.K.; Nunemaker, C.S. Circulating Levels of IL-1B+IL-6 Cause ER Stress and Dysfunction in Islets from Prediabetic Male Mice. *Endocrinology* **2013**, *154*, 3077–3088. [CrossRef]

metabolites

Article

Respiratory Parameters for the Classification of Dysfunctional Insulin Secretion by Pancreatic Islets

Uma D. Kabra [1,2], Charles Affourtit [3] and Martin Jastroch [4,5,*]

[1] Division of Metabolic Diseases, Technische Universität München, 80333 Munich, Germany; uma.kabra16205@paruluniversity.ac.in
[2] Department of Pharmaceutical Chemistry, Parul Institute of Pharmacy, Parul University, Vadodara 391760, India
[3] Faculty of Health, School of Biomedical Sciences, University of Plymouth, Plymouth Science Park, Plymouth PL6 8BU, UK; charles.affourtit@plymouth.ac.uk
[4] Department of Molecular Biosciences, The Wenner-Gren Institute, The Arrhenius Laboratories F3, Stockholm University, SE-106 91 Stockholm, Sweden
[5] Helmholtz Diabetes Center, Institute for Diabetes and Obesity, Helmholtz Zentrum München, German Research Center for Environmental Health (GmbH), 85764 Neuherberg, Germany
* Correspondence: martin.jastroch@su.se; Tel.: +46-720836997

Abstract: The development of obesity and type 2 diabetes (T2D) has been associated with impaired mitochondrial function. In pancreatic beta (β) cells, mitochondrial energy metabolism plays a central role in triggering and controlling glucose-stimulated insulin secretion (GSIS). Here, we have explored whether mitochondrial bioenergetic parameters assessed with Seahorse extracellular flux technology can quantitatively predict insulin secretion. We metabolically stressed male C57BL/6 mice by high-fat feeding (HFD) and measured the glucose sensitivity of islet respiration and insulin secretion. The diet-induced obese (DIO) mice developed hyperinsulinemia, but no pathological secretory differences were apparent between isolated DIO and chow islets. Real-time extracellular flux analysis, however, revealed a lower respiratory sensitivity to glucose in DIO islets. Correlation of insulin secretion with respiratory parameters uncovers compromised insulin secretion in DIO islets by oxidative power. Normalization to increased insulin contents during DIO improves the quantitative relation between GSIS and respiration, allowing to classify dysfunctional properties of pancreatic insulin secretion, and thereby serving as valuable biomarker for pancreatic islet glucose responsiveness and health.

Keywords: mitochondria; bioenergetics; glucose-stimulated insulin secretion; pancreatic islets; respiration

check for updates

Citation: Kabra, U.D.; Affourtit, C.; Jastroch, M. Respiratory Parameters for the Classification of Dysfunctional Insulin Secretion by Pancreatic Islets. *Metabolites* **2021**, *11*, 405. https://doi.org/10.3390/metabo11060405

Academic Editors: Melkam Kebede and Belinda Yau

Received: 3 June 2021
Accepted: 18 June 2021
Published: 21 June 2021

1. Introduction

Pancreatic β cells are specialized endocrine cells that integrate signals from glucose and other fuels to control the secretion of insulin [1]. Glucose induces insulin secretion via both triggering and amplifying pathways [2]. The triggering pathway involves oxidative glucose catabolism, a rise in the cytosolic adenosine triphosphate/adenosine diphosphate (ATP/ADP) ratio, closure of ATP-sensitive potassium (K_{ATP}) channels, depolarization of the plasma membrane, influx of calcium ions, and eventual exocytosis of insulin-containing granules [3–5]. The amplifying pathways boost this glucose-stimulated insulin secretion (GSIS) in a K_{ATP}-channel-independent (but Ca^{2+}-dependent) way [4,6]. Many steps of the β cell insulin secretory pathways can affect the efficiency of insulin secretion. Mitochondria are intimately involved in glucose catabolism because oxidative phosphorylation has control over the ATP/ADP ratio and, thus, GSIS [7]. Oxidative phosphorylation is ideally assessed by combining respiratory flux and mitochondrial membrane potential measurements [8,9]. While mitochondrial membrane potential measurements of β cell mitochondria are challenging in intact β cells because of glucose-evoked plasma membrane potential fluctuations [10], cellular respiratory flux measurements are relatively straightforward [11]. Several parameters are readily calculated from whole-cell oxygen uptake to give insight in the efficiency by which energy liberated from oxidative glucose breakdown is conserved as ATP. These parameters include glucose sensitivity (GS), i.e., the magnitude of respirational increase upon glucose stimulus, and coupling efficiency (CE) of oxidative phosphorylation, i.e., the part of mitochondrial respiration coupled to ATP synthesis [8]. Notably, these bioenergetic parameters are internally normalized and, thus, dimensionless, which renders them more sensitive indicators of oxidative phosphorylation than absolute oxygen uptake rates [8]. For example, CE has been instrumental for demonstrating a glucose sensitivity of oxidative phosphorylation in INS-1E insulinoma cells that is regulated by mitochondrial uncoupling protein-2 [12]. This bioenergetic regulation is reflected by insulin secretion activity, which suggests the possibility that CE has broader predictive power that may be exploited to forecast β cell dysfunction.

In this study, we correlated mitochondrial respiration and insulin secretion in pancreatic islets from chow-fed and diet-induced obese (DIO) mice, with the aim to identify parameters that allow classification of insulin secretion deficiencies. DIO mice represent a model of a well-established risk scenario in which obesity leads to insulin resistance, β cell dysfunction, and eventually type 2 diabetes mellitus [13]. Our findings suggest that loss of mitochondrial respiratory sensitivity to glucose is an early warning sign for compromised insulin secretion.

2. Results and Discussion

2.1. Inducing Metabolic Stress in Mice

We tested our hypothesis on the relation between respiration and insulin secretion in DIO mice, a typical model for the development of insulin resistance and hyperinsulinemia. Male C57BL/6 mice were kept on high-fat diet (HFD) for 16 weeks to induce obesity (~45.8 g HFD vs. ~28.8 g chow; Figure 1A). Plasma insulin levels were dramatically increased (Figure 1B). However, plasma glucose levels were unchanged (Figure 1C), suggesting that hyperinsulinemia was sufficient to compensate insulin resistance in HFD mice.

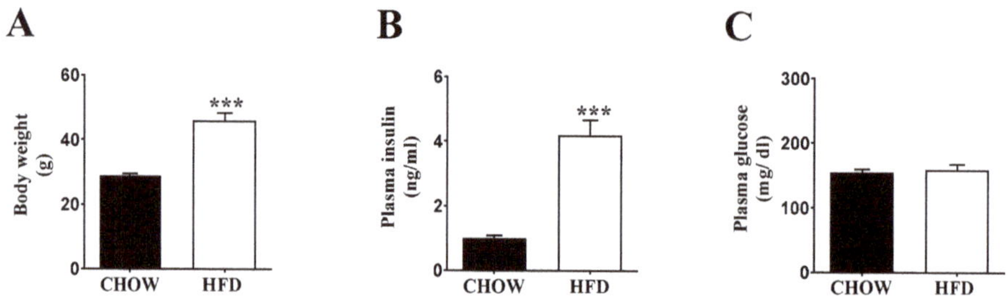

Figure 1. Characterization of chow (black bars) and HFD (white bars) fed C57BL/6 mice. (**A**) Body weight; (**B**) plasma insulin level; and (**C**) plasma glucose levels. The mice were 8 weeks old at the start of the feeding experiment. Data are represented as ± SEM ($n = 8$ mice per group). Statistical significance of mean differences was tested by unpaired two-tailed student t-test. $p < 0.001$ (***).

2.2. Respiratory Activity but Not Insulin Secretion Is Impaired in Islets of DIO Mice

Interestingly, the assessment of insulin secretion between chow and DIO islets revealed no differences (Figure 2A). In contrast, averaged real-time oxygen consumption traces in Figure 2B demonstrate clearly that respiration in DIO islets differs drastically from that exhibited by their control counterparts. After normalizing oxygen uptake to DNA content and correcting respiratory activity for non-mitochondrial (i.e., rotenone-and-antimycin-A-resistant) oxygen consumption, it transpired that there was no difference in basal respiration between the two systems (Figure 2C). However, the strong respiratory stimulation provoked by glucose in control islets was more than halved in DIO islets (Figure 2D,E), which was the combined result of attenuated respiration linked to mitochondrial proton leak (Figure 2F) and ATP synthesis (Figure 2G). In other words, both oligomycin-sensitive and -insensitive oxygen uptake was decreased by DIO, which explains why the CE of oxidative phosphorylation, which reflects the oligomycin sensitivity of overall glucose-stimulated respiration (GSR), was not affected to a statistically significant degree (Figure 2H). The reduction of both proton leak and ATP-linked respiration can be indicative of compromised substrate oxidation capacity [8]. Next, we plotted absolute secreted insulin values vs respiratory parameters related to insulin triggering (ATP-linked respiration and GSR), and assessed their relation by correlation analysis. Within chow islets, insulin values correlate significantly with ATP-linked respiration and GSR (Figure 2I,J). This relationship appeared to be shifted upwards in DIO islets (dotted regression line). In our experimental conditions in vitro, however, no secretagogues or other amplifying mechanisms are present, to the best of our knowledge, which could selectively act on DIO islets. However, we cannot formally exclude amplifying factors deriving from glucose catabolism [14], and we thus refer to altered secretory pathways of insulin secretion rather than to altered triggering of insulin secretion.

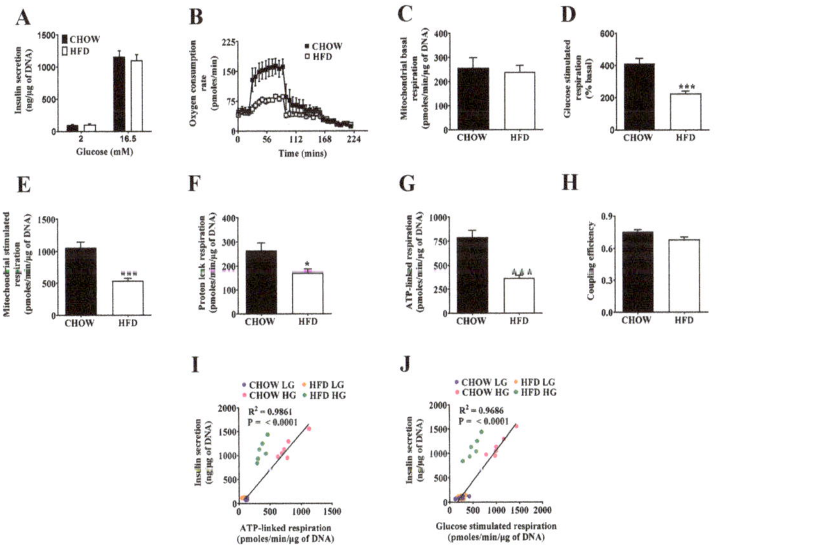

Figure 2. Glucose-stimulated insulin secretion and mitochondrial bioenergetics of pancreatic islets from chow and HFD-fed mice. (**A**) Insulin secretion. Batches of eight size-matched islets were exposed to either 2 or 16.5 mM glucose for 1 h before supernatants and lysates were collected. (**B**) Representative time-resolved oxygen consumption traces. Batches of 30 size-matched islets were exposed to 2 mM glucose to assess basal respiration (4 cycles), glucose-stimulated respiration (10 cycles), proton leak respiration by inhibition of ATP synthase using oligomycin (10 cycles), and non-mitochondrial respiration by final injection of rotenone/antimycin A. (**C**) Mitochondrial basal respiration at low glucose (2 mM). (**D**) Glucose-stimulated respiration expressed as percentage of basal (**E**) Mitochondrial stimulated respiration at high glucose (16.5 mM). (**F**) Proton leak respiration at high glucose. (**G**) ATP-linked respiration at high glucose. (**H**) Coupling efficiency at high glucose. (**I**) Correlation of insulin secretion (absolute values) and ATP-linked respiration. (**J**) Correlation of insulin secretion (absolute values) and glucose-stimulated respiration. Data are represented as means ± SEM (n = 6 mice per group, each mouse was considered as independent experiment and islets were plated in triplicate). Statistical significance of mean differences was tested by unpaired two-tailed student t-test. $p < 0.05$ (*), $p < 0.001$ (***).

2.3. Compensating Increase of Insulin Content in DIO Islets Masks Defects of the Triggering Pathway

Exploring compensatory responses to loss of insulin sensitivity in HFD mice (Figure 1), we found a 15% higher insulin content in DIO islets (Figure 3A), consistent with increases in β cell proliferation during DIO [15]. Normalizing insulin secretion to insulin content, GSIS of the DIO islets was lower than that from control islets (Figure 3B). Importantly, plotting GSIS values normalized to insulin content shifted DIO insulin values onto the regression of chow islets, with a better regression to glucose-stimulated than to ATP-linked respiration (Figure 3C,D). Insulin secretion correlates less well with coupling efficiency (CE, Figure 3E), but it should be noted that the internal standardization benefits data comparisons between different experimental settings and laboratories. The normalization-dependency of the GSIS phenotype suggests that DIO islets respond to an obesity-induced drop in glucose sensitivity by increasing insulin content, a response that appears to fix the secretory impairment.

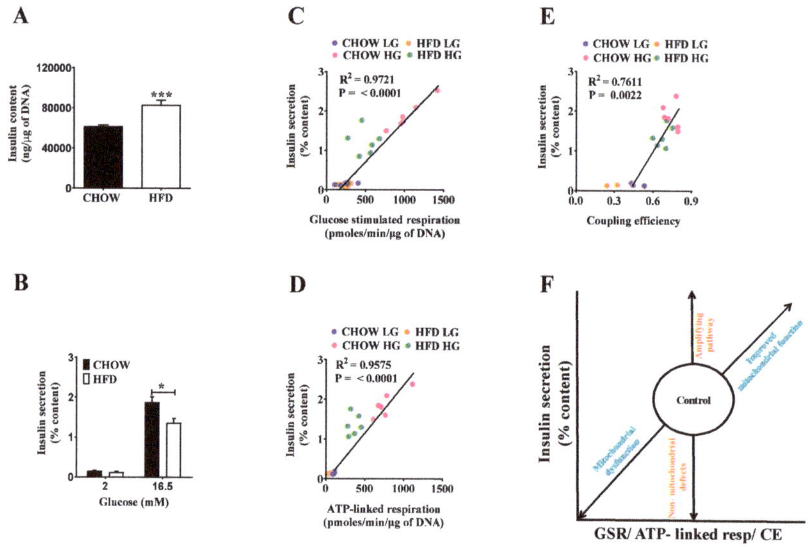

Figure 3. Glucose-stimulated insulin secretion in chow and HFD mice. Batches of eight size-matched islets were exposed to either 2 or 16.5 mM glucose for 1 h at which point supernatants and lysates were collected for (**A**) insulin content, and (**B**) secreted insulin measurements. Correlation between insulin secretion (% content) and (**C**) glucose-stimulated respiration, (**D**) ATP-linked respiration, and (**E**) coupling efficiency (**F**) correlation model classifying defects of insulin secretion. Data are represented as means ± SEM (n = 6 mice per group, each mouse was considered as independent experiment and islets were plated in triplicate). Statistical significance of mean differences was tested by unpaired two-tailed student t-test. $p < 0.05$ (*), $p < 0.001$ (***).

2.4. Classifying Defects in Insulin Secretion

From Figure 3C–E, it transpires that secreted insulin values require normalization to insulin content to establish a robust linear relationship between insulin and bioenergetics parameters. The relationship between oxidative phosphorylation and GSIS leads to a simple linear correlation model to classify defects of insulin secretion (Figure 3F). In relation to control values, descending along the regression line suggests reduced oxidative power by either compromised substrate delivery or respiratory dysfunction. In the case of DIO islets, the secretory pathway is compromised by reduced oxidative power, which could be mediated by impaired glucose uptake, glycolysis, or pyruvate oxidation—the latter being recently suggested in response to impaired mitochondrial dynamics [16] or inflammation [17]. In contrast, ascending values suggest improved substrate delivery or oxidative phosphorylation. Upwards deviation from regression is explained by amplifying pathways, while downwards deviation points towards secretory dysfunction downstream or no mitochondrial impact.

3. Materials and Methods

Animals—male C57BL/6 mice with an age of 8 to 10 weeks were purchased from Janvier Labs (Le Genest-Saint-Isle, France). The animals were maintained on a 12/12 h light/dark cycle in a temperature-controlled environment and allowed free access to standard chow diet (5.6% fat, LM-485, Harlan Teklad) or a high-fat diet (HFD) (58% kcal fat; Research Diets Inc., New Brunswick, NJ, USA) for 16 weeks. All in vivo procedures were conducted under the guidelines of the Institutional Animal Care Committee of the Helmholtz Center Munich, which approved all animal maintenance and experimental procedures. The animal experiments complied with all ethical regulations for animal testing and research, including animal maintenance and experimental procedures that the animal welfare authorities of the local animal ethics committee of the state of Bavaria (Regierung Oberbayern) approved in accordance with European guidelines.

Islet isolation and culture—mouse islets were isolated by digestion with collagenase as described elsewhere [12]. Around 150–200 islets were obtained per mouse. Islets were incubated overnight in RPMI 1640 culture medium supplemented with 10% (*v/v*) fetal calf serum (Life technologies) at 37 °C and 5% CO_2 before experimentation.

Insulin secretion—groups of eight size-matched islets were handpicked into individual wells of V-bottomed 96-well plates and incubated for 60 min at 37 °C in HEPES-balanced Krebs-Ringer (KRH) bicarbonate buffer containing 114 mM NaCl, 4.7 mM KCl, 2.5 mM $CaCl_2$, 1.16 mM $MgSO_4$, 1.2 mM KH_2PO_4, 25.5 mM $NaHCO_3$, 20 mM HEPES (pH 7.2–7.4), supplemented with 0.2% (*w/v*) bovine serum albumin (BSA), and 2 mM glucose. The islets were then incubated for a further 60 min in KRH-bicarbonate buffer containing 2 or 16.5 mM glucose. Subsequently, supernatants were collected to quantify secreted insulin. Following the secretion assay, islets were lysed with ice-cold RIPA buffer to allow total insulin content to be determined. Insulin was detected with the Ultra-Sensitive Mouse Insulin Elisa Kit (ALPCO, Salem, NH, USA). Data were normalized to DNA content measured with the Quant-it Pico Green DNA assay kit (Invitrogen, Darmstadt, Germany).

Mitochondrial respiration—oxygen consumption rates (OCR) were measured in islet-capture plates of the XF24 extracellular flux analyzer (Agilent, Seahorse Bioscience, Santa Clara, CA, USA). Briefly, groups of 30 size-matched islets were handpicked into individual wells of islet capture plates and incubated for 60 min at 37 °C without CO_2 in KRH-bicarbonate free buffer containing 2 mM glucose. Additional glucose (16.5 mM), oligomycin (10 µg/mL), and a mixture of rotenone and antimycin A (both 2 µM) were injected sequentially. Mitochondrial respiration was calculated by subtracting non-mitochondrial respiration from all other oxygen uptake rates. The individual bioenergetics parameters of OXPHOS parameters were calculated as follows: basal mitochondrial respiration = (last rate measured before glucose and/or other secretagogues injection) − (non-mitochondrial respiration rate). Glucose-stimulated mitochondrial respiration = (last rate measured before oligomycin injection) − (non-mitochondrial respiration rate). Proton-leak-linked respiration = (minimum rate measured after oligomycin injection) − (non-mitochondrial respiration rate). ATP-synthesis-linked respiration = (glucose-stimulated mitochondrial respiration) − (proton leak-linked respiration). Coupling efficiency = the fraction of respiration used to drive ATP synthesis for each run, calculated as CE = 1 − (proton-leak-linked respiration/mitochondrial glucose-stimulated respiration). Data were normalized to DNA content.

Metabolic tests—mice were fasted for six hours prior to blood collection. Blood samples taken from the tail vein were used to measure glycaemia with a glucometer (Abbott, Wiesbaden, Germany) and insulin by ELISA.

Statistical analysis—statistical analysis was performed with GraphPad Prism version 6.0. The data passed normality tests (Shapiro–Wilk and Kolmogorov–Smirnov test) and the groups were compared using unpaired Student's t test. All the data are shown as mean ± standard error of mean (S.E.M). p values < 0.05 were considered statistically significant.

4. Conclusions

In this study, we demonstrate that mitochondrial respiratory parameters have predictive value for insulin secretion from pancreatic β cells. In a typical model for metabolic disease, DIO mice, we found that defective insulin secretion is compensated by increased insulin content. Although absolute values of insulin secretion are not affected, mitochondrial respiration is severely compromised in DIO islets. Correlation of respiration and GSIS is firmly established by normalizing GSIS to insulin content, showing that mitochondrial respiratory parameters quantitatively predict changes in GSIS. Absolute values of ATP-linked respiration and GSR predict insulin release the best, while CE is less predictive. CE, however, is a powerful bioenergetic parameter that has been successfully applied to uncover molecular mechanisms in mitochondria, e.g., the role of UCP2 in β cells [8]. CE is based on the thermodynamic laws of energy conversion that mitochondrial oxidation energy is either converted to ATP or lost as heat due to the mitochondrial proton leak. CE is defined as the fraction of energy that is converted to ATP (thus ranging from 0 to 1). As internally standardized parameter, CE is not prone to variation in absolute values between independent experiments. Although CE correlates with GSIS, the relation is steeper and more variable concerning linear regression, as compared to ATP-linked respiration and GSR in our experimental setup. Thus, CE may only be used to compare independent studies. However, the relation of CE and GSIS suggests a "threshold" CE value that is required to trigger insulin secretion. Considering previous studies from our laboratory [16,18], we find a tight CE of about 0.4–0.6 as requirement for insulin triggering in islets and in β cell models. Based on the results we designed a model correlating GSR or ATP linked respiration vs. insulin secretion that allows classifying dysfunctional properties of pancreatic insulin secretion under pathological conditions. All these analyses suggest that mitochondrial bioenergetic parameters reflect insulin secretion in a quantifiable manner, and may thus serve as biomarkers for glucose responsiveness and pancreatic islet health.

Author Contributions: Conceptualization, M.J., C.A., and U.D.K.; investigation, U.D.K.; resources, M.J.; writing—original draft preparation, U.D.K., C.A., and M.J.; writing—review and editing, U.D.K., M.J., and C.A; visualization, U.D.K.; supervision, M.J.; project administration, M.J.; funding acquisition, M.J. All authors have read and agreed to the published version of the manuscript.

Funding: This research was partially supported by the German Center for Diabetes Research (DZD) and Novo Nordisk Fonden (to M. J. grant number 0059646).

Institutional Review Board Statement: All in vivo procedures were conducted under the guidelines of the Institutional Animal Care Committee of the Helmholtz Center Munich, which approved all animal maintenance and experimental procedures. The ethic code 5.1-568-Gas was granted by the Landratsamt München. The animal experiments complied with all ethical regulations for animal testing and research, including animal maintenance and experimental procedures, which the animal welfare authorities of the local animal ethics committee of the state of Bavaria (Regierung Oberbayern) approved, in accordance with European guidelines.

Informed Consent Statement: Not applicable.

Data Availability Statement: Not applicable.

Acknowledgments: The authors thank Katrin Pfuhlmann and Maria Kutschke at the Helmholtz Zentrum München for excellent technical assistance. We are grateful for excellent support by the central animal facility of the Helmholtz Center Munich and their animal caretakers.

Conflicts of Interest: The authors declare no conflict of interest. The funders had no role in the design of the study; in the collection, analyses, or interpretation of data; in the writing of the manuscript, or in the decision to publish the results.

References

1. Rutter, G.A.; Pullen, T.J.; Hodson, D.J.; Martinez-Sanchez, A. Pancreatic beta-cell identity, glucose sensing and the control of insulin secretion. *Biochem. J.* **2015**, *466*, 203–218. [CrossRef] [PubMed]
2. Henquin, J.C. Triggering and amplifying pathways of regulation of insulin secretion by glucose. *Diabetes* **2000**, *49*, 1751–1760. [CrossRef] [PubMed]
3. Ashcroft, F.M. ATP-sensitive potassium channelopathies: Focus on insulin secretion. *J. Clin. Investig.* **2005**, *115*, 2047–2058. [CrossRef] [PubMed]
4. Bergsten, P.; Lin, J.; Westerlund, J. Pulsatile insulin release: Role of cytoplasmic Ca^{2+} oscillations. *Diabetes Metab.* **1998**, *24*, 41–45. [PubMed]
5. Lewandowski, S.L.; Cardone, R.L.; Foster, H.R.; Ho, T.; Potapenko, E.; Poudel, C.; VanDeusen, H.R.; Sdao, S.M.; Alves, T.C.; Zhao, X.; et al. Pyruvate kinase controls signal strength in the insulin secretory pathway. *Cell Metab.* **2020**, *32*, 736–750. [CrossRef] [PubMed]
6. Gembal, M.; Detimary, P.; Gilon, P.; Gao, Z.Y.; Henquin, J.C. Mechanisms by which glucose can control insulin release independently from its action on adenosine triphosphate-sensitive K+ channels in mouse B cells. *J. Clin. Investig.* **1993**, *91*, 871–880. [CrossRef] [PubMed]
7. Maechler, P.; Li, N.; Casimir, M.; Vetterli, L.; Frigerio, F.; Brun, T. Role of mitochondria in beta-cell function and dysfunction. In *The Islets of Langerhans. Advances in Experimental Medicine and Biology*; Springer: Dordrecht, The Netherlands, 2010; Volume 654, pp. 193–216. [CrossRef]
8. Affourtit, C.; Brand, M.D. Measuring mitochondrial bioenergetics in INS-1E insulinoma cells. In *Methods in Enzymology*; William, S.A., Anne, N.M., Eds.; Elsevier: San Diego, CA, USA, 2009; Volume 457, pp. 405–424. [CrossRef]
9. Gerencser, A.A.; Mookerjee, S.A.; Jastroch, M.; Brand, M.D. Measurement of the Absolute Magnitude and Time Courses of Mitochondrial Membrane Potential in Primary and Clonal Pancreatic Beta-Cells. *PLoS ONE* **2016**, *11*, e0159199. [CrossRef] [PubMed]
10. Gerencer, A.A.; Chinopoulos, C.; Birket, M.J.; Jastroch, M.; Vitelli, C.; Nicholls, D.G.; Brand, M.D. Quantitative measurement of mitochondrial membrane potential in cultured cells: Calcium-induced de-and hyperpolarization of neuronal mitochondria. *J. Physiol.* **2012**, *590*, 2845–2871. [CrossRef] [PubMed]
11. Mookerjee, S.A.; Gerencser, A.A.; Nicholls, D.G.; Brand, M.D. Quantifying intracellular rates of glycolytic and oxidative ATP production and consumption using extracellular flux measurements. *J. Biol. Chem.* **2017**, *292*, 7189–7207. [CrossRef] [PubMed]
12. Affourtit, C.; Jastroch, M.; Brand, M.D. Uncoupling protein-2 attenuates glucose-stimulated insulin secretion in INS-1E insulinoma cells by lowering mitochondrial reactive oxygen species. *Free Radic. Biol. Med.* **2011**, *50*, 609–616. [CrossRef] [PubMed]
13. Nilsson, C.; Raun, K.; Yan, F.F.; Larsen, M.O.; Tang-Christensen, M. Laboratory animals as surrogate models of human obesity. *Acta Pharmacol. Sin.* **2012**, *33*, 173–181. [CrossRef] [PubMed]
14. Kalwat, M.A.; Cobb, M.H. Mechanisms of the amplifying pathway of insulin secretion in the β cell. *Pharmacol. Ther.* **2017**, *179*, 17–30. [CrossRef] [PubMed]
15. Mosser, R.E.; Maulis, M.F.; Moullé, V.S.; Dunn, J.C.; Carboneau, B.A.; Arasi, K.; Pappan, K.; Poitout, V.; Gannon, M. High-fat diet-induced β-cell proliferation occurs prior to insulin resistance in C57Bl/6J male mice. *Am. J. Physiol. Endocrinol. Metab.* **2015**, *308*, E573–E582. [CrossRef] [PubMed]
16. Kabra, U.D.; Pfuhlmann, K.; Migliorini, A.; Keipert, S.; Lamp, D.; Korsgren, O.; Gegg, M.; Woods, S.C.; Pfluger, P.T.; Lickert, H.; et al. Direct substrate delivery into mitochondrial fission–deficient pancreatic islets rescue insulin secretion. *Diabetes* **2017**, *66*, 1247–1257. [CrossRef] [PubMed]
17. Barlow, J.; Solomon, T.P.J.; Affourtit, C. Pro-inflammatory cytokines attenuate glucose-stimulated insulin secretion from INS-1E insulinoma cells by restricting mitochondrial pyruvate oxidation capacity—Novel mechanistic insight from real-time analysis of oxidative phosphorylation. *PLoS ONE* **2018**, *13*, e0199505. [CrossRef] [PubMed]
18. Schulz, N.; Kluth, O.; Jastroch, M.; Schürmann, A. Minor role of mitochondrial respiration for fatty-acid induced insulin secretion. *Int. J. Mol. Sci.* **2013**, *14*, 18989–18998. [CrossRef] [PubMed]

metabolites

MDPI

Article

Machine Learning Algorithms, Applied to Intact Islets of Langerhans, Demonstrate Significantly Enhanced Insulin Staining at the Capillary Interface of Human Pancreatic β Cells

Louise Cottle [1], Ian Gilroy [2], Kylie Deng [1], Thomas Loudovaris [3], Helen E. Thomas [3,4], Anthony J. Gill [5,6,7], Jaswinder S. Samra [5,8], Melkam A. Kebede [1], Jinman Kim [2] and Peter Thorn [1,*]

1 Charles Perkins Centre, School of Medical Sciences, University of Sydney, Camperdown 2006, Australia; louise.cottle@sydney.edu.au (L.C.); kden4987@uni.sydney.edu.au (K.D.); melkam.kebede@sydney.edu.au (M.A.K.)
2 School of Computer Science, University of Sydney, Camperdown 2006, Australia; ian.gilroy@sydney.edu.au (I.G.); jinman.kim@sydney.edu.au (J.K.)
3 St Vincent's Institute, Fitzroy 3065, Australia; tloudovaris@svi.edu.au (T.L.); hthomas@svi.edu.au (H.E.T.)
4 Department of Medicine, St Vincent's Hospital, University of Melbourne, Fitzroy 3065, Australia
5 Northern Clinical School, University of Sydney, St Leonards 2065, Australia; Anthony.Gill@health.nsw.gov.au (A.J.G.); jas.samra@bigpond.com (J.S.S.)
6 Department of Anatomical Pathology, Royal North Shore Hospital, St Leonards 2065, Australia
7 Cancer Diagnosis and Pathology Research Group, Kolling Institute of Medical Research, St Leonards 2065, Australia
8 Upper Gastrointestinal Surgical Unit, Royal North Shore Hospital, St Leonards 2065, Australia
* Correspondence: p.thorn@sydney.edu.au

Citation: Cottle, L.; Gilroy, I.; Deng, K.; Loudovaris, T.; Thomas, H.E.; Gill, A.J.; Samra, J.S.; Kebede, M.A.; Kim, J.; Thorn, P. Machine Learning Algorithms, Applied to Intact Islets of Langerhans, Demonstrate Significantly Enhanced Insulin Staining at the Capillary Interface of Human Pancreatic β Cells. *Metabolites* **2021**, *11*, 363. https://doi.org/10.3390/metabo11060363

Academic Editor: Amedeo Lonardo

Received: 3 May 2021
Accepted: 28 May 2021
Published: 7 June 2021

Abstract: Pancreatic β cells secrete the hormone insulin into the bloodstream and are critical in the control of blood glucose concentrations. β cells are clustered in the micro-organs of the islets of Langerhans, which have a rich capillary network. Recent work has highlighted the intimate spatial connections between β cells and these capillaries, which lead to the targeting of insulin secretion to the region where the β cells contact the capillary basement membrane. In addition, β cells orientate with respect to the capillary contact point and many proteins are differentially distributed at the capillary interface compared with the rest of the cell. Here, we set out to develop an automated image analysis approach to identify individual β cells within intact islets and to determine if the distribution of insulin across the cells was polarised. Our results show that a U-Net machine learning algorithm correctly identified β cells and their orientation with respect to the capillaries. Using this information, we then quantified insulin distribution across the β cells to show enrichment at the capillary interface. We conclude that machine learning is a useful analytical tool to interrogate large image datasets and analyse sub-cellular organisation.

Keywords: insulin; beta cell; human; islet; polarisation; machine learning; deep learning; cell segmentation; automation

1. Introduction

Defective insulin secretion from islet β cells is a characteristic feature of diabetes mellitus [1]. To better understand molecular mechanisms that regulate insulin secretion, we need to be able to image and study β cells and their subcellular structures [2], particularly their organisation within the native environment of the islets of Langerhans.

With the advancing capabilities of modern microscopy systems, the detailed visualisation of cells and their subcellular components is possible [2,3]. However, this has also led to the rapid generation of complex and ever-expanding image datasets [4,5]. The bottleneck facing researchers now is the extraction and quantification of valuable biological insights from these large image datasets [4,6]. Thus, the need for automated image analysis methods becomes increasingly important. Computational image processing methods have traditionally relied on static and predefined rules [3]. However, a major shortcoming of this approach emerges when these static algorithms are applied to different datasets or datasets of high variability, often requiring labour-intensive reprogramming and/or the manual adjustment of predefined parameters [3]. In contrast, machine learning seeks to detect patterns from training data, and then apply those patterns to new datasets [3,7]. With sufficient training data, encompassing wide variations in morphology, the same algorithm can be reused, even for different experimental setups, without the need for code tweaking [6,8]. This approach not only reduces human workload [9], but also offers significant advantages over conventional image processing methods in its ability to ensure objective, reproducible and timely analysis [3,6,10].

The last two decades have seen an expansion of machine learning applications in biological studies [4,5,11,12]. In particular, deep learning, a subtype of machine learning, has gained significant popularity in automated applications including image classification [5], tissue [13,14] and cell image segmentation [6,7], and nuclei identification and quantification [8]. A deep learning approach to image processing works by using neural network structures to extract features of a given image dataset in "layers" or levels of hierarchy [15]. Successive layers of representations are generated such that the higher levels of hierarchy are composed using the output of lower-level features [16]. Deep learning methods have demonstrated success at the cellular level in segmentation applications of a range of cell types, including bacteria and mammalian cells from phase contrast images [17], HeLa cells from DIC microscopy images [18], neuronal membranes in electron microscopy images [19], yeast cells [6], and circulating tumour cells [20]. At the subcellular level, deep learning algorithms have also precisely segmented the nuclei and cytoplasm in fibroblasts, HeLa, HepG2 cells [2,21,22].

Applied to the study of islets of Langerhans, automated analyses have been used for the segmentation of islets and pancreatic exocrine tissue [23], as well as the quantification of individual islets and islet cell density [24]. At the cellular level, however, only a few studies have applied machine learning methods to the study of islets. Human islets are composed of five endocrine cell types, insulin-secreting β cells (~65%), glucagon-secreting α-cells (~30%), somatostatin-secreting δ-cells (~5%), pancreatic polypeptide-secreting γ-cells and ghrelin-secreting ε-cells (<1%) [25,26]. The challenge of using machine learning for islet cell segmentation lies in the complex variation in structure and shape of these islet cells [27]. Not only is it difficult to distinguish between the different cell types, but cells are also often of irregular shape and closely packed together [27,28], leading to challenges in border detection between cells and the generation of labelled images for training models.

In the last decade, a small number of studies have highlighted the potential of automated image analysis methods for the segmentation of these islet cells. In 2012, an analytical software program, *Pancreas++*, was developed for the classification and positional quantification of α and β cells within islets in fluorescence microscopy images [29]. In another study, using the immunofluorescence staining of TMEM27 and BACE2 in islets, an automated image analysis pipeline was generated to determine β cell number, area and density per islet [23]. However, while these studies employ automated image analysis approaches, they largely focus on cellular arrangements within an islet, rather than individual β cells and their subcellular structures and protein distributions. Alternative computational approaches to study islets using mathematical modelling have generated three-dimensional reconstructions of pancreatic islets; however, these are not without limitations. For example, many models have not been able to accurately capture the heterogeneity of cell sizes and shapes within an islet [30,31]. In other models, the presence of various islet structures including vasculature have not been considered [32].

There are many aspects of the biology of β cells that could be advanced by machine learning approaches. For example, accumulating evidence indicates the presence of the structural and functional polarisation of β cells [33–36], reminiscent of cell polarity in epithelial cells [37]. Key regulators of cell polarity such as liver kinase B1 (LKB1) have been identified in rodent β cells [37], as well as cell polarity determinants including discs large (Dlg), partitioning defective 3 homologue (Par3) and scribble, showing consistent orientation with respect to islet vasculature in both human and rodent β cells [35,36]. Previous studies have also indicated β cell regional specialisations, such as the selective localisation of the GLUT2 transporter on the lateral membrane domain between adjacent β cells [38], as well as the targeting of insulin granule fusion at the vascular interface of the β cells [39]. The presynaptic scaffold proteins liprin, ELKS, Rab3-interacting protein (RIM2) and piccolo show enriched expression at the β cell–vasculature interface [36,39], suggesting that insulin secretion may be regulated by mechanisms similar to a neuronal synapse [39]. Furthermore, it has also been suggested that insulin content is asymmetrically distributed in the β cell, with an enrichment at the β cell–vasculature interface [36,40]. However, little is currently known about the mechanism linking β cell structural polarity and cell function.

Here, we use a deep learning approach to segment β cells, and subsequently investigate the subcellular organization of β cells within islets by analysing the distribution of insulin with respect to cell contacts with islet vasculature. We assessed two commonly used deep learning models for image segmentation applications, namely, the U-Net fully convolutional networks (FCN) and residual neural networks (RNN), for the automated segmentation of β cells from microscopy images of human pancreatic islet slices. We next applied the U-Net model to create β cell mask images, used to predict the location of β cells within islets. Analysis of insulin distribution in over 2000 β cell instances using computational techniques demonstrated an enrichment at the capillary interface of β cells.

2. Results

Human pancreas samples sourced from either partial pancreatectomy patients or cadaveric donors were processed using the pancreatic slice technique [41]. In this process, 150 μm sections were stained and imaged using 3D fluorescent microscopy. Deep learning approaches were undertaken on the resultant images, first to predict β cell locations and boundaries and then to assess subcellular fluorescent staining.

2.1. Manual Analysis Reveals Increased Insulin Staining at the Capillary Interface of β Cells

In situ analysis of β cells in islets within pancreatic slices provides evidence that β cells are polarised, and that both mouse and human β cells maintain a consistent orientation with respect to the vasculature [35,36]. The islet vasculature is composed of cells and secreted basement membrane, which is a complex mixture of proteins including laminin [42], and in this work we have used laminin-β-1 as a marker for the islet vasculature/capillaries.

β cell orientation has important functional consequences, such as the precise targeting of insulin secretion to the vasculature [35,36]. A recent study, in mice, suggested the presence of a population of β cells with an asymmetric distribution of insulin content, showing an enrichment of insulin in the regions adjoining the islet capillaries and an avascular location for insulin mRNA [40]. Therefore, we investigated whether vasculature contact influenced insulin distribution within human β cells. In this study, human islets were immunostained to visualise the β cells (insulin), their cell boundaries (syntaxin 1A) and the surrounding vasculature (laminin) (Figure 1a). To assess insulin distribution, manual analysis involved assessing one z-plane of the islet and drawing a perpendicular line across each β cell from the vascular face (laminin) to the avascular face (opposite the vasculature), and the fluorescence intensity was measured at each face using a line-scan (white lines, Figure 1b). The results showed that insulin distribution across each β cell was asymmetric and enriched towards the vasculature (Figure 1c, $n = 25$). This relationship was consistently observed in all islets analysed (Figure 1d, $n = 3$ donors, 1–2 islets per donor, $n = 83$ cells) [36]. This analysis, whilst informative, was performed manually, and so was relatively labour-intensive and sampled only a subset of β cells in the islet.

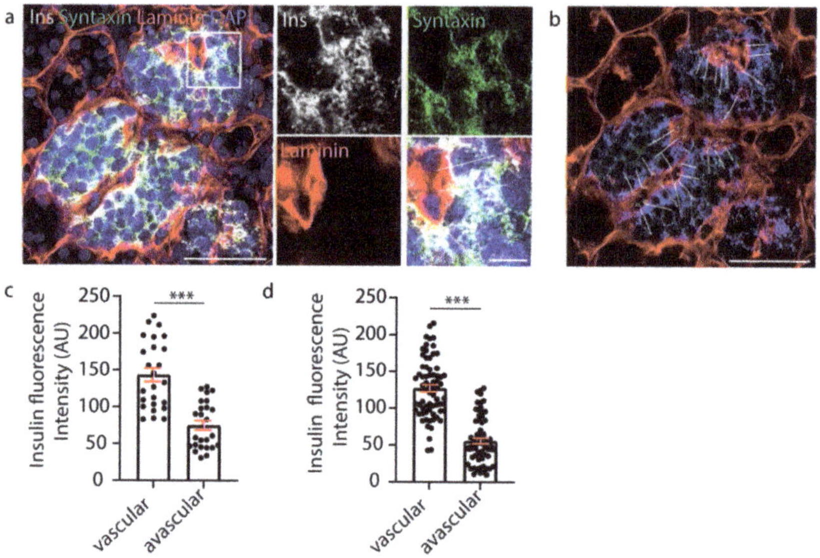

Figure 1. Manual analysis of β cell insulin intensity. (**a**) Representative image of a whole human islet with inset. Ins-grey, Syntaxin 1A-green, Laminin-red, DAPI-blue. (**b**) Whole islet image with line-scans (white) over β cells contacting the vasculature. Ins-blue, Syntaxin 1A-green, Laminin-red. (**c**) Graph of insulin intensity analysis for the islet in a-b, cells $n = 25$ *** $p < 0.0001$. (**d**) Graph of insulin intensity analysis for all β cells analysed. Data are representative of $n = 3$ donors (1–2 islets per donor), cells $n = 83$ *** $p < 0.0001$. Scale bar 50 μm on whole islet images and 10 μm on insets.

2.2. *U-Net-Based Deep Learning Was the Most Efficient for β Cell Segmentation*

We set out to develop a new automated approach to provide a more objective, time-efficient analysis that would allow the inclusion of the majority of β cells within the islet volumes imaged. Previous studies have used automated approaches to assess islet cell density and islet cell proportions (α and β cells) with islet 3D reconstructions [24,29,32,43]. Here, we aimed to create an automated model capable of identifying islet cells (insulin-labelled β cells) to then refine it for further downstream analyses to assess the subcellular distribution of key β cell proteins.

Firstly, we determined the most suitable approach for use in cell segmentation of human pancreatic islet images. We evaluated the performance of two deep learning methodologies, the Fully Convolutional Network (FCN) and Residual Neural Network (RNN). Testing involved the use of publicly available cell image data (670 labelled training images, and 65 test images of segmented nuclei) [44], pancreatic islet data (855 training images and 606 test images) [36] and a transfer learning approach involving pre-training using the public cell image data then pancreatic islet images.

The datasets were divided into training and validation datasets by *K* partitions where the model is trained on *K-1* and evaluated on the remaining data [45]. The models were tested using 10-fold cross validation with the results listed in Table 1. The FCN-based U-Net model trained using pancreatic islet images only was determined to be the most effective, with accuracy 0.9773, loss 0.0586 and precision 0.5920 (Table 1); therefore, this model was implemented for downstream analyses. The U-Net model was able to correctly identify cells in the original training images. In addition, with new images the U-Net model confirmed the cells that had been manually identified as beta cells (Figure 2). The U-Net model also identified additional insulin-labelled β cells present within the images (Figure 2), suggesting the ability of the model to learn and then predict cells.

Table 1. Testing results from deep learning methodologies [1].

Model	Data	Accuracy	Loss	Precision	Recall	F1	Epoch
U-Net	Public	0.9750	0.0628	0.9207	0.9072	0.9125	54.6
	Islet	0.9773	**0.0586**	**0.5920**	0.1308	0.2012	19.5
	Transfer/ Islet	**0.9777**	0.0594	0.5828	0.1407	0.2189	22.3
ResNet	Public	0.9640	0.0933	0.9022	0.8664	0.8821	37.0
	Islet	0.9764	0.0624	0.5267	0.2081	0.2852	19.0
	Transfer/ Islet	**0.9765**	**0.0622**	**0.5288**	0.1688	0.2442	15.8

[1] The best values achieved for each model are displayed in bold text. Table descriptors defined in Supplementary Table S1.

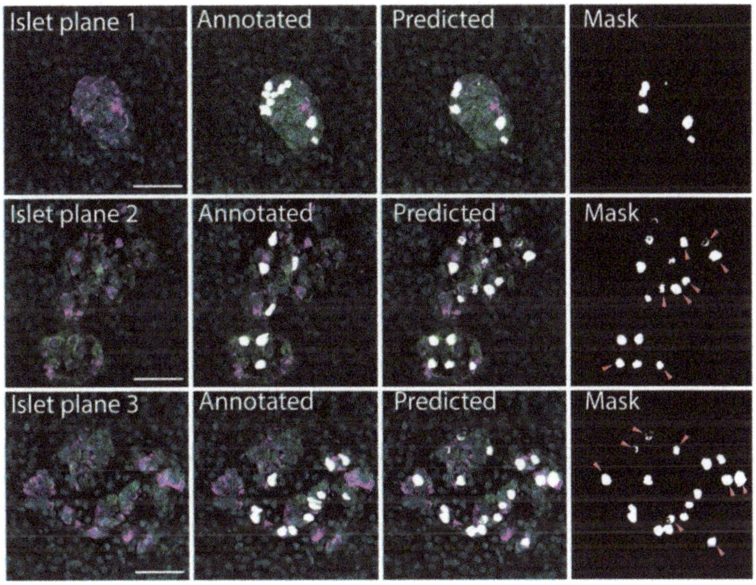

Figure 2. β cell prediction using U-Net deep learning approaches. Examples of microscopy images (Islet plane), the related manually segmented mask overlayed on the original image (Annotated), and the predicted mask image overlayed on the original image (Predicted). The predicted mask (Mask) shows predicted β cells that were not labelled in the annotated image with red arrows. Insulin—magenta, Syntaxin 1A—green, DAPI—cyan, β cell masks—white. Scale bar represents 50 μm.

2.3. *Using Machine Learning to Model β Cells within Islets in 3D*

The 3D modelling of islets is important to assess cell–cell and cell–vasculature relationships, which have recently been demonstrated as important to islet function [36]. To create 3D models of islets, the β cells are predicted using the U-Net machine learning approach for every z-plane that was imaged. These files were then loaded into ImageJ as a sequence, scaled and projected using the 3D Viewer plugin (Figure 3a). This allows a comprehensive view of cell size and shape (Figure 3b). To assess the spatial relationship of the β cells to the islet vasculature, the laminin images were added to the 2D images or 3D reconstructions (3c, purple). Whilst 3D computational modelling of islets has been performed previously [32,43], our data validate the modelling capacity of our approaches, which we then used in downstream applications.

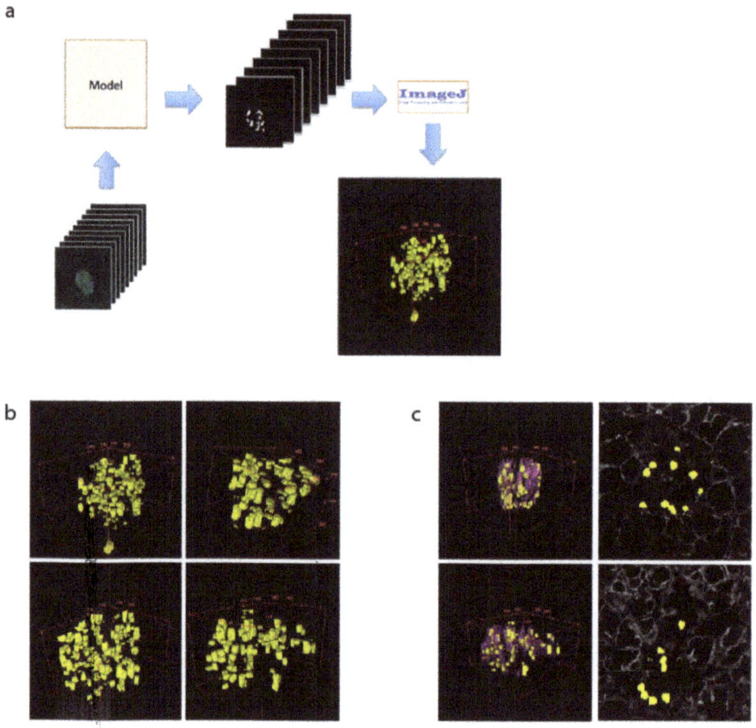

Figure 3. The 3D modelling of β cells within pancreatic islets. (**a**) Schematic of workflow used to create 3D representation of islets using cell masks. (**b**) Examples of 3D models of β cells (yellow) within islets. (**c**) The 3D models showing β cells and the vasculature (as labelled with laminin-magenta) and a single plane showing β cell location (yellow) and vasculature (grey).

2.4. Using Machine Learning to Assess Subcellular Proteins within β Cells

After demonstrating that our U-Net-based deep learning approach can successfully identify insulin-positive β cells within islets from image files, we then wanted to investigate the subcellular staining profiles of proteins of interest within β cells. Firstly, to identify individual β cells, instance segmentation was performed on the semantic segmentation of the predicted β cells. Once individual cell boundaries were identified, the fluorescence values were extracted and presented as a heat map (blue to red; low to high) (Figure 4a). An overlay of the β cell boundary heat map image onto the laminin channel image can be used to assess whether proteins are polarised towards the islet cell vasculature.

Extracting the fluorescent values of insulin staining from the whole cell mask gives insight into the subcellular localisation and concentration of insulin with respect to the vasculature (Figure 4b). To develop an unbiased, automated approach involving a line-scan originating from regions of high and low laminin concentration near the β cell boundary representing the vascular and avascular faces, respectively, software was developed to automatically identify a 10-pixel boundary region (Figure 4c, grey) around the predicted β cells (Figure 4c, white), and locate points of high and low laminin concentration in this boundary region by scanning using a 9 × 9 pixel window (Figure 8) representing the cell circumference. The software then determined the pixel locations of a line-scan from these high and low laminin points towards the cell centre (Figure 4d, vascular region: pink line; avascular region: white line).

The mean fluorescence intensities of insulin staining along the line-scans were determined and this produced statistical data for 2365 predicted β cell instances (cells within individual planes). The vascular face was identified with significantly higher laminin signal than the avascular face (Figure 4e, *** $p < 0.001$; vascular laminin 47.41 ± 30.89 avascular laminin 5.807 ± 4.264). At this computationally identified vascular face a significantly higher concentration of insulin was observed than at the avascular face (Figure 4f, *** $p < 0.001$; vascular insulin 90.78 ± 46.57, avascular insulin 65.22 ± 42.3). These data recapitulate the results from the manually analysed data in Figure 1. However, now the deep learning approach was able to generate statistical data for over 2000 β cell instances to investigate the polarisation of β cells in an unbiased and timely manner.

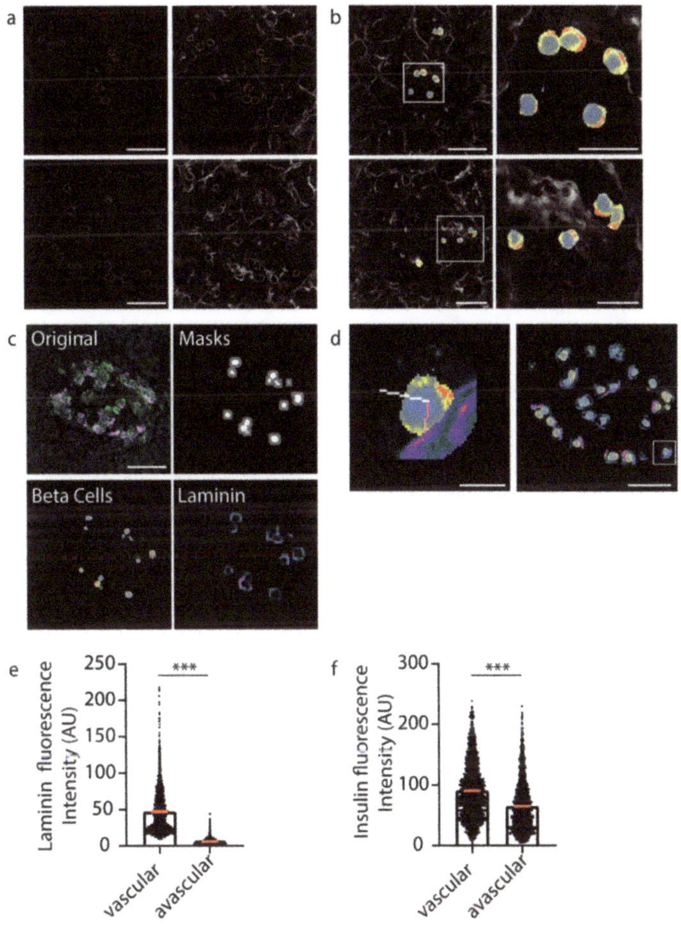

Figure 4. Insulin intensity analyses using computational techniques. (**a**) Representative β cell boundaries with insulin fluoresce represented using a defined heatmap LUT in ImageJ with vasculature (laminin-grey). Heat map fluorescence low to high; blue to red. (**b**) Heat maps applied to whole β cells within islets (**c**) Original islet image with masks for cells (white) and cell boundaries (grey), heatmap for insulin (Beta cells; fluorescence low to high; blue to red) and laminin (Laminin; fluorescence low to high; teal to orange). (**d**) A 10 pixel-wide scan line from the vasculature (laminin high intensity) in pink was used to determine insulin florescence intensity. In white is the line used to measure insulin intensity from the avascular cell boundary (low laminin intensity). The location of the example cell in the whole islet image is shown using a white box. (**e**) Graph of laminin intensity at the vascular and avascular regions. (**f**) Graph of insulin intensity at the vascular and avascular regions. *** $p < 0.001$. Scale bar represents 50 μm for whole islets, 20 μm for insets in (**b**) and 5 μm for the inset in (**d**).

3. Discussion

We found that machine learning approaches are useful in the analysis of large datasets and can be applied to facilitate an understanding the organisation of sub-cellular structures. In this example, we show that the conclusion reached by machine learning algorithms is coincident with that from manual analysis, and both methods show that insulin contents within individual β cells are enriched at the β cell–vascular interface.

The importance of the automated approach is that it is unbiased and drawn from a much bigger dataset than is reasonably feasible with a manual approach. This not only is useful in terms of time efficiency and the increase in the number of example cells that are analysed, but it also demonstrates that an automated approach, where only a few initial quantitative constraints are placed on the model, can confirm the results from a manual approach, which is driven by user expertise. This is important for complex structures, such as islets of Langerhans, where there can be ambiguous images where we currently rely on experts for interpretation. If, through machine learning, we do not need such expert input, then this further underscores the robustness and reproducibility of the findings. It is interesting that the U-Net approach identified β cells that were not found in the manual approach, which suggests either inaccuracies in the algorithm or in the expert identification of the cells. In this context, it is important to note that in any single image plane β cells will be sectioned randomly. Those cells sectioned through the middle will show the clearest, most obvious insulin staining, whereas cells sectioned at their periphery could show fragmented insulin staining, making their identification problematic. Thus, while we expect that the U-Net learning approach might misidentify some β cells (as would an expert), there are still advantages of being able to sample across large volumes and include large numbers of cell instances.

The U-Net modelling approach achieved very high precision with the public learning image data. The images used in the training were DAPI-stained nuclei that have a very consistent morphology both within a single image and across datasets. In contrast, insulin-stained β cells within islets of Langerhans have quite a different morphology, reflecting the close-packing of the cells around the complex network of capillaries. We believe that the diversity in β cell morphology is the basis of the reduced precision in the images of the islets.

The image segmentation applied in our approach is applicable to the identification of the organisation of any subcellular compartment. In the example used here, β cells orientate consistently with respect to capillaries, and therefore identification of the capillary contact provides a spatial point of reference around which the distribution of other compartments or proteins can be mapped. However, it is common for most tissues to show a characteristic organisation, and therefore, with an appropriate external point of reference, such as a lumen or contact with basement membrane, exactly the same approach we use here will be applicable.

We conclude that machine learning is a valuable approach to the analysis of sub-cellular structures within the complex architecture of an organ. In the example here, the approach has enabled a far larger dataset than is practical through manual segmentation, and the results add further evidence for the polarisation of β cells.

4. Materials and Methods

4.1. Human Pancreas Samples

Human pancreatic samples were processed via methods previously described [36]. In brief, tissue was sourced from pancreatic tumour resections (with patient consent, approved by the Northern Sydney Local Heath District Human Research Ethics Committee) or cadaveric donors (study approved by the Human Research Ethics Committee at the University of Sydney). Tissue samples were fixed in 4% paraformaldehyde then mounted in 1.5% low-melting point agarose and 150 μm sections were cut on a vibratome as described by Marciniak et al. [41]. Free-floating sections were stained as described by Meneghel-Rozzo et al. [46]; this involved incubations in blocking buffer (3% BSA, 3% donkey serum,

0.3% Triton X-100) for 4 h at room temperature, and then in primary antibody at 4 °C for 16 h. Sections were washed in 1X PBS and secondary antibody with DAPI incubations were for 5 h at room temperature. After washing, the sections were mounted using ProLong Diamond Antifade (Thermo Fisher Scientific) and imaged on a Leica SP8 confocal microscope using the 63X objective (Leica Microsystems, Wetzlar, Germany).

4.2. Quantification of Insulin Intensity

Image analysis was performed using Fiji (ImageJ) [47] and Metamorph 7.8 (Molecular Devices, San Jose, CA, USA). Graphs were produced using GraphPad Prism v7.02 (GraphPad Software, San Diego, CA, USA). We identified β cells (insulin staining) making contact with the vasculature (laminin staining) and β cell boundaries were identified using Syntaxin 1A staining. To analyse insulin intensity, a line-scan was drawn from the vascular face to the avascular face of the cell and the average intensity across the line extending from the Syntaxin-labelled cell boundary for 2 μm into the cell was measured from each face.

4.3. Statistical Analyses

Statistical analyses were performed using GraphPad Prism. A paired two-tailed student's t test was used to analyse insulin intensity (Figure 1). A paired two-tailed student's t test was used to analyse insulin and laminin intensity (Figure 4). Data are expressed in the text as mean ± SEM.

4.4. Imaging Datasets

Human pancreatic islet images: The dataset consisted of confocal microscopy images stained with insulin, syntaxin 1A, laminin and DAPI [36]. The data were produced with the Leica SP8 confocal microscope using the 63X objective. Each islet consisted of between 50 and 90 z-stacked images. The images were 2048 × 2048 pixels in size, and the dimensions of each voxel (x, y, z) were 0.0901 × 0.0901 × 0.3362 micron^3. The data are 8-bit grey scale. The training images were selected from each of the series spaced five slices apart, sufficient to encompass large variations in the images. These data consisted of 855 training images and 606 test images.

Segmented nuclei images: Image set BBBC038v1 from the Broad Bioimage Benchmark Collection Cacicedo et al. [44]. This dataset consisted of 675 training images and 65 test images.

4.5. Image Format Conversion

The human pancreatic islet images were originally stored in Leica image file (LIF) format (.lif files). The LIF files were opened in ImageJ/Fiji and the required channels were combined into a composite image that was saved as a PNG file. An ImageJ/Fiji macro was developed to quickly generate PNG files for all image slices to be used in the deep learning model (Supplementary Information S1). Images were resized to 512 × 512 pixels as this size was determined to be optimal to maintain accuracy, while fitting within the limitations of computing resources available.

4.6. Training Data-Manual Annotation

Training data were manually labelled to produce mask files for use in training and for evaluating a supervised learning deep learning model. The cell boundary was manually traced in ImageJ/Fiji using insulin staining to identify β cells, syntaxin staining to identify the cell boundary and DAPI staining to identify the cell nucleus; the annotation was then used to create a cell mask (Figure 5). A basic ImageJ macro to improve efficiency was developed (Supplementary Information S2). The image annotation was performed by a student working under the supervision of PhD-level cell biologists.

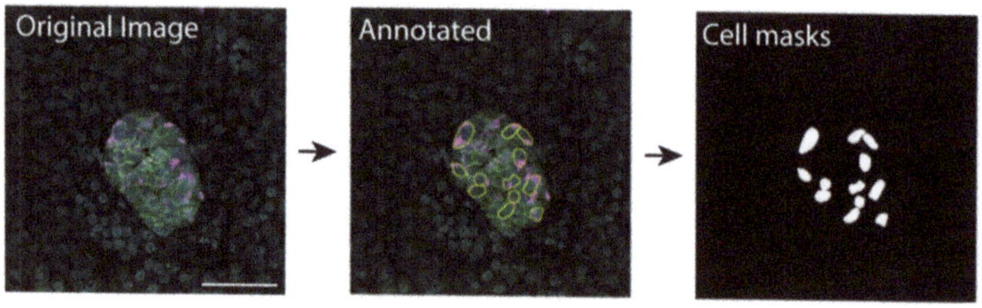

Figure 5. Manual annotation of images using ImageJ and resulting masks used for training the machine learning model. Scale bar represents 50 μm.

4.7. Training Data-Image Augmentation

To create additional training data, we used image augmentation using the Python Keras image data generator. Each of the 95 manually created training images was subject to eight iterations of random transformations, including horizontal and vertical mirroring (or flipping), shearing and shifting (horizontally and vertically), and rotation (Figure 6). This resulted in the creation of 760 additional images, which, along with the original 95, is sufficient for training the deep learning models developed.

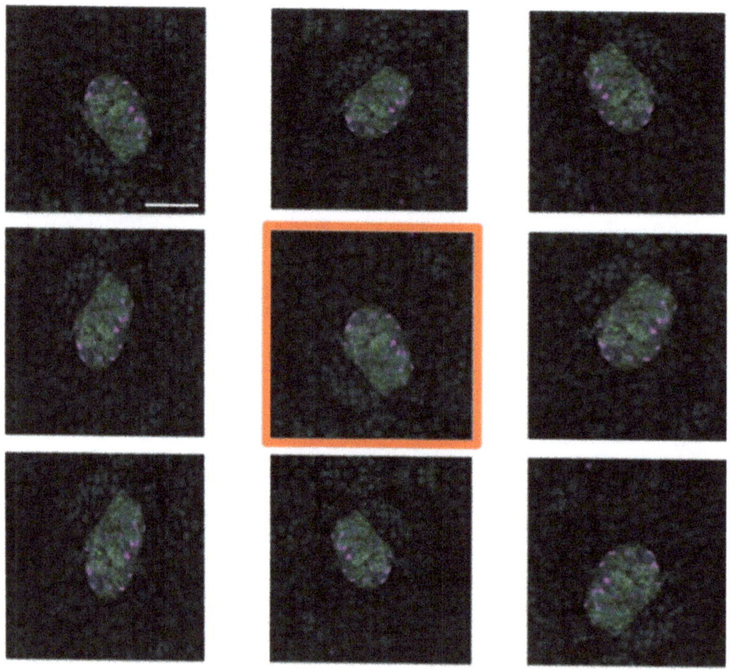

Figure 6. Example of image augmentation used to increase image numbers in the training dataset. The original image has an orange border. Scale bar represents 50 μm.

4.8. Model Development and Testing

An FCN model, based on U-Net [48], was built using the Python Keras high-level neural network API library, based on an example that used publicly available cell image data consisting of 670 labelled segmented nuclei images [19]. The ResNet RNN model was built in Python using the Keras library and was based on example implementations [49,50]. ResNet was computationally expensive with a scaling up of CPU and required memory in comparison to U-Net.

4.9. 3D Models of β Cells within Islets

The best performing FCN model that had been trained on 855 training images was then applied to the test images to make predictions of β cell locations (semantic segmentation). The predicted β cell mask images generated for each series were then loaded into ImageJ/Fiji as a single image sequence. The 3D representations were created using the 3D Viewer plugin in ImageJ/Fiji. The image stack was scaled to match the original voxel ratios (1:1:3.7) to create a realistic depiction of the islet in three dimensions. The voxel dimension from the imaging files was 0.0902 × 0.0902 × 0.3362 microns (x, y, z).

4.10. Instance Segmentation of β Cells

Instance segmentation was performed using the marker-controlled watershed transform [51,52]. This was performed using the Python library Scikit-Image library. A Gaussian filter was applied to reduce noise in cell detection. The resulting output was an image with each cell boundary assigned a label representing its unique cell instance number (Figure 7).

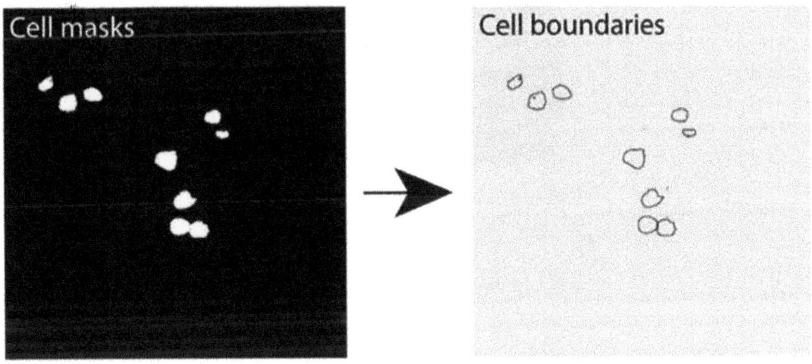

Figure 7. Instance segmentation output showing β cell boundaries. Example image outputs showing cell masks (white) and the resulting cell boundaries predicted using instance segmentation in Python.

4.11. Identifying the Vascular and Avascular Regions and Assessing β Cell Subcellular Insulin Fluorescence Values

After predicting cell location and boundary via instance segmentation, we then created a boundary region mask to extract protein intensity values from the appropriate channel in the microscopy images. Python image processing library routines ("scikit-image") were used to perform a binary dilation of each predicted β cell mask, and then subtraction of the original mask.

A 9 × 9-pixel window was scanned horizontally and then vertically across the boundary region to identify the vascular (high laminin) and avascular (low laminin) regions (Figure 8a). The high or low concentration point is taken to be the centre of the 9 by 9 grid. Once the regions of interest are determined, lines to the cell centre are drawn (Figure 8b). For each coordinate on each scan-line within the β cell, the mean insulin value is calculated for a 10-pixel by 10-pixel region around that point, and the maximum value is recorded for analysis (Figure 8c,d). Only insulin values for pixels that were within the predicted β cell were used in determining the maximum mean insulin concentration. The resulting fluorescence values were exported into csv files. The data were filtered to exclude β cells instances with a radius of less than 5 μm; therefore, all cell instances greater than 490 pixels were used in the analyses.

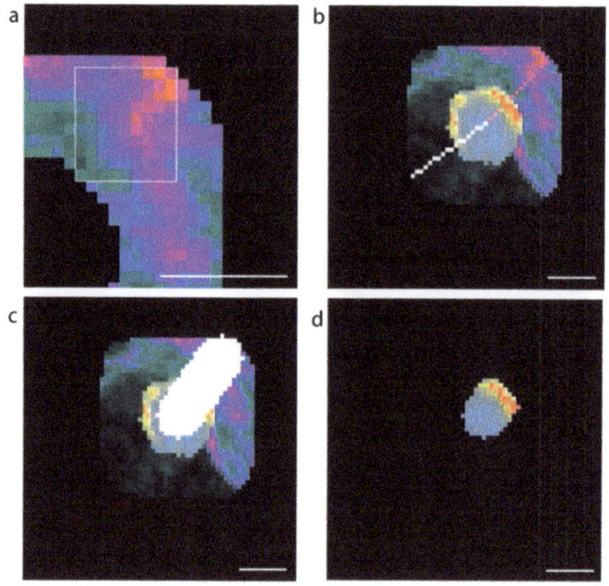

Figure 8. Methods used to computationally assess vascular and avascular regions. (**a**) The 9 × 9-pixel window used for determining the high and low mean laminin points in the cell boundary region, (**b**) the software generated scan lines from the high (pink) and low (white) laminin concentration points to the cell centre. (**c**) The 10-pixel-wide scan-line and (**d**) the insulin concentrations along the 10-pixel-wide scan line within the β cell (blue low to red high insulin staining intensity). Scale bar represents 5 μm.

Supplementary Materials: The following are available online at https://www.mdpi.com/article/10.3390/metabo11060363/s1, Table S1: Definitions of the descriptors used to explain the performance of deep learning models, Information S1: Image Processing Macros: Image format conversion, Information S2: Image Processing Macros: Processing Cell Masks.

Author Contributions: Conceptualization, L.C., I.G., J.K. and P.T.; methodology, L.C., I.G., J.K. and P.T.; software, I.G.; investigation, L.C. and I.G.; formal analysis, L.C. and I.G.; visualization, L.C. and I.G.; resources, T.L., H.E.T., A.J.G., J.S.S. and M.A.K.; writing—original draft preparation, L.C., I.G., K.D. and P.T.; writing—review and editing, L.C., I.G., K.D., T.L., H.E.T., A.J.G., J.S.S. and M.A.K., J.K. and P.T.; supervision, L.C. and P.T.; project administration, P.T.; funding acquisition, J.S.S. and P.T. All authors have read and agreed to the published version of the manuscript.

Funding: This project was funded by the NHMRC (APP1146788) to P.T. and J.S., (APP1128273) to P.T., a Diabetes Australia (DART Y18G_THOP) grant to P.T.

Institutional Review Board Statement: The study was conducted according to the guidelines of the Declaration of Helsinki, and approved by the Human Research Ethics Committee of Northern Sydney Local Heath District (HREC/16/HAWKE/105) and certified by the NHMRC under the National model for Harmonization of Multicentre Ethical Review (HoMER), and the Human Research Ethics Committee of the University of Sydney (2017/042).

Informed Consent Statement: Informed consent was obtained from all subjects involved in the study.

Data Availability Statement: The data presented in this study are available in the supplementary material.

Acknowledgments: The authors acknowledge the facilities and the scientific and technical assistance of Microscopy Australia at the Australian Centre for Microscopy & Microanalysis at the University of Sydney. We thank all organ donors, patients and their families for their generosity and for enabling this work. Thanks to Donatelife for obtaining research consent and providing the human pancreases.

Conflicts of Interest: The authors declare no conflict of interest.

References

1. Ashcroft, F.M.; Rorsman, P. Diabetes mellitus and the beta cell: The last ten years. *Cell* **2012**, *148*, 1160–1171. [CrossRef] [PubMed]
2. Al-Kofahi, Y.; Zaltsman, A.; Graves, R.; Marshall, W.; Rusu, M. A deep learning-based algorithm for 2-D cell segmentation in microscopy images. *BMC Bioinform.* **2018**, *19*, 365. [CrossRef] [PubMed]
3. Sommer, C.; Gerlich, D.W. Machine learning in cell biology—Teaching computers to recognize phenotypes. *J. Cell Sci.* **2013**, *126*, 5529–5539. [CrossRef] [PubMed]
4. Xu, C.; Jackson, S.A. Machine learning and complex biological data. *Genome Biol.* **2019**, *20*, 76. [CrossRef]
5. Kraus, O.Z.; Ba, J.L.; Frey, B.J. Classifying and segmenting microscopy images with deep multiple instance learning. *Bioinformatics* **2016**, *32*, i52–i59. [CrossRef] [PubMed]
6. Lugagne, J.B.; Lin, H.; Dunlop, M.J. DeLTA: Automated cell segmentation, tracking, and lineage reconstruction using deep learning. *PLoS Comput. Biol.* **2020**, *16*, e1007673. [CrossRef]
7. Falk, T.; Mai, D.; Bensch, R.; Cicek, O.; Abdulkadir, A.; Marrakchi, Y.; Bohm, A.; Deubner, J.; Jackel, Z.; Seiwald, K.; et al. U-Net: Deep learning for cell counting, detection, and morphometry. *Nat. Methods* **2019**, *16*, 67–70. [CrossRef] [PubMed]
8. Caicedo, J.C.; Roth, J.; Goodman, A.; Becker, T.; Karhohs, K.W.; Broisin, M.; Molnar, C.; McQuin, C.; Singh, S.; Theis, F.J.; et al. Evaluation of Deep Learning Strategies for Nucleus Segmentation in Fluorescence Images. *Cytom. A* **2019**, *95*, 952–965. [CrossRef] [PubMed]
9. Zhang, X.; Peng, X.; Han, C.; Zhu, W.; Wei, L.; Zhang, Y.; Wang, Y.; Zhang, X.; Tang, H.; Zhang, J.; et al. A unified deep-learning network to accurately segment insulin granules of different animal models imaged under different electron microscopy methodologies. *Protein Cell* **2019**, *10*, 306–311. [CrossRef] [PubMed]
10. Glory, E.; Murphy, R.F. Automated subcellular location determination and high-throughput microscopy. *Dev. Cell* **2007**, *12*, 7–16. [CrossRef]
11. Boland, M.V.; Markey, M.K.; Murphy, R.F. Automated recognition of patterns characteristic of subcellular structures in fluorescence microscopy images. *Cytometry* **1998**, *33*, 366–375. [CrossRef]
12. Huang, K.; Murphy, R.F. Boosting accuracy of automated classification of fluorescence microscope images for location proteomics. *BMC Bioinform.* **2004**, *5*, 78. [CrossRef]
13. Vu, Q.D.; Graham, S.; Kurc, T.; To, M.N.N.; Shaban, M.; Qaiser, T.; Koohbanani, N.A.; Khurram, S.A.; Kalpathy-Cramer, J.; Zhao, T.; et al. Methods for Segmentation and Classification of Digital Microscopy Tissue Images. *Front. Bioeng Biotechnol.* **2019**, *7*, 53. [CrossRef]
14. Kong, Z.; Li, T.; Luo, J.; Xu, S. Automatic Tissue Image Segmentation Based on Image Processing and Deep Learning. *J. Healthc. Eng.* **2019**, *2019*, 2912458. [CrossRef]
15. Rizwan, I.; Haque, I.; Neubert, J. Deep learning approaches to biomedical image segmentation. *Inform. Med. Unlocked* **2020**, *18*, 100297. [CrossRef]
16. Bengio, Y. Learning Deep Architectures for AI. Found. *Trends Mach. Learn.* **2009**, *2*, 1–127. [CrossRef]
17. Van Valen, D.A.; Kudo, T.; Lane, K.M.; Macklin, D.N.; Quach, N.T.; DeFelice, M.M.; Maayan, I.; Tanouchi, Y.; Ashley, E.A.; Covert, M.W. Deep Learning Automates the Quantitative Analysis of Individual Cells in Live-Cell Imaging Experiments. *PLoS Comput. Biol.* **2016**, *12*, e1005177. [CrossRef]
18. Essa, E.; Xie, X. Phase contrast cell detection using multilevel classification. *Int. J. Numer. Methods Biomed. Eng.* **2018**, *34*, e2916. [CrossRef]
19. Ronneberger, O.; Fischer, P.; Brox, T. U-Net: Convolutional Networks for Biomedical Image Segmentation. In Proceedings of the 18th International Conference on Medical Image Computing and Computer-Assisted Intervention—MICCAI 2015, Munich, Germany, 5–9 October 2015; Lecture Notes in Computer Science. pp. 234–241.
20. He, B.; Lu, Q.; Lang, J.; Yu, H.; Peng, C.; Bing, P.; Li, S.; Zhou, Q.; Liang, Y.; Tian, G. A New Method for CTC Images Recognition Based on Machine Learning. *Front. Bioeng Biotechnol.* **2020**, *8*, 897. [CrossRef]
21. Jung, H.; Lodhi, B.; Kang, J. An automatic nuclei segmentation method based on deep convolutional neural networks for histopathology images. *BMC Biomed. Eng.* **2019**, *1*, 1–12. [CrossRef]
22. Naylor, P.; Lae, M.; Reyal, F.; Walter, T. Nuclei segmentation in histopathology images using deep neural networks. In Proceedings of the 2017 IEEE 14th International Symposium on Biomedical Imaging (ISBI 2017), Melbourne, Australia, 18–21 April 2017; pp. 933–936.
23. Rechsteiner, M.P.; Floros, X.; Boehm, B.O.; Marselli, L.; Marchetti, P.; Stoffel, M.; Moch, H.; Spinas, G.A. Automated assessment of beta-cell area and density per islet and patient using TMEM27 and BACE2 immunofluorescence staining in human pancreatic beta-cells. *PLoS ONE* **2014**, *9*, e98932. [CrossRef] [PubMed]
24. Huang, Y.; Liu, C.; Eisses, J.F.; Husain, S.Z.; Rohde, G.K. A supervised learning framework for pancreatic islet segmentation with multi-scale color-texture features and rolling guidance filters. *Cytom. A* **2016**, *89*, 893–902. [CrossRef] [PubMed]
25. Brissova, M.; Fowler, M.J.; Nicholson, W.E.; Chu, A.; Hirshberg, B.; Harlan, D.M.; Powers, A.C. Assessment of Human Pancreatic Islet Architecture and Composition by Laser Scanning Confocal Microscopy. *J. Histochem. Cytochem.* **2005**, *53*, 1087–1097. [CrossRef] [PubMed]

26. Barbieux, C.; Parnaud, G.; Lavallard, V.; Brioudes, E.; Meyer, J.; Alibashe Ahmed, M.; Berishvili, E.; Berney, T.; Bosco, D. Asymmetrical distribution of δ and PP cells in human pancreatic islets. *J. Endocrinol.* **2016**, *229*, 123–132. [CrossRef]
27. Zimny, M.L.; Blackard, W.G. The surface structure of isolated pancreatic islet cells. *Cell Tissue Res.* **1975**, *164*, 467–471. [CrossRef]
28. Mikami, S.i.; Mutoh, K. Light- and electron-microscopic studies of the pancreatic islet cells in the chicken under normal and experimental conditions. *Z. Zellforsch. Mikrosk. Anat.* **1971**, *116*, 205–227. [CrossRef]
29. Chen, H.; Martin, B.; Cai, H.; Fiori, J.L.; Egan, J.M.; Siddiqui, S.; Maudsley, S. Pancreas++: Automated quantification of pancreatic islet cells in microscopy images. *Front. Physiol.* **2012**, *3*, 482. [CrossRef]
30. Lei, C.-L.; Kellard, J.A.; Hara, M.; Johnson, J.D.; Rodriguez, B.; Briant, L.J.B. Beta-cell hubs maintain Ca2+ oscillations in human and mouse islet simulations. *Islets* **2018**, *10*, 151–167. [CrossRef]
31. von Herrath, M.G.; Hoang, D.-T.; Hara, M.; Jo, J. Design Principles of Pancreatic Islets: Glucose-Dependent Coordination of Hormone Pulses. *PLoS ONE* **2016**, *11*, e0152446. [CrossRef]
32. Félix-Martínez, G.J.; Mata, A.N.; Godínez-Fernández, J.R. Reconstructing human pancreatic islet architectures using computational optimization. *Islets* **2020**, *12*, 121–133. [CrossRef]
33. Bonner-Weir, S. Morphological Evidence for Pancreatic Polarity of -Cell Within Islets of Langerhans. *Diabetes* **1988**, *37*, 616–621. [CrossRef]
34. Lombardi, T.; Montesano, R.; Wohlwend, A.; Amherdt, M.; Vassalli, J.D.; Orci, L. Evidence for polarization of plasma membrane domains in pancreatic endocrine cells. *Nature* **1985**, *313*, 694–696. [CrossRef] [PubMed]
35. Gan, W.J.; Zavortink, M.; Ludick, C.; Templin, R.; Webb, R.; Webb, R.; Ma, W.; Poronnik, P.; Parton, R.G.; Gaisano, H.Y.; et al. Cell polarity defines three distinct domains in pancreatic β-cells. *J. Cell Sci.* **2017**, *130*, 143–151. [CrossRef]
36. Cottle, L.; Gan, W.J.; Gilroy, I.; Samra, J.S.; Gill, A.J.; Loudovaris, T.; Thomas, H.E.; Hawthorne, W.J.; Kebede, M.A.; Thorn, P. Structural and functional polarisation of human pancreatic beta cells in islets from organ donors with and without type 2 diabetes. *Diabetologia* **2021**, *64*, 618–629. [CrossRef]
37. Granot, Z.; Swisa, A.; Magenheim, J.; Stolovich-Rain, M.; Fujimoto, W.; Manduchi, E.; Miki, T.; Lennerz, J.K.; Stoeckert, C.J., Jr.; Meyuhas, O.; et al. LKB1 regulates pancreatic beta cell size, polarity, and function. *Cell Metab.* **2009**, *10*, 296–308. [CrossRef]
38. Orci, L.; Thorens, B.; Ravazzola, M.; Lodish, H. Localization of the pancreatic beta cell glucose transporter to specific plasma membrane domains. *Science* **1989**, *245*, 295–297. [CrossRef]
39. Low, J.T.; Zavortink, M.; Mitchell, J.M.; Gan, W.J.; Do, O.H.; Schwiening, C.J.; Gaisano, H.Y.; Thorn, P. Insulin secretion from beta cells in intact mouse islets is targeted towards the vasculature. *Diabetologia* **2014**, *57*, 1655–1663. [CrossRef]
40. Farack, L.; Golan, M.; Egozi, A.; Dezorella, N.; Bahar Halpern, K.; Ben-Moshe, S.; Garzilli, I.; Tóth, B.; Roitman, L.; Krizhanovsky, V.; et al. Transcriptional Heterogeneity of Beta Cells in the Intact Pancreas. *Dev. Cell* **2019**, *48*, 115–125.e114. [CrossRef]
41. Marciniak, A.; Cohrs, C.M.; Tsata, V.; Chouinard, J.A.; Selck, C.; Stertmann, J.; Reichelt, S.; Rose, T.; Ehehalt, F.; Weitz, J.; et al. Using pancreas tissue slices for in situ studies of islet of Langerhans and acinar cell biology. *Nat. Protoc.* **2014**, *9*, 2809–2822. [CrossRef]
42. Virtanen, I.; Banerjee, M.; Palgi, J.; Korsgren, O.; Lukinius, A.; Thornell, L.E.; Kikkawa, Y.; Sekiguchi, K.; Hukkanen, M.; Konttinen, Y.T.; et al. Blood vessels of human islets of Langerhans are surrounded by a double basement membrane. *Diabetologia* **2008**, *51*, 1181–1191. [CrossRef]
43. Dybala, M.P.; Hara, M. Heterogeneity of the Human Pancreatic Islet. *Diabetes* **2019**, *68*, 1230–1239. [CrossRef] [PubMed]
44. Caicedo, J.C.; Goodman, A.; Karhohs, K.W.; Cimini, B.A.; Ackerman, J.; Haghighi, M.; Heng, C.; Becker, T.; Doan, M.; McQuin, C.; et al. Nucleus segmentation across imaging experiments: The 2018 Data Science Bowl. *Nat. Methods* **2019**, *16*, 1247–1253. [CrossRef] [PubMed]
45. Chollet, F. *Deep Learning with Python*; Manning Publications: Shelter Island, NY, USA, 2018.
46. Meneghel-Rozzo, T.; Rozzo, A.; Poppi, L.; Rupnik, M. In vivo and in vitro development of mouse pancreatic beta-cells in organotypic slices. *Cell Tissue Res.* **2004**, *316*, 295–303. [CrossRef]
47. Schindelin, J.; Arganda-Carreras, I.; Frise, E.; Kaynig, V.; Longair, M.; Pietzsch, T.; Preibisch, S.; Rueden, C.; Saalfeld, S.; Schmid, B.; et al. Fiji: An open-source platform for biological-image analysis. *Nat. Methods* **2012**, *9*, 676–682. [CrossRef]
48. Navab, N.; Hornegger, J.; Wells, W.M.; Frangi, A. *Medical Image Computing and Computer-Assisted Intervention—MICCAI 2015, 18th International Conference, Munich, Germany, 5–9 October 2015, Proceedings, Part I*, 1st ed.; Springer International Publishing: Cham, Switzerland, 2015.
49. Liciotti, D.; Paolanti, M.; Pietrini, R.; Frontoni, E.; Zingaretti, P. Convolutional Networks for Semantic Heads Segmentation using Top-View Depth Data in Crowded Environment. In Proceedings of the 2018 24th International Conference on Pattern Recognition (ICPR), Beijing, China, 20–24 August 2018; pp. 1384–1389.
50. He, K.; Zhang, X.; Ren, S.; Sun, J. Deep Residual Learning for Image Recognition. *arXiv* **2015**, arXiv:1512.03385.
51. Meyer, F.; Beucher, S. Morphological segmentation. *J. Vis. Commun. Image Represent.* **1990**, *1*, 21–46. [CrossRef]
52. Beucher, S.; Lantuejoul, C. Use of Watersheds in Contour Detection. In Proceedings of the International Workshop on Image Processing, Real-Time Edge and Motion Detection/Estimation, Rennes, France, 17–21 September 1979.

metabolites

Article

Successful Islet Outcomes Using Australia-Wide Donors: A National Centre Experience

Wayne J Hawthorne [1,2,*], Sussan Davies [1], Hee-chang Mun [1], Yi Vee Chew [1], Lindy Williams [1], Patricia Anderson [1], Natasha Rogers [1,2] and Philip J O'Connell [1]

1 The Centre for Transplant & Renal Research, Westmead Institute for Medical Research, Westmead Hospital, Westmead, NSW 2145, Australia; sussan.davies@sydney.edu.au (S.D.); hee.mun@sydney.edu.au (H.-c.M.); yi.chew@sydney.edu.au (Y.V.C.); lindy.williams@sydney.edu.au (L.W.); Patricia.Anderson@health.nsw.gov.au (P.A.); natasha.rogers@health.nsw.gov.au (N.R.); philip.oconnell@sydney.edu.au (P.J.O.)

2 Westmead Clinical School, University of Sydney, Westmead Hospital, Westmead, NSW 2145, Australia

* Correspondence: Wayne.Hawthorne@Sydney.edu.au

Abstract: Cold ischemia and hence travel time can adversely affect outcomes of islet isolation. The aim of this study was to compare the isolation and transplant outcomes of donor pancreata according to the distance from islet isolation facility. Principally, those within a 50 km radius of the centre were compared with those from regional areas within the state and those from interstate donors within Australia. Organ donors were categorised according to distance from National Pancreas Transplant Unit Westmead (NPTU). Donor characteristics were analysed statistically against islet isolation outcomes. These were age, BMI, cause and mechanism of death, days in ICU, gender, inotrope and steroid use, cold ischemia time (CIT) and retrieval surgical team. Between March 2007 and December 2020, 297 islet isolations were performed at our centre. A total of 149 donor pancreata were local area, and 148 non-local regions. Mean distance from the isolation facility was 780.05 km. Mean pancreas CIT was 401.07 ± 137.71 min and was significantly different between local and non-local groups (297.2 vs. 487.5 min, $p < 0.01$). Mean age of donors was 45.22 years, mean BMI was 28.82, sex ratio was 48:52 F:M and mean time in ICU was 3.07 days. There was no significant difference between local and non-local for these characteristics. The mean CIT resulting in islet transplantation was 297.1 ± 91.5 min and longest CIT resulting in transplantation was 676 min. There was no significant difference in islet isolation outcomes between local and non-local donors for characteristics other than CIT. There was also no significant effect of distance from the isolation facility on positive islet transplant outcomes (C-peptide > 0.2 at 1 month post-transplant). *Conclusions:* Distance from the isolation centre did not impact on isolation or transplant outcomes supporting the ongoing nationwide use of shipping pancreata for islet isolation and transplantation.

Keywords: diabetes; hypoglycemic unaware; ischaemia; islet cell transplantation; organ donation; pancreas

check for updates

Citation: Hawthorne, W.J; Davies, S.; Mun, H.-c.; Chew, Y.V.; Williams, L.; Anderson, P.; Rogers, N.; O'Connell, P.J Successful Islet Outcomes Using Australia-Wide Donors: A National Centre Experience. *Metabolites* **2021**, *11*, 360. https://doi.org/10.3390/metabo11060360

Academic Editor: Amedeo Lonardo

Received: 21 April 2021
Accepted: 3 June 2021
Published: 5 June 2021

1. Introduction

Pancreatic islet cell transplantation has become a successful modality of treatment for a select group of patients with type 1 diabetes. [1–4]. In order to provide this service, centralized islet isolation centres need to overcome a number of unique logistical problems, in particular retrieving donor pancreata and transplanting patients from distant areas. In Australia, the problem is particularly acute [5,6] as the service covers an area of 7,692,024 square kilometers, which is approximately twice the size of Europe or three-quarters the size of the United States. The National Islet Consortium comprises two isolation centres—Westmead Hospital in Sydney and St Vincent's Hospital in Melbourne; and three transplantation Hospitals—Westmead, St Vincent's and Queen Elizabeth in Adelaide. [6] Almost one-third of the Australian population lives outside these major urban centres and patients from regional and rural areas face a number of barriers to accessing medical services (Figure 1). The aim of this study was to assess whether pancreas donors accessed outside the local region of the isolation facility provided equivalent outcomes to those accessed from the local region of the isolation facility.

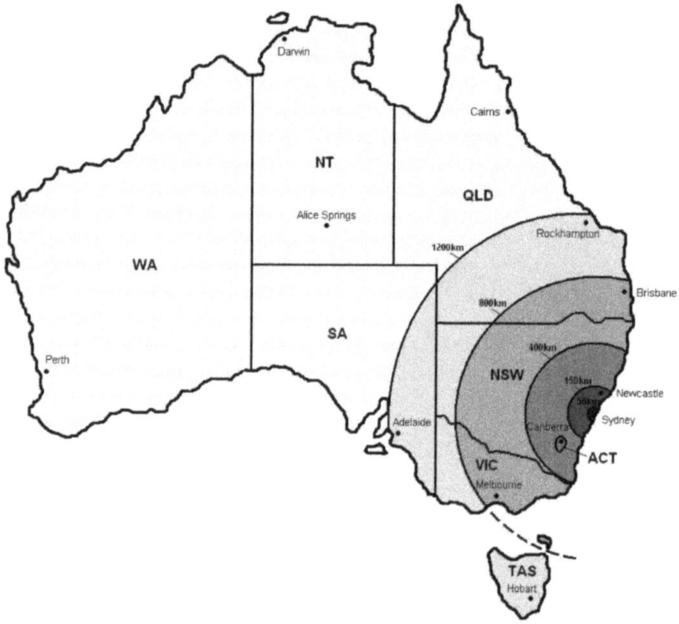

Figure 1. Distance zones from Westmead Hospital. Schematic map of Australia with states and major cities denoted. Distance rings (different areas of shading) are shown radiating from NPTU in Westmead Sydney, NSW.

2. Results

2.1. The Effect of Distance on Isolation Outcome

Between March 2007 and December 2021, 297 islet isolations were performed and analysed for this study. Donor pancreata came from multiorgan donors that were accessed from multiple hospitals across Australia (149 were from the local region, 148 were from non-local regions). The mean distance from isolation centre was 780.05 km (range 0 to 3278.7 km) for all donors. The distance from the isolation facility provided significant impact on the islet isolation outcome the closer local donors provided a positive outcome at 11.34 km ± 10.22 vs. 883.04 km ± 625.45 for a positive outcome from non-local donors, $p =< 0.01$) (Figure 2A). The chance of obtaining an isolation outcome that achieved release criteria was greater from local donors with a positive outcome occurring 32% of the time from local donors as compared to only 24% of isolations from non-local donors ($p < 0.01$).

The mean pancreas cold ischemic time for all donors was 401.07 ± 137.71 min (range 78 to 870). CIT correlated closely with distance from isolation facility (Figure 2B). From local (zone 1) to zone 2, the mean CIT increased from 279.6 ± 89.2 to 337.6 ± 74.4 min ($p = 0.011$). As one moves out from zone 2 to zone 6, there was a progressive increase in cold ischemia times. (The mean CIT for each zone is shown on Table 1.) When all non-local donors were compared, there was a significant increase in mean CIT when compared to local CIT (mean 487.5 ± 103.4 vs. 297.2 ± 95.7 min, Figure 2C, $p = 0.12$). However, regardless of the region, there was a negative association between CIT and chances of a positive isolation outcome ($p = 0.55$). Although isolations with a positive outcome tended to have a shorter CIT compared to those with a negative outcome (297.1 ± 91.5 min vs. 297.6 ± 105.5 min) (Figure 2C) this was not statistically significant ($p = 0.73$) (Figure 2C).

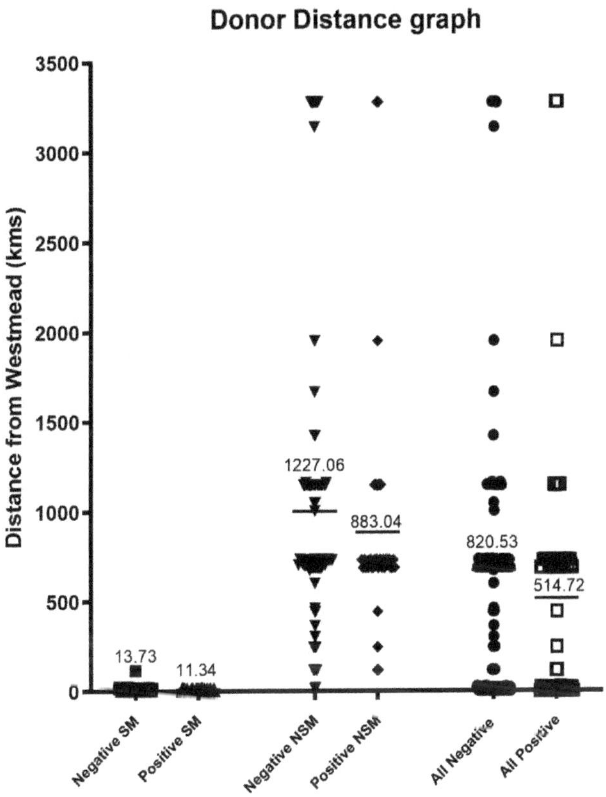

(A)

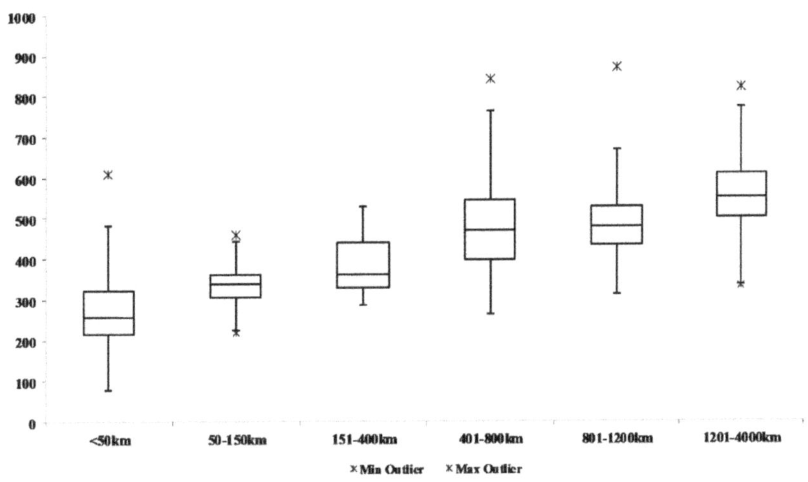

(B)

Figure 2. *Cont.*

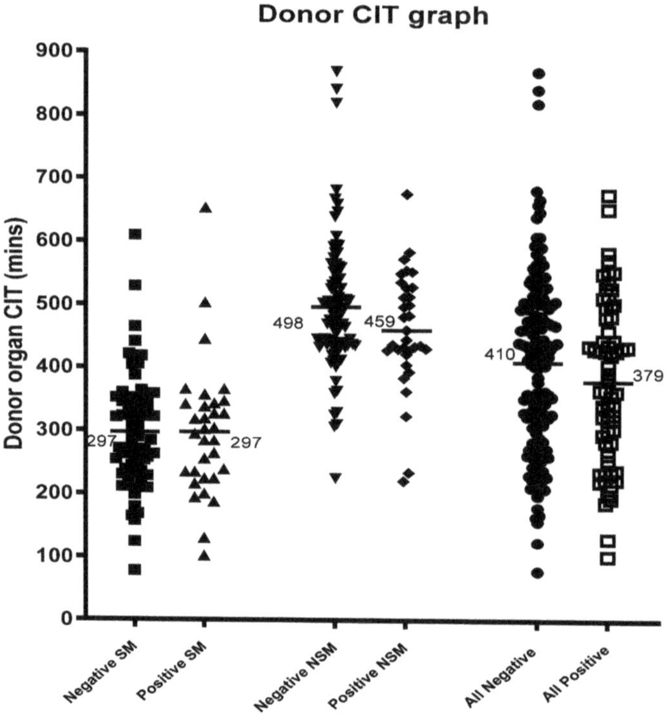

(C)

Figure 2. (**A**). Scatter plot of donor distance outcomes. Distance from Westmead (NPTU) in kilometers is shown for each analysis group, and the mean of each group is marked and labelled. Negative local (■ $n = 65$), positive local (▲ $n = 30$), negative non-local (▼ $n = 129$), positive non-local (◆ $n = 41$), all negative (• $n = 194$), all positive (□ $n = 71$). (**B**). Box plot of donor distance zone vs. CIT. The CIT is shown for each distance zone, and outliers are represented by ✱ on the graph. <50 km $n = 71$, 50–150 km $n = 12$, 151–400 km $n = 9$, 401–800 km $n = 76$, 801–1200 km $n = 31$, 1201–4000 $n = 17$. From local (zone 1) to zone 2, the mean CIT increases from 279.6 to 337.6 min which was statistically different ($p = 0.03$). There was no significant difference in the mean CIT from zone 2 to zone 3 (385.1 min) ($p = 0.182$), from zone 3 to zone 4 (474.8 min) ($p = 0.009$) and from zone 4 to zone 5 (504.9 min) ($p = 0.171$) (each zone representing a distance of 250 to 400 km). There was also a significant increase in CIT when a donor pancreas was received from zone 6 compared to zone 5 ($p = 0.109$), as zone 6 represents a distance of 2800 km further than zone 5 (mean CIT 560.6 min). (**C**). Scatter plot of CIT Outcomes. CIT in minutes is shown for each analysis group, the mean of each group is marked and labelled. Negative local (■ $n = 67$), positive local (▲ $n = 32$), negative non-local (▼ $n = 86$), positive non-local (◆ $n = 33$), all negative (• $n = 153$), all positive (□ $n = 65$). CIT was statistically significant between local and non-local groups ($p < 0.01$).

Table 1. Descriptive statistics of the data collected, grouped into donor-related factors, those that are Edmonton score donor-related factors and recipient-related factors. For non-continuous data, a number was assigned to each set of data.

	N	Min	Max	Mean	SD
	DONOR-RELATED FACTORS				
Outcome	263	0.00	1.00	0.27	0.44
Donor Distance (kms)	263	0.00	3278.70	780.06	2195.2
Donor Distance Zone	263	1.00	6.00	3.09	1.73

Table 1. *Cont.*

	N	Min	Max	Mean	SD
EDMONTON SCORE DONOR-RELATED FACTORS					
CIT (mins)	215	124.00	870.00	404.35	140.21
Donor Age	296	10.00	71.00	45.22	13.03
Donor BMI	298	19.80	57.46	28.82	6.01
Donor Gender	308	1.00	2.00	1.49	0.49
Donor Days in ICU	252	1.00	13.00	3.07	2.51
Retrieval Team	268	1.00	2.00	1.56	0.50
Cause of Death	246	1.00	6.00	2.64	0.84
Mechanism of Death	246	1.00	12.00	7.78	1.98
Inotrope Usage	244	0.00	3.00	1.38	1.00
Steroid Usage	244	0.00	1.00	0.02	0.13
RECIPIENT-RELATED FACTORS					
C-Peptide at 1 Month Post-Transplant	23	0.00	0.91	0.28	0.23
C-Peptide level	73	0.00	1.00	0.55	0.36

2.2. *Other Donor Factors Affecting Isolation Outcome*

Forty-eight percent of the donors were female and patient gender had no effect on isolation outcomes (p = 0.554) and there were no differences in the gender balance between local and non-local donors (p = 0.6493, data not shown). There was no difference in age between donors accepted from local versus those from non-local regions (44.5 ± 12.3 vs. 45.7 ± 12.2, p = 0.46). The mean age of all donors was 45.22 years (range 10 to 71), and there was no difference in age of the donor between those with a positive isolation outcome and those that did not (44.34 (range 23 to 69) vs. 45.49 (range 14 to 71), p = 0.46, Figure 3A). The mean BMI for all donors was 28.82 kg/m^2 (range 19.80 to 57.46) and the mean BMI was greater in isolations with a positive isolation outcome with those that did not (31.01 (range 20.09 to 51.05) vs. 28.12 (range 17.90 to 57.46), p = 0.0004, Figure 3B). However, there was no difference in the BMI between local compared with non-local groups (28.81 (range 17.48 to 50.50) vs. 28.90 (range 19.81 to 57.46), p = 0.927). The mean time spent in ICU for all donors was 3.07 days (range 1 to 13) and no impact on isolation outcome, (mean time for a positive outcome 2.8 ± 1.7 days versus 3.1 ± 2.7 days for a negative outcome and this was not significantly different (p = 0.451, Figure 3C), nor was it significant between local and non-local groups (2.85 days vs. 3.10 days). Cause and mechanism of death demonstrated that 'other causes' or 'other mechanisms' have the highest positive results with a significant p value (p= 0.042) (Table 2), whereas inotrope dose, steroid use and retrieval team had no significant effect on the islet isolation outcome (Table 1).

The clinical outcome of those isolations that were transplanted were assessed and evaluated for their effect of donor-related factor on outcome using the non-parametric Spearman's rank correlation. Donor distance, CIT, days in ICU, donor BMI had no effect on C-peptide levels achieved (Table 3).

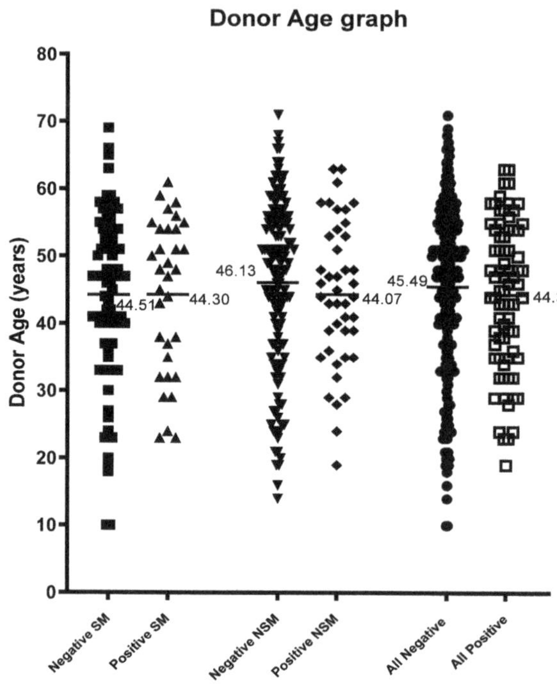

(A)

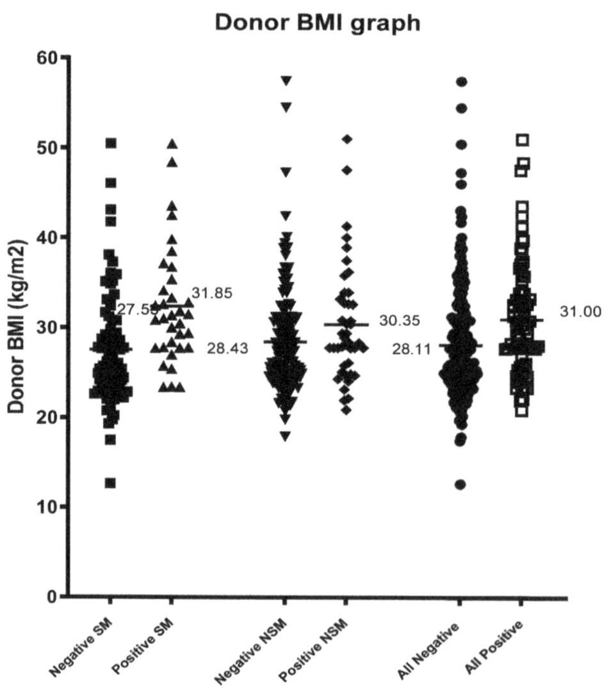

(B)

Figure 3. *Cont.*

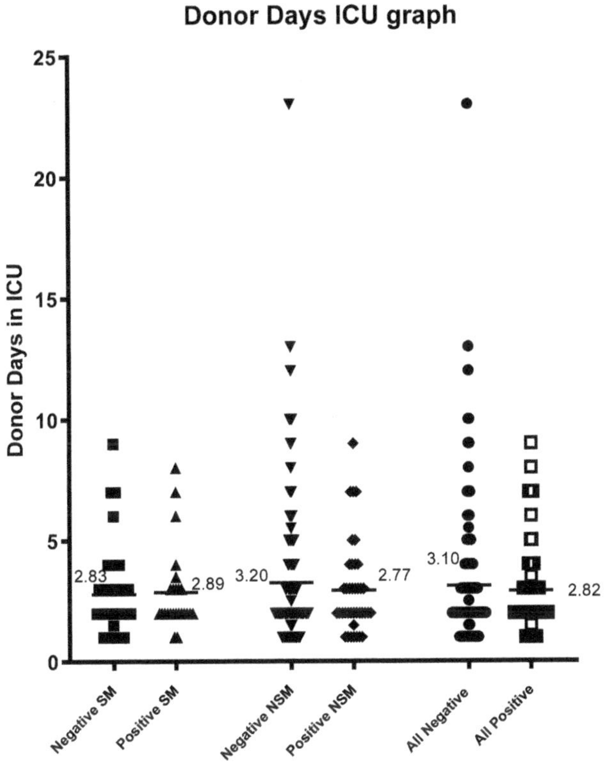

(C)

Figure 3. (**A**). Scatter plot of donor age outcomes. The age of the donor at time of donation in years is shown for each analysis group, the mean of each group is marked and labelled. Negative local (■ *n* = 79), positive local (▲ *n* = 33), negative non-local (▼ *n* = 142), positive non-local (◆ *n* = 43), all negative (• *n* = 221), all positive (□ *n*= 76). (**B**). scatter plot of donor BMI outcomes. The BMI as calculated from height and weight of the patient is shown for each analysis group, the mean of each group is marked and labelled. Negative local (■ *n* = 79), positive local (▲ *n* = 33), negative non-local (▼ *n* = 142), positive non-local (◆ *n* = 43), all negative (• *n* = 221), all positive (□ *n* = 76). BMI was statistically significant between negative and positive groups ($p = 0.0004$). (**C**). Scatter plot of donor ICU outcomes. The amount of time in days that the donor spent in ICU is shown for each analysis group, the mean of each group is marked and labelled. Negative local (■ *n* = 52), positive local (▲ *n* = 25), negative non-local (▼ *n* = 129), positive non-local (◆ *n* = 39), all negative (• *n* = 181), all positive (□ *n* = 64).

Table 2. Cause and mechanism of donor death statistics. The number of each cause or mechanism that had a negative or positive isolation outcome is compared with the percent of each total. 'Other causes' or 'other mechanisms' have the highest positive results with a significant *p* value, while those with anoxia or cardiac arrest have the least. Sharp injury and seizure as a mechanism of death have the lowest numbers. However, total numbers in these groups are very low.

Cause of Donor Death	Outcome (N)		% Positive	*p* Value
	Negative	Positive		
Anoxia/Cardiac Arrest	22	4	15.38	0.113
Head Trauma	40	17	29.82	0.829

Table 2. *Cont.*

Cause of Donor Death	Outcome (N)		% Positive	*p* Value
	Negative	Positive		
Cerebrovascular/Stroke	108	44	28.95	0.909
Other	4	5	55.56	0.124
Total	174	70	28.69	-
Mechanism of Donor Death	**Outcome (N)**		**% Positive**	***p* Value**
	Negative	**Positive**		
Asphyxiation	8	2	20.00	0.728
Blunt Injury	2	1	33.33	1.000
Cardiovascular	4	1	20.00	1.000
Sharp Injury	4	0	0.00	0.325
Intracranial Haemorrhage/Stroke	148	58	28.16	0.396
Seizure	1	0	0.00	1.000
Other	7	8	53.33	0.042
Total	170	70	28.69	-

Table 3. Factors affecting C-peptide levels. The correlation coefficient of each factor in relation to C-peptide level as a positive transplant outcome are shown with their significance. Donor distance, CIT, donor BMI, donor days in ICU and recipient distance all showed a weak positive correlation to C-peptide levels (as one increases, C-peptide level increases).

	C-Peptide Level	
	Correlation Coefficient	Significance (2-Tailed)
Donor Distance (kms)	0.311	0.832
CIT (mins)	−0.308	0.065
Donor Age	0.354	0.977
Donor BMI	−0.171	0.770
Donor Days in ICU	0.135	0.659

3. Discussion

Availability of suitable donor pancreata is a major limiting factor for islet transplant activity. Hence, the ability to access pancreata that are retrieved at a distance from the isolation centre is essential if more patients are to be transplanted. However, it is essential to ascertain whether pancreata retrieved from more distant centres provide equivalent outcomes to those retrieved locally. This study evaluated the differences in donor characteristics between locally retrieved local donor organs with those retrieved from non-local regions. Because of the distances between major cities and states within Australia there was a large variation in distances between the donor hospital and the isolation centre. Because of the logistics of transport, there were substantial differences in CIT between local and non-local donors. This did impact on the chances of achieving a positive isolation resulting in reaching release criteria. However, provided the release criteria were met there were no differences in transplant outcomes between local and non-local donors. The one caveat to this conclusion was that no successful isolation was achieved if the CIT was greater than 676 min.

The contributing factors that affect the logistics of the transport of the organ included-time from organ retrieval to shipping. These included the time to separate the pancreas from the liver after removal en-bloc during organ retrieval and the availability of a commercial flight to Sydney. These variables tended to have a substantial impact on CIT. These logistical issues had a major impact on organs retrieved from regions 3 to 5 where flight times range from 44 min to 2 h. Pancreata transported from zone 6 had flight times ranging from 3 to 6 h. By contrast those retrieved within the local region were transported to the isolation facility by car which meant that it tended to travel with the retrieval team and did not suffer from delays caused by transfer to couriers and delays with airport transfers. In most instances though the careful planning of organ donor retrieval surgical times around flight times and urgency to meet assigned flights made significant differences to flight times as seen between zone 6 and those undertaken in zone 4 where similar urgency to make flight times was not as critical or there were significant distances driven from a distant regional hospital to make the flight interstate.

Other studies have looked at the impact of distance on isolation outcome including an initial report of 3 patients with a follow up of 11 patients, transplanted with islets isolated from pancreata obtained in Houston that were transported by air to the isolation centre in Miami, before being transported back to be implanted into patients in Houston [7–9]. However, the major focus was on recipient outcomes with transplanted preparations and the analysis of the factors affecting isolation were limited. More recently, established consortiums have been actively recruiting remote sites, with the UK and GRAGIL being notable examples [10]. In the GRAGIL consortium, pancreata are transported to a single isolation centre in Geneva and the distances travelled were all within 300 min driving time.

The study reported by the GRAGIL study is a relatively small number of isolations performed and the consequent small number of preparations suitable for transplantation. Whilst we did not identify an impact of CIT and days in ICU on isolation outcome, these have been identified as important variables in other studies. In this setting, CIT is a good surrogate for distance travelled but only when taken in the right context and in relation to the exceptionally long distances travelled across Australia that can only occur by air travel. CIT has been shown to have an adverse outcome on islet isolation in both single-centre and registry studies [1,9].

A unique feature of this study was the distance travelled between the donor hospital and the islet isolation facility. Donor pancreata were transported up to 3290 km from the city of Perth situated on the west coast to Sydney on the east coast of Australia for islet isolation. Whilst these large distances did impact on the chances of a successful outcome, we show that it is possible to achieve release criteria and good outcomes from those organs that come from distant regions.

The BMI of a donor was also significant in determining a positive isolation outcome meeting release criteria and thus we must continue to consider this when deciding whether we should perform a donor pancreas isolation. The BMI of a donor has been found to correlate with the size of their pancreas and bigger pancreata often result in greater islet numbers [11]. This result is consistent with other islet groups' isolation and transplant outcomes around the world [12,13].

All other factors were not significantly different between all comparison groups. The data we used for this analysis were retrospective and donor selection criteria were based on the Edmonton Score [14]. If possible, we choose organs that are in an accepted age group, larger BMI, limited hypoxia and minimal or no steroid use as these things have been shown to impact on isolation results [11,14–16]. This may have prevented us identifying factors other than distance and donor BMI as important criteria for achieving release criteria.

In conclusion, excellent islet transplant outcomes can be achieved from pancreata retrieved at distant centres, despite substantial logistical issues involved. The nationally funded program provides a fair and equitable use for donor organs regardless of the state they are retrieved in and will provide outcomes for patients equivalent to the best units in the world despite major logistical hurdles compared to other units. Whilst pancreata retrieved from distant sites are less likely to achieve release criteria, those that do achieve release criteria have comparable outcomes to locally retrieved outcomes. Distant pancreas retrieval does pose challenges and careful selection of appropriate donors and minimisation of CIT by improved logistics is essential to ensure success.

4. Materials and Methods

4.1. Donor Selection

All multiorgan heart-beating brain dead donors were accepted for donation based upon their suitability as described previously [15,17]. Organ donor characteristics that could influence islet isolation and transplantation outcomes were recorded including cold ischemia time (CIT), transport time, donor age, BMI, admission blood glucose levels, hypotension, use of vasopressors prior to death, and cause of death [18].

4.2. Islet Preparation

Islets were separated as described previously using a variation of the closed-loop method described by Ricordi et al. [5,18]. Pancreata were disaggregated by infusing the ducts with cold collagenase NB1 GMP grade (SERVA, Heidelberg, Germany). Dissociated islet and acinar tissue were separated on a continuous Biocoll (Biochrom AG, Berlin, Germany) density gradient (polysucrose 400 and amidotrizoic acid) on a refrigerated apheresis system (Model 2991, COBE Laboratories, Lakewood, Colorado).

4.3. Release Criteria

Purified islets were counted and islet number and mass were expressed in terms of islet equivalents (IEQ) [19]. Islet preparations underwent pre-culture quality assurance, which included purity and viability assessment, packed cell volume measurement and evaluation of islet morphology to exclude excessive fragmentation. Islets were cultured in Miami media in 95% room air and 5% CO_2 at 37 °C for up to 24 h with quality assurance, including beta cell viability index, oxygen consumption rate, endotoxin and Gram stain, being repeated prior to release of the islets for transplant. Islets were deemed suitable for transplantation if they reached the appropriate release criteria defined as greater than 5000 IEQ per kg of recipient body weight, a negative Gram stain, less than 5 EU/kg endotoxin and the total tissue volume less than 10 mL as based on CIT release criteria published previously [20].

4.4. Recipient Patients

The patient selection criteria and outcomes of the trial have been published previously. Eligible patients had type 1 diabetes mellitus for more than 5 years and were aged between 18 and 65 years. Additionally, they had recurrent severe hypoglycaemia unawareness that required constant monitoring or regular intervention by a third party with a hyposcore as assessed by the Edmonton criteria of greater than 1000 [21]. All patients gave informed consent, and the protocol was approved by the Human Research Ethics Committee of the Western Sydney Local Health District.

4.5. Islet Transplantation

For the purposes of this evaluation, all islets were isolated, and all patients were transplanted at Westmead Hospital. The islets were resuspended in 120 mL of medium 199 (ThermoTrace, Melbourne, Australia) containing 5000 U heparin and 20% human albumin. Patients received a general anaesthetic, and a mini-laparotomy was performed to access a mesenteric vein. An arterial angiographic catheter was inserted and threaded into the main portal vein with the assistance of image intensification and the islets infused under gravity.

4.6. Data Collection and Grouping

Donor information was collected at the time of retrieval on national organ donor data sheets supplied with the organ by the organ donor agency. Additional information was also requested from the relevant organ donor agency if this was missing. Distance from islet isolation facility was calculated as-the-crow-flies and then grouped into six zones:

1 =< 50 km, 2 = 50 to 150 km, 3 = 151 to 400 km, 4 = 401 to 800 km, 5 = 801 to 1200 km and 6 = 1201 to 4000 km (Figure 1). The distances were divided into zones to incorporate the local Sydney Metro region in zone 1, the wider Sydney region in zone 2, the rest of highly populated NSW and the Australian Capital Territory in zone 3, major cities Melbourne and Brisbane in zone 4, major cities Adelaide, Hobart, and Rockhampton in zone 5 and remote areas of Australia more than 1200 km from the Isolation facility including the city of Perth some 3290 km away in zone 6.

For CIT, age and days in ICU, absolute values were used for analysis. For gender, females were assigned to group 1 and males to group 2. Retrieval team was split into group 1—our local team and group 2—all other retrieval teams.

Cause and mechanism of death were classified into CITR structured groups[13] by agreement between two members of our team (Surgeon and Islet Operations Manager). Inotrope use was classified into 4 groups: 1 = none, 2 = normal/low (4–6 mg/100 mls at <6 mls/h), 3 = moderate (4–6 mg/100 mls at 6–10 mls/h) and 4 = high (4–6 mg/100 mls at >10 mls/h). Steroid use was assigned a 0 for none and 1 for some and C-peptide in recipients 0 for a negative isolation (<0.2 at one month) and 1 for a positive isolation (>0.2 at one month).

A positive isolation outcome was defined as 'preparation of transplant quality islets meeting all release criteria allowing for transplantation of the islets'. A positive transplant outcome was defined as 'recipient C-peptide >0.2 at 1 month post-transplant'.

4.7. Statistical Analysis

The statistical software package S-PLUS v8 was used to analyse the data. Ranges (or minimum/maximum values), means and standard deviations (SD) were calculated for all data. For non-continuous data, a value was assigned to each group for analysis (Table 1), and Chi-square and Fisher's Exact tests were used to test significance (including Table 2). Non-parametric Spearman's rank correlations were used for transplant outcome analysis. A *p*-value of less than 0.05 was considered significant.

Author Contributions: W.J.H. designed this study, interpreted the data and wrote this paper. S.D. participated in design of the experiments, collated data, interpreted the data and helped write this paper. H.-c.M. collated data, interpreted the data and helped write this paper. Y.V.C. provided management of islet isolation and collection and collation of data. L.W. provided management of islet isolation and collection of data. P.A. coordinated patient treatments and collection of data. N.R. Helped in review of the paper. P.J.O. helped in writing this paper. All authors have read and agreed to the published version of the manuscript.

Funding: These studies were supported by grants from the National Health & Medical Research Council of Australia the Juvenile Diabetes Research Foundation International, and the Nationally Funded Centres Program, and the Nationally Funded Centres approved by the Australian Health Ministers' Advisory Council (AHMAC) funded by states and territories.

Institutional Review Board Statement: These studies were undertaken after approval by the Western Sydney Local Health District Ethics Committee, HREC2003/10/MH/25 and HREC2006/3/4.5(2305). All research was carried out in compliance with the Helsinki Declaration.

Informed Consent Statement: Informed consent was obtained for all human subjects involved in this study.

Data Availability Statement: The data presented in this study are available in article.

Acknowledgments: These studies were supported by grants from the National Health & Medical Research Council of Australia and the Juvenile Diabetes Research Foundation International. Thank you to the Donate Life Network—Australia's tissue and organ donation service—for their help in obtaining donor pancreata and relevant donor information. Thank you also to Karen Byth, biostatistician, for data analysis.

Conflicts of Interest: The authors declare no conflict of interest.

Abbreviations

BMI	Body mass index
CIT	Cold ischemia time
ICU	Intensive care unit
IEQ	Islet equivalents
SD	Standard deviation

References

1. Markmann, J.F.; Deng, S.; Huang, X.; Desai, N.M.; Velidedeoglu, E.H.; Lui, C.; Frank, A.; Markmann, E.; Palanjian, M.; Brayman, K.; et al. Insulin Independence Following Isolated Islet Transplantation and Single Islet Infusions. *Ann. Surg.* **2003**, *237*, 741–750. [CrossRef]
2. Robertson, R.P. Islet Transplantation as a Treatment for Diabetes—A Work in Progress. *N. Engl. J. Med.* **2004**, *350*, 694–705. [CrossRef]
3. Ryan, E.A.; Lakey, J.R.; Paty, B.W.; Imes, S.; Korbutt, G.S.; Kneteman, N.M.; Bigam, D.; Rajotte, R.V.; Shapiro, A.J. Successful islet transplantation: Continued insulin reserve provides long-term glycemic control. *Diabetes* **2002**, *51*, 2148–2157. [CrossRef]
4. Shapiro, A.M.J.; Ricordi, C.; Hering, B.J.; Auchincloss, H.; Lindblad, R.; Robertson, R.P.; Secchi, A.; Brendel, M.D.; Berney, T.; Brennan, D.C.; et al. International Trial of the Edmonton Protocol for Islet Transplantation. *N. Engl. J. Med.* **2006**, *355*, 1318–1330. [CrossRef]
5. O'Connell, P.J.; Hawthorne, W.J.; Nankivell, B.J.; Patel, A.T.; Walters, S.N.; Pleass, H.C.; Allen, R.D.; Chapman, J.R.; Holmes-Walker, D.J.; Gunton, J.E. Clinical islet transplantation in type 1 diabetes mellitus: Results of Australia's first trial. *Med. J. Aust.* **2006**, *184*, 221–225. [CrossRef] [PubMed]
6. O'Connell, P.J.; Holmes-Walker, D.J.; Goodman, D.; Hawthorne, W.J.; Loudovaris, T.; Gunton, J.E.; Thomas, H.E.; Grey, S.T.; Drogemuller, C.J.; Ward, G.M.; et al. Multicenter Australian Trial of Islet Transplantation: Improving Accessibility and Outcomes. *Arab. Archaeol. Epigr.* **2013**, *13*, 1850–1858. [CrossRef]
7. Alejandro, R.; Barton, F.B.; Hering, B.J.; Wease, S. 2008 Update From the Collaborative Islet Transplant Registry. *Transplantation* **2008**, *86*, 1783–1788. [CrossRef] [PubMed]
8. Goss, J.A.; Schock, A.P.; Brunicardi, F.C.; Goodpastor, S.E.; Garber, A.J.; Soltes, G.; Barth, M.; Froud, T.; Alejandro, R.; Ricordi, C. Achievement of insulin independence in three consecutive type-1 diabetic patients via pancreatic islet transplantation using islets isolated at a remote islet isolation center. *Transplantation* **2002**, *74*, 1761–1766. [CrossRef] [PubMed]
9. Goss, J.A.; Goodpastor, S.E.; Brunicardi, F.C.; Barth, M.H.; Soltes, G.D.; Garber, A.J.; Hamilton, D.J.; Alejandro, R.; Ricordi, C. Development of a human pancreatic islet-transplant program through a collaborative relationship with a remote islet-isolation center. *Transplantation* **2004**, *77*, 462–466. [CrossRef]
10. Kempf, M.-C.; Andres, A.; Morel, P.; Benhamou, P.-Y.; Bayle, F.; Kessler, L.; Badet, L.; Thivolet, C.; Penfornis, A.; Renoult, E.; et al. Logistics and Transplant Coordination Activity in the GRAGIL Swiss-French Multicenter Network of Islet Transplantation. *Transplantation* **2005**, *79*, 1200–1205. [CrossRef]
11. Kin, T.; Murdoch, T.B.; Shapiro, A.M.J.; Lakey, J.R.T. Estimation of Pancreas Weight from Donor Variables. *Cell Transplant.* **2006**, *15*, 181–185. [CrossRef]
12. Briones, R.; Miranda, J.; Mellado-Gil, J.; Castro, M.; Gonzalez-Molina, M.; Cuesta-Munoz, A.; Alonso, A.; Frutos, M. Differential Analysis of Donor Characteristics for Pancreas and Islet Transplantation. *Transplant. Proc.* **2006**, *38*, 2579–2581. [CrossRef] [PubMed]
13. Sakuma, Y.; Ricordi, C.; Miki, A.; Yamamoto, T.; Pileggi, A.; Khan, A.; Alejandro, R.; Inverardi, L.; Ichii, H. Factors That Affect Human Islet Isolation. *Transplant. Proc.* **2008**, *40*, 343–345. [CrossRef] [PubMed]
14. O'Gorman, D.; Kin, T.; Murdoch, T.; Richer, B.; McGhee-Wilson, D.; Ryan, E.A.; Shapiro, J.A.; Lakey, J.R. The Standardization of Pancreatic Donors for Islet Isolations. *Transplantation* **2005**, *80*, 801–806. [CrossRef] [PubMed]
15. Lakey, J.R.; Warnock, G.L.; Rajotte, R.V.; Suarez-Almazor, M.E.; Ao, Z.; Shapiro, A.J.; Kneteman, N.M. Variables in organ donors that affect the recovery of human islets of Langerhans. *Transplantation* **1996**, *61*, 1047–1053. [CrossRef]
16. Ponte, G.M.; Pileggi, A.; Messinger, S.; Alejandro, A.; Ichii, H.; Baidal, D.A.; Khan, A.; Ricordi, C.; Goss, J.A.; Alejandro, R. Toward Maximizing the Success Rates of Human Islet Isolation: Influence of Donor and Isolation Factors. *Cell Transplant.* **2007**, *16*, 595–607. [CrossRef]
17. Thwaites, S.E.; Gurung, B.; Yao, J.; Kable, K.; Robertson, P.; Ryan, B.J.; Lam, V.W.; Pleass, H.C.; Chapman, J.R.; Hawthorne, W.J.; et al. Excellent outcomes of simultaneous pancreas kidney transplantation in patients from rural and urban Australia: A national service experience. *Transplantation* **2012**, *94*, 1230–1235. [CrossRef]
18. Ricordi, C.; Lacy, P.E.; Scharp, D.W. Automated Islet Isolation from Human Pancreas. *Diabetes* **1989**, *38*, 140–142. [CrossRef]
19. Ricordi, C.; Gray, D.W.R.; Hering, B.J.; Kaufman, D.B.; Warnock, G.L.; Kneteman, N.M.; Lake, S.P.; London, N.J.M.; Socci, C.; Alejandro, R.; et al. Islet isolation assessment in man and large animals. *Acta Diabetol.* **1990**, *27*, 185–195. [CrossRef] [PubMed]

20. Yamamoto, T.; Horiguchi, A.; Ito, M.; Nagata, H.; Ichii, H.; Ricordi, C.; Miyakawa, S. Quality control for clinical islet transplantation: Organ procurement and preservation, the islet processing facility, isolation, and potency tests. *J. Hepato-Biliary-Pancreat. Surg.* **2009**, *16*, 131–136. [CrossRef]
21. Ryan, E.A.; Shandro, T.; Green, K.; Paty, B.W.; Senior, P.A.; Bigam, D.; Shapiro, A.J.; Vantyghem, M.-C. Assessment of the Severity of Hypoglycemia and Glycemic Lability in Type 1 Diabetic Subjects Undergoing Islet Transplantation. *Diabetes* **2004**, *53*, 955–962. [CrossRef] [PubMed]

metabolites

MDPI

Article

Ectopic Leptin Production by Intraocular Pancreatic Islet Organoids Ameliorates the Metabolic Phenotype of *ob/ob* Mice

Barbara Leibiger *, Tilo Moede, Ismael Valladolid-Acebes, Meike Paschen, Montse Visa, Ingo B. Leibiger and Per-Olof Berggren *

The Rolf Luft Research Center for Diabetes and Endocrinology, Department of Molecular Medicine and Surgery, Karolinska Institutet, Karolinska Sjukhuset L1:03, 17176 Stockholm, Sweden; tilo.moede@ki.se (T.M.); ismael.valladolid.acebes@ki.se (I.V.-A.); paschen.meike@gmail.com (M.P.); montserrat.visa.majoral@ki.se (M.V.); ingo.leibiger@ki.se (I.B.L.)

* Correspondence: barbara.leibiger@ki.se (B.L.); per-olof.berggren@ki.se (P.-O.B.)

Abstract: The pancreatic islets of Langerhans consist of endocrine cells that secrete peptide hormones into the blood circulation in response to metabolic stimuli. When transplanted into the anterior chamber of the eye (ACE), pancreatic islets engraft and maintain morphological features of native islets as well as islet-specific vascularization and innervation patterns. In sufficient amounts, intraocular islets are able to maintain glucose homeostasis in diabetic mice. Islet organoids (pseudo-islets), which are formed by self-reassembly of islet cells following disaggregation and genetic manipulation, behave similarly to native islets. Here, we tested the hypothesis that genetically engineered intraocular islet organoids can serve as production sites for leptin. To test this hypothesis, we chose the leptin-deficient *ob/ob* mouse as a model system, which becomes severely obese, hyperinsulinemic, hyperglycemic, and insulin resistant. We generated a Tet-OFF-based beta-cell-specific adenoviral expression construct for mouse leptin, which allowed efficient transduction of native beta-cells, optical monitoring of leptin expression by co-expressed fluorescent proteins, and the possibility to switch-off leptin expression by treatment with doxycycline. Intraocular transplantation of islet organoids formed from transduced islet cells, which lack functional leptin receptors, to *ob/ob* mice allowed optical monitoring of leptin expression and ameliorated their metabolic phenotype by improving bodyweight, glucose tolerance, serum insulin, and C-peptide levels.

Keywords: pancreatic islets; leptin; exocytosis; tissue-engineering; in vivo imaging; metabolism; diabetes; β-cell; viral transduction; transplantation

check for updates

Citation: Leibiger, B.; Moede, T.; Valladolid-Acebes, I.; Paschen, M.; Visa, M.; Leibiger, I.B.; Berggren, P.-O. Ectopic Leptin Production by Intraocular Pancreatic Islet Organoids Ameliorates the Metabolic Phenotype of *ob/ob* Mice. *Metabolites* **2021**, *11*, 387. https://doi.org/10.3390/metabo11060387

Academic Editors: Melkam Kebede, Belinda Yau and Dawn Belt Davis

Received: 13 April 2021
Accepted: 11 June 2021
Published: 14 June 2021

Publisher's Note: MDPI stays neutral with regard to jurisdictional claims in published maps and institutional affiliations.

1. Introduction

Pancreatic islets of Langerhans are micro-organs that form the endocrine part of the pancreas. They consist of endocrine alpha-, beta-, delta-, epsilon- and PP-cells, that produce and secrete glucagon, insulin, somatostatin, ghrelin, and pancreatic polypeptide, respectively. The fenestrated islet blood vessels allow for an efficient exchange of blood-derived factors stimulating these cells as well as the immediate release of the secreted hormones into the blood circulation. Importantly, these characteristics are maintained when pancreatic islets are transplanted into the anterior chamber of the eye, ACE [1]. In fact, transplantation of a sufficient number of islets to the ACE (75–300 islets in mice) showed that these islet grafts are capable of maintaining glycaemia in streptozotocin-treated diabetic mice ([1], reviewed in [2]). Islet organoids (also called pseudo-islets) that are formed by self-reassembly of islet cells following disaggregation and genetic manipulation, for example by adenoviral-mediated ectopic gene expression [3,4], behave similarly to native islets. Because of these features, we wanted to test the hypothesis that genetically engineered intraocular islet organoids can serve as production sites for blood-born proteins/peptides as a novel treatment strategy. To test this hypothesis, we chose the *ob/ob* mouse as a disease model. *Ob/ob* mice lack functional leptin, which is an adipocyte-produced peptide hormone, and develop a very pronounced phenotype based, among others, on their extreme hyperphagia. After weaning, they become severely obese, hyperinsulinemic, hyperglycemic, and insulin resistant (reviewed in [5]). Treating these mice with peripherally administered leptin reverses [6,7] or ameliorates this phenotype in a dose-dependent manner [8]. Ectopic expression of leptin via gene therapy approaches by either injecting virus-encoded leptin constructs [9,10] or transplanting engineered leptin-expressing cells [11,12] improves the metabolic phenotype of *ob/ob* mice even further. Because leptin acts in concert with insulin as anorexic stimuli in the central nervous system (reviewed in [13]), we decided to genetically engineer insulin-producing beta-cells to become production sites for leptin. This allowed us to test the hypothesis whether ectopic leptin production by islet organoids in the ACE ameliorates the metabolic phenotype of *ob/ob* mice.

2. Results

2.1. Generation of a Tet-OFF-Based Beta-Cell Specific Adenoviral Expression Construct for Mouse Leptin and Its In Vitro Assessment

When designing the expression construct for mouse leptin, we considered the following points: We wanted (1) leptin to be expressed in pancreatic beta-cells, (2) to be able to reduce/turn-off leptin expression if necessary, (3) to optically monitor the activity of the expression construct by fluorescence microscopy, and (4) an adenoviral-based expression construct for efficient transduction of native beta-cells. As a result, we generated an adenoviral vector, vAd-RIP-leptin-OFF, that contains two expression cassettes that are positioned in opposite directions and are separated by a 'transcription-block' sequence to allow independent expression (Figure 1a). The first expression cassette allows rat insulin-1 promoter-driven expression of the synthetic transcription factor tTA (Tet-OFF) and the green fluorescent protein ZsGreen in pancreatic beta-cells. The second expression cassette consists of a TRE-tight promoter-driven mouse leptin-IRES-mCherry cassette. Binding of tTA to the TRE-tight promoter induces the expression of leptin and the red fluorescent protein mCherry in beta-cells. Addition of doxycycline inhibits binding of tTA and turns-off the expression of the two proteins. The IRES-element in the two expression cassettes allows the co-expression of tTA with ZsGreen and leptin with mCherry. Hence, 'green' and 'red' serve as visual read-outs for the expression of tTA and leptin in beta-cells, respectively. For in vitro assessment of the expression construct, we used organoids created from transduced islets cells of B6.BKS(D)-Lepr^{db}/J (*db/db*) mice. Organoids from the same preparations were also used for the transplantations. The detection of green and red fluorescence by laser-scanning confocal microscopy in beta-cells of the islet organoids verified that both expression cassettes were active in the same cells (Figure 1b,c). The islet organoids secreted leptin into the culture medium, the secretion was glucose-dependent: 7.02 $\pm$ 1.76 pg/organoid/h at 3 mM glucose vs. 12.94 $\pm$ 2.83 pg/organoid/h at 16 mM glucose (stimulation index 16 versus 3 mM glucose: 1.93 $\pm$ 0.14).

2.2. Ectopic Leptin Production by Islet Organoid Grafts in the ACE Ameliorates the Metabolic Phenotype of ob/ob Mice

To test whether intraocular leptin production by islet organoids affects the metabolic phenotype of *ob/ob* mice, we considered the following points. Because leptin deficiency after weaning leads to a rapid development of the *ob/ob* phenotype, including strong beta-cell proliferation in both the native in situ islets as well as in islets transplanted to the ACE and leptin treatment decelerates this process [14], we decided to treat *ob/ob* mice with daily intraperitoneal injections of leptin (1.5 µg/g bodyweight/day) immediately after their arrival from the vendor until 4 weeks after transplantation of the islet organoid grafts, i.e., until their full engraftment. Moreover, because leptin is discussed in the literature to have a negative effect on insulin secretion [15], we decided to avoid the potential negative feed-back of beta-cell produced leptin on the organoids by generating them from islets of leptin receptor-deficient mice, i.e., islets from *db/db* mice with a B6 genetic background (B6.BKS(D)-Lepr^{db}/J) to match the B6 background of the *ob/ob* recipient mice (B6.Cg-

Lep^{ob}/J). Hence, we isolated islets from *db/db* mice, disaggregated them, transduced the islets cells with vAd-RIP-leptin-OFF and generated islet organoids by self-reassembly.

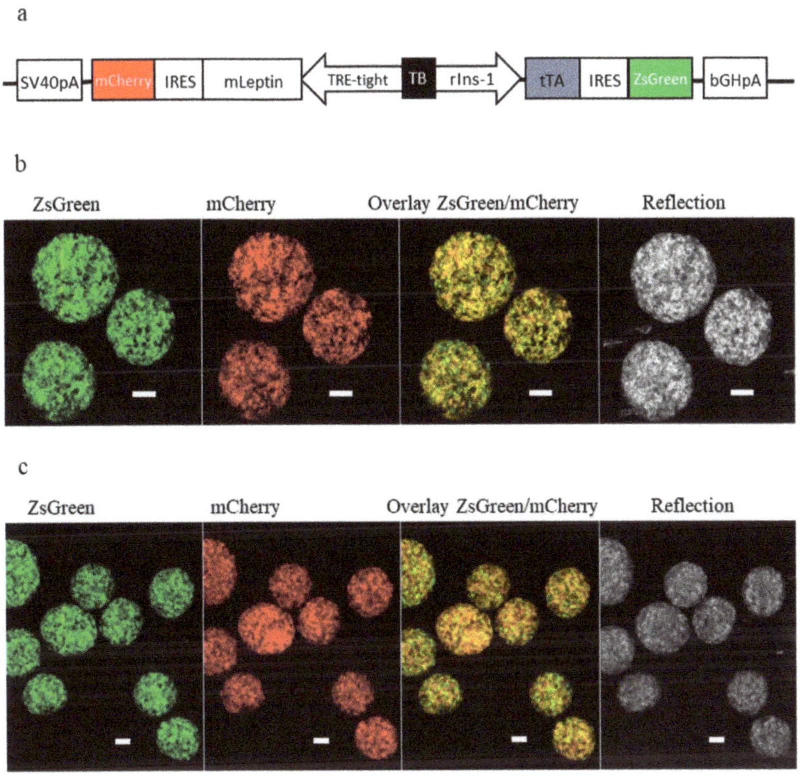

Figure 1. In vitro characterization of leptin-expressing islet organoids. (**a**) Schematic illustration of vAd-RIP-leptin-OFF. The rat insulin-1 promoter (rIns-1) drives expression of the synthetic transcription factor tTA and the green fluorescent protein ZsGreen in pancreatic beta-cells. The TRE-tight promoter drives expression of mouse leptin (mLeptin) and the red fluorescent protein mCherry. The two expression cassettes are separated by a transcription blocker sequence (TB). Binding of tTA to the TRE-tight promoter induces in the absence of doxycycline the expression of leptin and mCherry, while addition of doxycycline turns-off the expression of the two proteins. IRES-elements in the cassettes ensures stoichiometric expression of the two proteins under the same promoter. (**b**) Representative maximum projection of a 3D-stack of leptin-expressing islet organoids, obtained by confocal imaging, cultured for 4 weeks in vitro (1st set of experiments), showing reflection, expression of ZsGreen, mCherry and their overlay. Scale bar: 50 μm. (**c**) Representative maximum projection of a 3D-stack of leptin-expressing islet organoids, obtained by confocal imaging, cultured for 4 weeks in vitro (2nd set of experiments), showing reflection, expression of ZsGreen, mCherry and their overlay. Scale bar: 50 μm.

In a first set of experiments (Figure A1a) we transplanted 130 leptin-expressing *db/db* islet organoids into the ACE of female *ob/ob* mice ($n = 3$, see Figure 2a,b) that were treated with leptin from 15 d before transplantation until 28 d after transplantation and used leptin-treated (for the same time period) female *ob/ob* mice that were not transplanted with islet organoids as a control group ($n = 3$). To use non-transplanted leptin-treated *ob/ob* mice as controls was based on previous observations showing extreme growth of non-leptin expressing intraocular islet grafts in *ob/ob* mice, which required premature termination of experiments for ethical reasons. Inspection of islet organoids kept in vitro (Figure 1b) as well as organoids transplanted to the ACE in vivo (Figure 2b) showed expression of both green and red fluorescent proteins, indicating that both the tTA-IRES-ZsGreen and mLeptin-IRES-mCherry expression cassettes were expressed in beta-cells. While we did not observe a difference in the bodyweight of mice between the two groups (Figure 2c), we observed an improvement in glucose tolerance starting 10 d after stopping leptin treatment, which became significant 25 d after stopping leptin treatment in the transplanted group (Figures 2d and A1b).

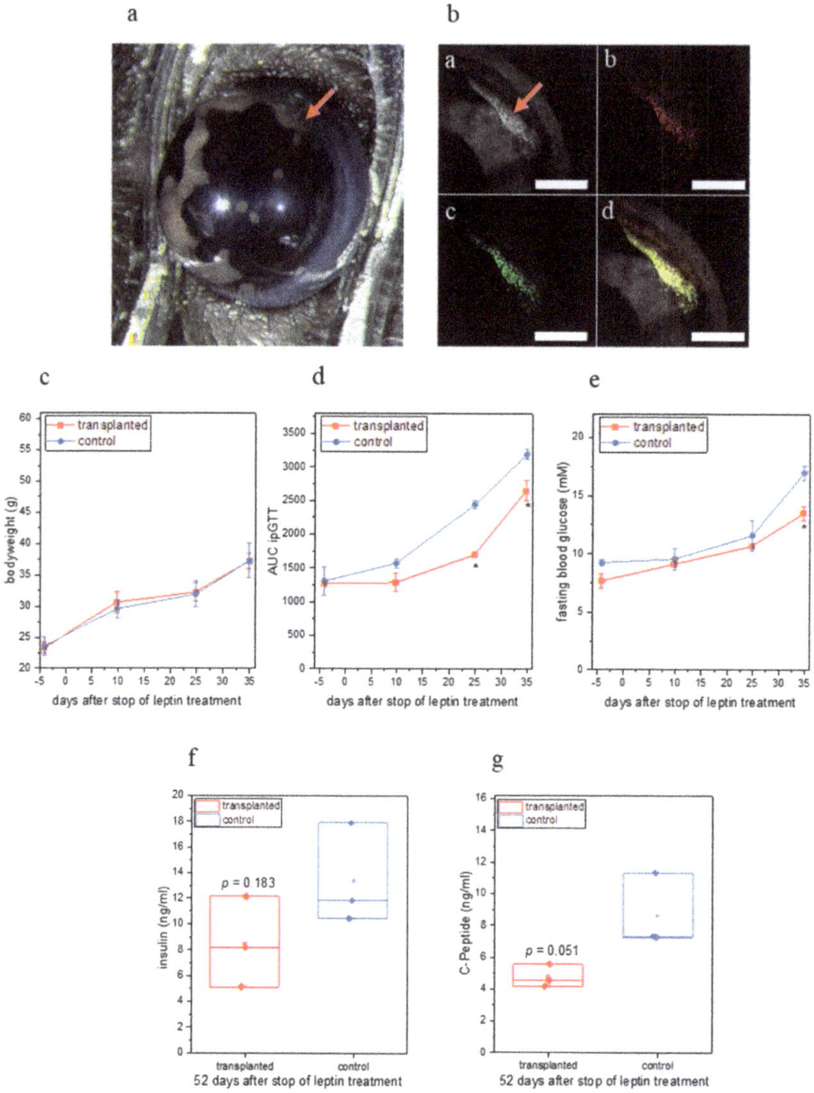

Figure 2. In vivo characterization of leptin-expressing islet organoids from the first set of experiments. (**a**) Photograph of the eye containing the leptin-expressing islet organoid graft (red arrow) 9 weeks after transplantation. (**b**) Maximum projection of a 3D-stack of a leptin-expressing islet organoid graft (red arrow) obtained by confocal imaging 9 weeks after transplantation; (**ba**) reflection, (**bb**) mCherry fluorescence, (**bc**) ZsGreen fluorescence, (**bd**) overlay image; scale bar 500 µm. (**c**) Bodyweight of transplanted and control mice over the entire period of the experiment. (**d**) Area under the curve (AUC) for ipGTT of transplanted and control mice over the entire period of the experiment. (**e**) Fasting blood glucose of transplanted and control mice over the entire period of the experiment. (**f**) Plasma insulin levels of transplanted and control mice at the end of the experiment. (**g**) Plasma C-peptide levels of transplanted and control mice at the end of the experiment. (**d**,**e**) * $p < 0.05$. (**c**–**g**) $n = 3$.

Moreover, in the group transplanted with leptin-producing organoids we observed a trend towards a decrease in fasting blood glucose (Figure 2e), which was significant at 35 d after stop of leptin treatment, a trend towards lower plasma insulin (Figure 2f) and insulin C-peptide (Figure 2g) levels, both measured at the end of the experiment. Finally, we detected leptin in the aqueous humor (11.375 ± 3.211 ng/mL) at the end of the experiment in the transplanted group, while no leptin was detectable in the control group. No leptin was detectable in the blood plasma from both groups, as the levels in the transplanted group were likely below the detection limit of the ELISA kit used.

In a second set of experiments (Figure A2a) we transplanted 200 leptin-expressing *db/db* islet organoids into the ACE of female *ob/ob* mice that were treated with leptin from 7 d before until 28 d after transplantation ($n = 7$, Figure 3a,b) and used leptin-treated female *ob/ob* mice that were not transplanted with islet organoids as a control group ($n = 5$).

Inspection of islet organoids kept in vitro (Figure 1c) as well organoids transplanted to the ACE in vivo (Figure 3b) showed expression of both green and red fluorescent proteins, indicating that both the tTA-IRES-ZsGreen and mLeptin-IRES-mCherry expression cassettes were expressed in beta-cells. Importantly in this experiment, mice that were transplanted with leptin-producing organoids showed a significant difference in bodyweight from 15 d after stop of leptin treatment (Figure 3c). We observed a significant improvement in ipGTT starting from 19 d after stop of leptin treatment (Figures 3d and A2b). Moreover, we observed a decrease in fasting blood glucose (Figure 3e, significant on days 19 and 68 after stop of leptin treatment), a significant decrease in plasma insulin (from day 14 after stop of leptin, Figure 3f) and a significant decrease in plasma C-peptide levels (from day 8 after stop of leptin, Figure 3g) in the group transplanted with leptin-producing organoids. Finally, we detected leptin in the aqueous humor (6.05 ± 2.91 ng/mL) of the transplanted group at the end of the experiment while no leptin was detectable in the control group. The values of leptin in plasma of the transplanted group were 185.52 ± 10.73 pg/mL between day 8 after stop of leptin treatment and the end of the experiment (Figure A2c). No leptin was detectable in plasma from the control group.

2.3. Doxycycline Treatment Stops Ectopic Leptin Expression

Three animals of the transplanted group were treated with doxycycline (five intraperitoneal injections of 50 µg dox/kg/mouse over 10 days) to switch-off leptin production from the transplanted organoids. In grafts of these animals no mCherry fluorescence was detectable (Figure 4(bc)) and no leptin expression could be detected by immunohistochemistry (Figure 4(dc)) at the end of the experiment. Expression of only green (Figure 4(bb)), but not red (Figure 4(bc)), fluorescent protein in the graft of doxycycline-treated animals indicates that only the tTA-IRES-ZsGreen cassette was expressed in beta-cells and the expression of the mLeptin-IRES-mCherry cassette was switched-off. In grafts of non-doxycycline treated animals both mCherry and leptin expression were detectable (Figure 4(ac,cc)). More importantly, no leptin was measurable in the aqueous humor of doxycycline-treated animals ($n = 3$).

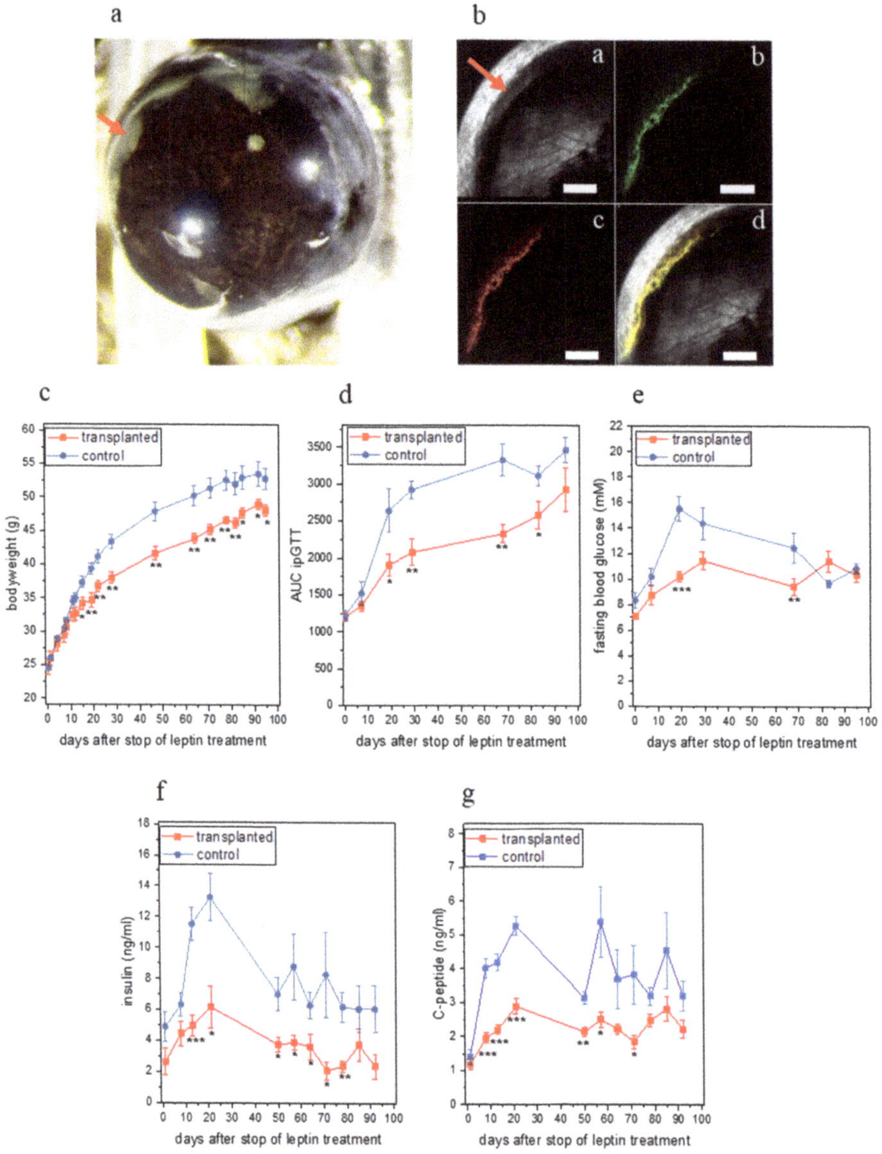

Figure 3. In vivo characterization of leptin-expressing islet organoids from the second set of experiments. (**a**) Photograph of the eye containing the leptin-expressing islet organoid graft (red arrow) 6 weeks after transplantation. (**b**) Maximum projection of a 3D-stack of a leptin-expressing islet organoid graft (red arrow) obtained by confocal imaging 12 weeks after transplantation; (**ba**) reflection, (**bb**) mCherry fluorescence, (**bc**) ZsGreen fluorescence, (**bd**) overlay image; scale bar 300 µm. (**c**) Bodyweight, (**d**) Area under the curve (AUC) for ipGTT, (**e**) Fasting blood glucose, (**f**) Plasma insulin levels and (**g**) Plasma C-peptide levels of transplanted and control mice over the period of the experiment. (**c–g**) * $p < 0.05$, ** $p < 0.01$, *** $p < 0.001$; $n = 7$ for transplanted and 5 for control group.

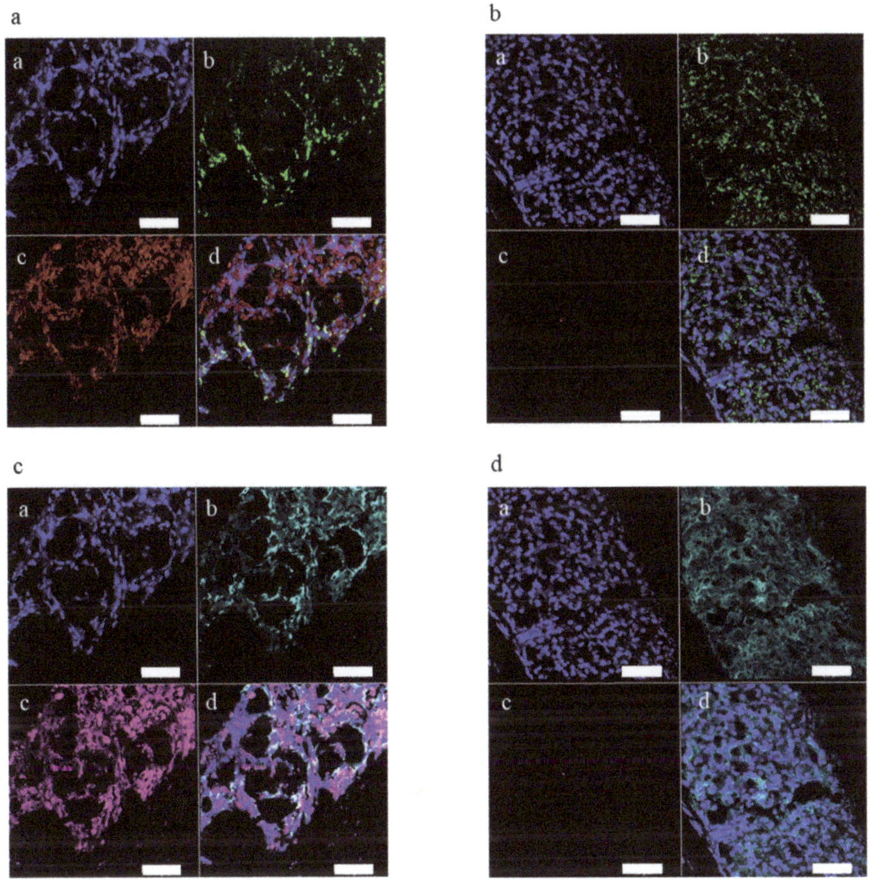

Figure 4. Characterization of leptin-expressing islet organoid grafts from animals treated with doxycycline, removed at the end of the experiment. (**a**) Fluorescent microscopy image of a leptin-expressing islet organoid graft from transplanted non-doxycycline-treated animal; (**aa**) DAPI staining; (**ab**) ZsGreen fluorescence; (**ac**) mCherry fluorescence; (**ad**) overlay. (**b**) Fluorescent microscopy image of a leptin-expressing islet organoid graft from transplanted doxycycline-treated animal; (**ba**) DAPI staining, (**bb**) ZsGreen fluorescence, (**bc**) mCherry fluorescence, (**bd**) overlay. (**c**) Immunohistochemistry image of a leptin-expressing islet organoid graft from transplanted non-doxycycline treated animal; (**ca**) DAPI staining, (**cb**) C-peptide staining, (**cc**) leptin staining, (**cd**) overlay. (**d**) Immunohistochemistry image of a leptin-expressing islet organoid graft from transplanted doxycycline treated animal; (**da**) DAPI staining, (**db**) C-peptide staining, (**dc**) leptin staining, (**dd**) overlay. (**a–d**) All scale bars 50 µm.

3. Discussion

We have previously shown that the ACE as a transplantation site allows (i) optical/microscopic monitoring of graft vascularization and function), (ii) monitoring of potential pharmacological intervention (systemic or ocular/intraocular treatment), and (iii) the possibility to remove the graft if needed (e.g., by iridectomy). Moreover, hormones secreted by intraocular pancreatic islets are capable of maintaining glucose homeostasis in a diabetic recipient (reviewed in [2]). In this proof-of-concept study we aimed to test whether genetically engineered islet organoids can serve as production sites for non-islet blood-born peptides/proteins, here leptin, and take advantage of the above-listed properties of the ACE as a transplantation site. We designed an adenoviral vector for beta-cell specific expression of mouse leptin that in addition allowed to monitor the activity of the expression cassettes by fluorescence microscopy and to switch-off leptin expression by doxycycline treatment. Expression of leptin in beta-cells brought the potential advantage of glucose-regulated leptin secretion. In addition to the glucose-regulated route of secretion, pancreatic beta-cells are also equipped for continuous release of proteins [16]. We used leptin-expressing islet organoids generated from *db/db* mouse pancreatic islets. Here, all islet cells lack the expression of the functional/signaling leptin receptor and thus do not allow a potential feedback action of secreted leptin on the organoid graft. We chose *ob/ob* mice as recipients, because these mice lack expression of functional leptin and thus develop a severe metabolic phenotype that is easy to monitor, i.e., increased bodyweight, hyperglycemia, and hyperinsulinemia. By transplanting leptin-expressing islet organoids to the ACE we could show that leptin, secreted by the organoids, ameliorates the metabolic phenotype of the *ob/ob* mice. Transplanted mice showed significantly reduced bodyweight, improved glucose tolerance, decreased fasting blood glucose and significant decreased plasma insulin and C-peptide levels. The food intake of these animals was not measured because it would require single animal housing to get reliable results, which was not possible due to the limited number of experimental animals and limitations set by our ethical permits. IpITTs could not be performed due to restrictions of our ethical permits not allowing us to fast the animals more than two times per month. Ectopically expressed leptin was detectable in the aqueous humor and plasma of transplanted mice, whereas no leptin was detected in non-transplanted mice. It is noteworthy that despite the additional beta-cell volume due to the islet organoid grafts, we observed a decrease in serum insulin and C-peptide levels as a result of the ectopic leptin expression. When some of the animals in our study were treated with doxycycline, the co-expression of leptin and mCherry was stopped, demonstrating that the ectopic leptin expression can be switched-off in vivo by the Tet-OFF part of the adenoviral construct.

Our results are comparable to those obtained from *ob/ob* mice that were transplanted with engineered and encapsulated gut-derived K-cells to the peritoneal cavity [11] or with engineered leptin-secreting skin grafts [12]. Additionally, leptin, produced by encapsulated engineered adipocytes and transplanted into visceral fat depots of *ob/ob* mice, was shown to reduce bodyweight and improve glucose tolerance [17].

Although we achieved long-term action of leptin, secreted from the transplanted islet organoids, the effect on glucose metabolism started to fade by the end of the experiment. This can be due to multiple reasons including among others immune response to adenoviral-transduced cells or the transgene [9,18], leading to development of anti-leptin antibodies [9] and increased demand for insulin from transplanted islet organoids, which can lead to hypertrophy and cell proliferation, thereby "diluting out" the leptin-producing cells [6,19]. The limited longevity of adenovirus-mediated expression has been discussed in the past [20,21].

Our study with relative low numbers of animals shows encouraging results. Future work is needed to confirm our preliminary conclusions. In the future, it will be possible to generate islet organoids from genetically engineered stem-cell derived material. This will not only help to overcome the shortage of supply of primary pancreatic islets but will allow engineering the cellular components for an optimized protein production site.

In summary, we have demonstrated that ectopic leptin production by genetic engineered islet organoids, transplanted into the ACE, ameliorates the obese and diabetic phenotype in leptin-deficient *ob/ob* mice. Using the ACE as the transplantation site allowed for optical monitoring of ectopic gene expression by the use of co-expressed fluorescent proteins and gives the safety potential to remove the intraocular graft by for example iridectomy.

In conclusion, our data provide a proof-of-concept that genetically engineered islet organoids can serve, as entire micro-organs, as factories for non-islet blood-born peptides in high enough concentrations to have a physiological function, here exemplified by the production of leptin.

4. Materials and Methods

4.1. Animals

Female B6.Cg-Lep^{ob}/J (*ob/ob*) mice at 4 weeks of age and B6.BKS(D)-Lepr^{db}/J (*db/db*) mice at 8 weeks of age were purchased from Jackson Laboratories (Bar Harbour, ME, USA). Both mouse strains share a C57BL/6J (B6) background. After delivery, the mice were allowed to adapt to the animal facility for 1 week before the start of the experiment. All mice were group-housed on a 12/12-h dark/light cycle with free access to food (chow diet R70 from Lantmännen, Kimstad, Sweden) and water. All experiments were performed in accordance with the Karolinska Institutet's guidelines for the care and use of animals in research and were approved by the institute's Animal Ethics Committee.

4.2. Leptin Treatment of ob/ob Mice

Ob/ob mice were treated daily with one intraperitoneal injection of recombinant human leptin protein (1.5 µg/g bodyweight, R&D Systems, Minneapolis, MN, USA) from the day of arrival until 4 weeks after transplantation.

4.3. Expression Vector Construction

The mouse leptin cDNA was obtained from pCMV6.mouse leptin (#MC208876, OriGene, Rockville, MD, USA) and subcloned into pTRE-tight (Clontech, Takara, Mountain View, CA, USA) to create pTRE-tight.mLeptin. Next, we inserted an IRES-mCherry sequence downstream of the mLeptin cDNA to obtain pTRE-tight.mLeptin-IRES-mCherry. To create pENTR.TRE-tight.mLeptin-IRES-mCherry/RIP1.DsRed2 we exchanged in pENTR.-rbGK.EGFP/RIP1.DsRed2 [22] the rbGK.EGFP cassette by TRE-tight.mLeptin-IRES-mCherry. Next, we exchanged in pENTR.TRE-tight.mLeptin-IRES-mCherry/RIP1.DsRed2 the DsRed cDNA by the TetOFF-IRES-ZsGreen cassette obtained from pTetOFF-Dual (Green) (Clontech, Takara, Mountain View, CA, USA), thus generating pENTR.TRE-tight.mLeptin-IRES-mCherry/RIP1.TetOFF-IRES-ZsGreen. All constructs were verified by DNA sequence analysis and leptin expression was verified by Simple Western. The TRE-tight.mLeptin-IRES-mCherry/RIP1.TetOFF-IRES-ZsGreen cassette was transferred into the promoter-less adenovirus plasmid pAd/PL-DEST (Thermo Fisher Scientific, Waltham, MA, USA) by the Gateway technique. The ViraPower Adenoviral Expression System (Thermo Fisher Scientific, Waltham, MA, USA) was used to generate a replication-deficient adenovirus called vAd-RIP-leptin-OFF, which was used for transduction of cells.

4.4. Isolation of Pancreatic Islets

Islets were isolated from B6.BKS(D)-Lepr^{db}/J (*db/db*) mice. Islets were prepared from mice by duct injection of collagenase (F. Hoffmann-La Roche, Basel, Switzerland) and were handpicked under a stereomicroscope MZ6 (Leica Microsystems, Wetzlar, Germany) after digestion. Thereafter, islets were cultured in RPMI-1640 medium (RPMI medium, Gibco, Thermo Fisher Scientific, Waltham, MA, USA), with a final concentration of 10% heat-inactivated fetal bovine serum, 2 mM glutamine, 100 U/mL penicillin, and 100 mg/mL streptomycin (all from Gibco, Thermo Fisher Scientific, Waltham, MA, USA) at 5% CO_2 and 37 °C.

4.5. Islet Organoid Production

The islets were collected into 1.5 mL Eppendorf tubes (250 islets/tube), digested with accutase (Sigma-Aldrich, St. Louis, MO, USA) for 10 min at 37 °C and centrifuged at 500 rpm. The islet cells were transduced with 4×10^6 plaque forming units/mL of the leptin encoding adenovirus in suspension culture dishes for 1h in RPMI medium at 5% CO_2 and 37 °C and then washed two times with an excess of RPMI medium to get rid of the adenovirus. 2500 islet cells were seeded into each well of Nunclon Sphera 96U Bottom plates (Thermo Scientific, Leicestershire, UK). Islet organoids formed over a time of 5 days, after which the organoids were transferred to suspension culture dishes.

4.6. In Vitro Imaging of Intraocular Islet Organoids

An inverted laser scanning confocal microscope (TCS SP8, Leica Microsystems, Wetzlar, Germany) was used to image islet organoids as 3-dimensional stacks with a 3-µm step size. ZsGreen fluorescent protein was excited at 488 nm, and the fluorescence was detected at 505–535 nm, mCherry was excited at 561 nm, and fluorescence was detected at 580–650 nm. Backscatter signal (reflection) from the 561-nm excitation was collected at 555–565 nm.

4.7. Leptin Secretion from Islet Organoids In Vitro

Islet organoids were cultured in vitro in suspension culture dishes in RPMI medium. For glucose-stimulated leptin secretion measurements the organoids were placed into 4-well-plate (Nunc, Thermo Fisher Scientific, Waltham, MA, USA) suspension dishes containing RPMI medium with 3 mM glucose for 1.5 h. After that, the organoids were first placed into new wells containing medium with 3 mM glucose for 30 min and then later into new wells containing 16 mM glucose for 30 min. Then the organoids were returned into suspension culture dishes containing normal RPMI medium for continued culture. The medium from the 4-well-plate wells was collected into Eppendorf tubes, centrifuged for 1 min to pellet possible organoid fragments. The supernatants were collected into fresh tubes and kept at −20 °C until analysis.

4.8. Transplantation of Islet Organoids into the ACE

The islet organoids were transplanted into the ACE of three (first set of experiments) and seven (second set of experiments) *ob/ob* recipients, using a technique previously described in [23]. Briefly, under anesthesia, organoids were transplanted into the ACE with a glass cannula after generating a puncture in the cornea with a 27-gauge needle. Great care was taken to avoid bleeding and damage to the iris. Each mouse of the transplanted group received 100 organoids/eye. Mice were injected subcutaneous with Temgesic (0.1 mL/kg; RB Pharmaceuticals, Berkshire, UK) for postoperative analgesia. Three *ob/ob* mice (first set of experiments) and five *ob/ob* mice (second set of experiments) were used as controls and were not transplanted.

4.9. In Vivo Imaging of Intraocular Islet Organoid Grafts

Islet organoid grafts were imaged in vivo, beginning 9 weeks after transplantation. An upright laser scanning confocal microscope (TCS SP5, Leica Microsystems, Wetzlar, Germany), equipped with a long-distance, water-dipping objective (HXC-APO10x/0.30 NA, Leica Microsystems, Wetzlar, Germany) and a custom-built stereotaxic head holder, allowing positioning of the mouse eye containing the engrafted islets toward the objective, was used. Viscotears (Thea Nordic, Örebro, Sweden) was used as an immersion liquid between the eye and the objective, and isoflurane was used to anesthetize the mice during in vivo imaging. Grafts were imaged as 3-dimensional stacks with 3-µm step size. ZsGreen fluorescent protein was excited at 488 nm, and the fluorescence was detected at 505–535 nm. mCherry was excited at 561 nm, and fluorescence was detected at 580–650 nm. Backscatter signal (reflection) from the 561-nm excitation was collected at 555–565 nm. After imaging, the mice were allowed to recover from anesthesia. Additionally, beginning 6 weeks after transplantation, overview images of the grafts were obtained using a digital camera connected to a Leica M60 stereomicroscope (Leica Microsystems, Wetzlar, Germany) while the mice were under anesthesia.

4.10. Doxycycline Treatment of Animals

Three animals of the transplanted group were treated with doxycycline (Clontech, Takara, Mountain View, CA, USA) in order to stop leptin production from the islet organoid grafts. Sterile doxycycline hydrochloride, dissolved in PBS, was administered intraperitoneally 5 times over 10 days (50 µg/kg/mouse), starting on day 53 after stopping leptin treatment.

4.11. Body Weight and Fasting and Non-Fasting Blood Glucose

Body weight was determined before 9 am with full access to food. Fasting blood glucose was measured after 6 h denial of food. Non-fasting blood glucose was measured at 4 pm with full access to food.

4.12. Intraperitoneal Glucose Tolerance Test (ipGTT)

To determine glucose tolerance, blood glucose levels were measured in mice that were unfed for 6 h at basal state (0 min) and at 10, 30, 60, and 120 min after glucose injection (2 g/kg bodyweight intraperitoneally, dissolved in PBS). The results were depicted as the area under the curve (AUC) of the ipGTT. Glucose concentrations were measured with the Accu-Chek Aviva monitoring system (F. Hoffmann-La Roche, Basel, Switzerland). If blood glucose concentrations exceeded the detection limit of the glucose meter, the limit value of 33.3 mM was used as value for the determination of the AUC of the ipGTT.

4.13. Plasma and Aqueous Humor Samples

Blood samples were taken in the mornings from non-fasted animals and collected into Microvette CB300 EDTA/PK100 tubes (Sarstedt, Nürnbrecht, Germany), centrifuged to gain blood plasma, and preserved at −20 °C until use. Aqueous humor samples were obtained at the end of the experiment and kept at −20 °C until use.

4.14. Insulin and C-Peptide Measurements

Ultrasensitive mouse ELISA kits (Crystal Chem, Elk Grove Village, IL, USA) were used to analyze insulin and C-peptide levels in the plasma.

4.15. Leptin Measurements

Leptin was measured in cell culture medium, blood plasma, and aqueous humor samples using the Mouse/Rat Leptin Quantikine ELISA Kit (R&D Systems, Minneapolis, MN, USA).

4.16. Tissue Extraction and Sectioning

Mice were anesthetized with isoflurane and sacrificed by cervical dislocation. Eyes were extracted and fixed with 4% paraformaldehyde for 1 week. Before cryopreservation, the eyes were processed with a sucrose gradient (10–30% (wt/vol) sucrose in PBS containing 0.01% (wt/vol) sodium azide and 0.02% (wt/vol) bacitracin, embedded in OCT-Compound (Tissue-Tek, Sakura Finetek, Torrance, CA, USA), frozen on dry ice, and preserved at −80 °C until use. Then, 20 μm thick cryosections of the anterior part of the eye were collected on SuperFrost Plus microscope slides (VWR International, Radnor, PA, USA) and kept at −20 °C until use.

4.17. Immunofluorescence in Eye Sections

For immunostaining, eye sections were equilibrated to room temperature, washed, blocked, and incubated with the primary antibodies, goat anti m-leptin (R&D Systems, Minneapolis, MN, USA) and rabbit anti-C-peptide (Cell Signaling, Danvers, MA, USA) in the presence of 0.1% Triton X-100 and 10% serum. After washing, secondary antibodies, anti-goat Alexa633, and anti-rabbit Alexa594, respectively (Thermo Fisher Scientific, Waltham, MA, USA), were applied, and mounting with medium containing DAPI for nuclear counterstaining (Thermo Fisher Scientific, Waltham, MA, USA) was performed after repeated washing. Imaging was performed using a confocal laser scanning microscope (Leica TCS SP8, Leica Microsystems, Wetzlar, Germany) with the following settings: DAPI excitation 405 nm, detection 450–470 nm; ZsGreen excitation 488 nm, detection 500–525 nm; mCherry excitation 548 nm, detection 560–580 nm; Alexa 594 (C-peptide staining) excitation 594 nm, detection 600–620 nm and Alexa 633 (leptin staining) excitation 633 nm, detection 640–680 nm. To avoid spectral overlap imaging was performed using in-between-frames sequential imaging.

4.18. Statistics

The values are expressed as means ± SEM. A 2-sided, unpaired t test was used to determine statistical significance among different treatment groups. Statistical significance was determined as follows: * $p < 0.05$, ** $p < 0.01$ and *** $p < 0.001$. Origin 2015 64-bit (OriginLab, Northampton, MA, USA) and Excel (Microsoft, Redmond, WA, USA) were used for statistical analyses.

5. Conclusions

Our data provide a proof-of-concept that genetically engineered islet organoids can serve as production sites for blood-born proteins/peptides.

Author Contributions: The authors contributed to the study in the following matter: conceptualization, P.-O.B., I.B.L., B.L. and T.M.; methodology, T.M., M.P., B.L., M.V. and I.V.-A.; validation, T.M., I.V.-A., B.L. and I.B.L.; formal analysis, T.M. and I.V.-A.; investigation, I.B.L., B.L. and T.M.; resources, P.-O.B. and I.B.L.; data curation, T.M. and I.V.-A.; writing—original draft preparation, B.L. and I.B.L.; writing—review and editing, P.-O.B., I.B.L. and B.L.; visualization, T.M. and B.L.; supervision, B.L. and I.B.L.; project administration, I.B.L.; funding acquisition: P.-O.B. and I.B.L. All authors have read and agreed to the published version of the manuscript.

Funding: This research was funded by Karolinska Institutet, Swedish Research Council, The Erling-Persson Family Foundation, the Novo Nordisk Foundation, the Stichting af Jochnick Foundation, Diabetesfonden (Sweden), Scandia Insurance Company Limited, the Diabetes Research and Wellness Foundation, the Berth von Kantzows Foundation, the Strategic Research Program in Diabetes at Karolinska Institutet, the European Research Council grant number ERC-2018-AdG 834860 EYELETS, the Swedish Foundation for Strategic Research and the Knut and Alice Wallenbergs Foundation.

Institutional Review Board Statement: The study was conducted according to the guidelines of the Declaration of Helsinki, and the animal experiments were approved by the Institutional Review Board (or Ethics Committee) of KAROLINSKA INSTITUTET (ethical permit numbers N34/16, approved on 23 February 2016 and 19462–2017, approved on 8 February 2018).

Informed Consent Statement: Not applicable.

Conflicts of Interest: P.-O.B. is cofounder and CEO of the biotech company Biocrine AB. B.L. and I.B.L. are consultants for the same company. T.M., M.P., M.V. and I.V.A. have no relationships or activities that might bias, or be perceived to bias, this work. The funders had no role in the design of the study; in the collection, analyses, or interpretation of data; in the writing of the manuscript, or in the decision to publish the results.

Appendix A

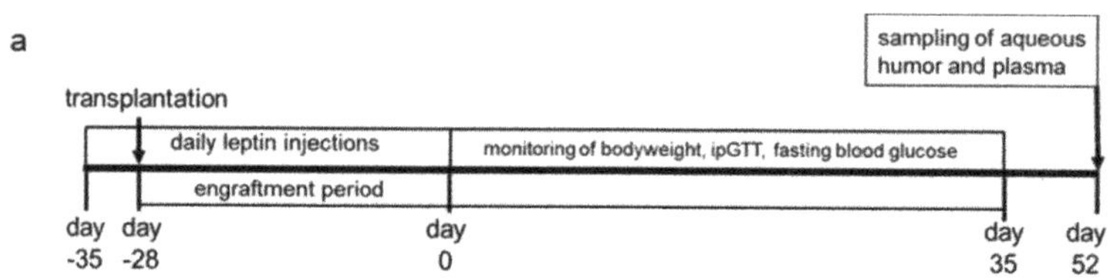

Figure A1. *Cont.*

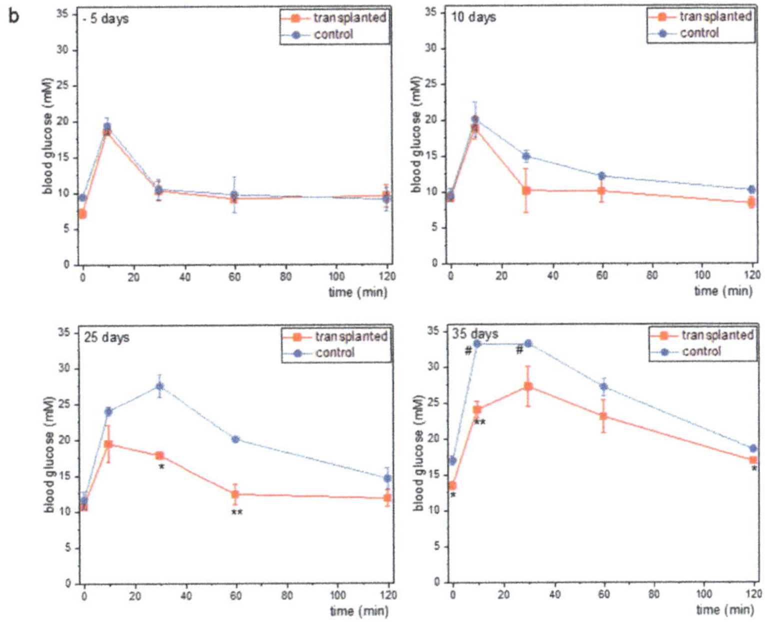

Figure A1. Additional information related to Figure 2. (**a**) Timeline of the first set of experiments. (**b**) Individual traces from the ipGTTs summarized as AUC in Figure 2d. Data are expressed as mean ± SEM. Data points that include saturated glucose measurements that were set to 33.3 mM are labeled with #. * $p < 0.05$; ** $p < 0.01$; $n = 3$.

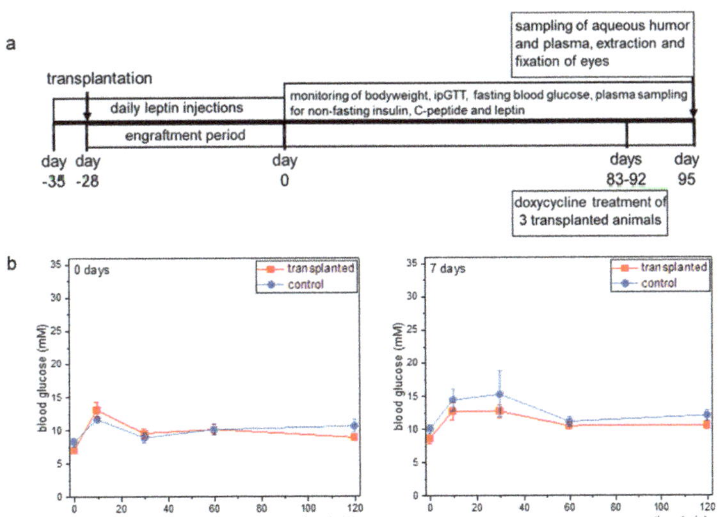

Figure A2. *Cont.*

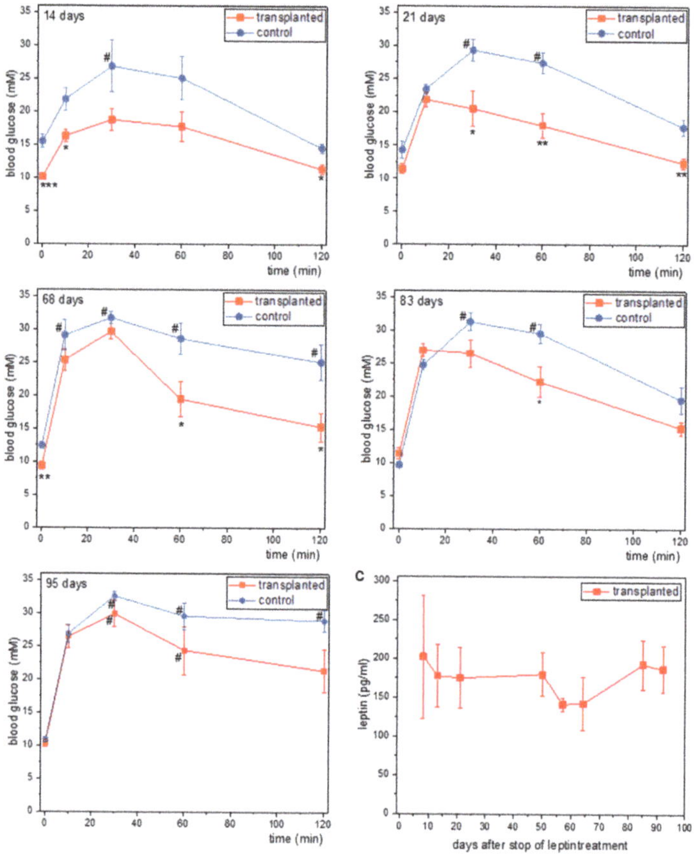

Figure A2. Additional information related to Figure 3. (**a**) Timeline of the second set of experiments. (**b**) Individual traces from the ipGTTs summarized as AUC in Figure 3d. Data are expressed as mean ± SEM. Data points that include saturated glucose measurements are labeled with #. * $p < 0.05$; ** $p < 0.01$; *** $p < 0.001$; $n = 7$ for transplanted and 5 for control group. (**c**) Plasma leptin levels of transplanted mice over the period of the experiment, $n = 3$–7.

References

1. Speier, S.; Nyqvist, D.; Cabrera, O.; Yu, J.; Molano, R.D.; Pileggi, A.; Moede, T.; Köhler, M.; Wilbertz, J.; Leibiger, B.; et al. Noninvasive in vivo imaging of pancreatic islet cell biology. *Nat. Med.* **2008**, *14*, 574–578. [CrossRef] [PubMed]
2. Leibiger, I.B.; Berggren, P.-O. Intraocular in vivo imaging of pancreatic islet cell physiology/pathology. *Mol. Metab.* **2017**, *6*, 1002–1009. [CrossRef] [PubMed]
3. Van Krieken, P.P.; Voznesenskaya, A.; Dicker, A.; Xiong, Y.; Park, J.H.; Lee, J.I.; Ilegems, E.; Berggren, P.-O. Translational assessment of a genetic engineering methodology to improve islet function for transplantation. *EBioMedicine* **2019**, *45*, 529–541. [CrossRef] [PubMed]
4. Mir-Coll, J.; Moede, T.; Paschen, M.; Neelakandhan, A.; Valladolid-Acebes, I.; Leibiger, B.; Biernath, A.; Ämmälä, C.; Leibiger, I.; Yesildag, B.; et al. Human Islet Microtissues as an In Vitro and an In Vivo Model System for Diabetes. *Int. J. Mol. Sci.* **2021**, *22*, 1813. [CrossRef] [PubMed]
5. Lindström, P. The Physiology of Obese-Hyperglycemic Mice [*ob/ob*Mice]. *Sci. World J.* **2007**, *7*, 666–685. [CrossRef]
6. Pelleymounter, M.A.; Cullen, M.J.; Baker, M.B.; Hecht, R.; Winters, D.; Boone, T.; Collins, F. Effects of the obese gene product on body weight regulation in ob/ob mice. *Science* **1995**, *269*, 540–543. [CrossRef]
7. Halaas, J.L.; Boozer, C.; Blair-West, J.; Fidahusein, N.; Denton, D.; Friedman, J.M. Physiological response to long-term peripheral and central leptin infusion in lean and obese mice. *Proc. Natl. Acad. Sci. USA* **1997**, *94*, 8878–8883. [CrossRef]
8. Harris, R.B.S.; Zhou, J.; Redmann, S.M.; Smagin, G.N.; Smith, S.R.; Rodgers, E.; Zachwieja, J.J. A Leptin Dose-Response Study in Obese (*ob/ob*) and Lean (+/?) Mice. *Endocrinology* **1998**, *139*, 8–19. [CrossRef] [PubMed]

9. Morsy, M.A.; Gu, M.C.; Zhao, J.Z.; Holder, D.J.; Rogers, I.T.; Pouch, W.J.; Motzel, S.L.; Klein, H.J.; Gupta, S.K.; Liang, X.; et al. Leptin gene therapy and daily protein administration: A comparative study in the *ob/ob* mouse. *Gene Ther.* **1998**, *5*, 8–18. [CrossRef] [PubMed]
10. Dhillon, H.; Ge, Y.-L.; Minter, R.; Prima, V.; Moldawer, L.; Muzyczka, N.; Zolotukhin, S.; Kalra, P.; Kalra, S. Long-term differential modulation of genes encoding orexigenic and anorexigenic peptides by leptin delivered by rAAV vector in *ob/ob* mice. *Regul. Pept.* **2000**, *92*, 97–105. [CrossRef]
11. Oosman, S.N.; Lam, A.W.; Harb, G.; Unniappan, S.; Lam, N.T.; Webber, T.; Bruch, D.; Zhang, Q.-X.; Korbutt, G.S.; Kieffer, T.J. Treatment of Obesity and Diabetes in Mice by Transplant of Gut Cells Engineered to Produce Leptin. *Mol. Ther.* **2008**, *16*, 1138–1145. [CrossRef] [PubMed]
12. Larcher, F.; Del Rio, M.; Serrano, F.; Segovia, J.C.; Ramirez, A.; Meana, A.; Page, A.; Abad, J.L.; González, M.A.; Bueren, J.; et al. A cutaneous gene therapy approach to human leptin deficiencies: Correction of the murine *ob/ob* phenotype using leptin-targeted keratinocyte grafts. *FASEB J.* **2001**, *15*, 1529–1538. [CrossRef] [PubMed]
13. D'Souza, A.M.; Neumann, U.H.; Glavas, M.M.; Kieffer, T.J. The glucoregulatory actions of leptin. *Mol. Metab.* **2017**, *6*, 1052–1065. [CrossRef] [PubMed]
14. Ilegems, E.; Dicker, A.; Speier, S.; Sharma, A.; Bahow, A.; Edlund, P.K.; Leibiger, I.B.; Berggren, P.-O. Reporter islets in the eye reveal the plasticity of the endocrine pancreas. *Proc. Natl. Acad. Sci. USA* **2013**, *110*, 20581–20586. [CrossRef]
15. Kulkarni, R.N.; Wang, Z.L.; Wang, R.M.; Hurley, J.D.; Smith, D.M.; Ghatei, M.A.; Withers, D.J.; Gardiner, J.V.; Bailey, C.J.; Bloom, S.R. Leptin rapidly suppresses insulin release from insulinoma cells, rat and human islets and, in vivo, in mice. *J. Clin. Investig.* **1997**, *100*, 2729–2736. [CrossRef]
16. Åvall, K.; Ali, Y.; Leibiger, I.B.; Leibiger, B.; Moede, T.; Paschen, M.; Dicker, A.; Daré, E.; Köhler, M.; Ilegems, E.; et al. Apolipoprotein CIII links islet insulin resistance to β-cell failure in diabetes. *Proc. Natl. Acad. Sci. USA* **2015**, *112*, E2611–E2619. [CrossRef] [PubMed]
17. DiSilvestro, D.J.; Melgar-Bermudez, E.; Yasmeen, R.; Fadda, P.; Lee, L.J.; Kalyanasundaram, A.; Gilor, C.L.; Ziouzenkova, O. Leptin Production by Encapsulated Adipocytes Increases Brown Fat, Decreases Resistin, and Improves Glucose Intolerance in Obese Mice. *PLoS ONE* **2016**, *11*, e0153198. [CrossRef]
18. Muzzin, P.; Eisensmith, R.C.; Copeland, K.C.; Woo, S.L.C. Correction of obesity and diabetes in genetically obese mice by leptin gene therapy. *Proc. Natl. Acad. Sci. USA* **1996**, *93*, 14804–14808. [CrossRef]
19. Kim, A.; Miller, K.; Jo, J.; Kilimnik, G.; Wojcik, P.; Hara, M. Islet architecture: A comparative study. *ISLETS* **2009**, *1*, 129–136. [CrossRef]
20. Yang, Y.; Nunes, F.A.; Berencsi, K.; Furth, E.E.; Gonczol, E.; Wilson, J. Cellular immunity to viral antigens limits E1-deleted adenoviruses for gene therapy. *Proc. Natl. Acad. Sci. USA* **1994**, *91*, 4407–4411. [CrossRef]
21. Tripathy, S.K.; Black, H.B.; Goldwasser, E.; Leiden, J.M. Immune responses to transgene–encoded proteins limit the stability of gene expression after injection of replication–defective adenovirus vectors. *Nat. Med.* **1996**, *2*, 545–550. [CrossRef]
22. Paschen, M.; Moede, T.; Valladolid-Acebes, I.; Leibiger, B.; Moruzzi, N.; Jacob, S.; García-Prieto, C.F.; Brismar, K.; Leibiger, I.B.; Berggren, P.-O. Diet-induced β-cell insulin resistance results in reversible loss of functional β-cell mass. *FASEB J.* **2019**, *33*, 204–218. [CrossRef] [PubMed]
23. Speier, S.; Nyqvist, D.; Köhler, M.; Caicedo, A.; Leibiger, I.B.; Berggren, P.-O. Noninvasive high-resolution in vivo imaging of cell biology in the anterior chamber of the mouse eye. *Nat. Protoc.* **2008**, *3*, 1278–1286. [CrossRef] [PubMed]

metabolites
MDPI

Article

Knockout of the Amino Acid Transporter SLC6A19 and Autoimmune Diabetes Incidence in Female Non-Obese Diabetic (NOD) Mice

Matthew F. Waters [1,2], Viviane Delghingaro-Augusto [1,2], Kiran Javed [3], Jane E. Dahlstrom [1,2,4], Gaetan Burgio [2], Stefan Bröer [3] and Christopher J. Nolan [1,2,5,*]

1 Australian National University Medical School, Australian National University, Acton, ACT 2601, Australia; matt.waters@anu.edu.au (M.F.W.); viviane.augusto@anu.edu.au (V.D.-A.); jane.dahlstrom@act.gov.au (J.E.D.)
2 John Curtin School of Medical Research, Australian National University, Acton, ACT 2601, Australia; gaetan.burgio@anu.edu.au
3 Research School of Biology, Australian National University, Acton, ACT 2601, Australia; kiran.javed@anu.edu.au (K.J.); stefan.broeer@anu.edu.au (S.B.)
4 ACT Pathology, The Canberra Hospital, Canberra Health Services, Garran, ACT 2605, Australia
5 Department of Endocrinology, The Canberra Hospital, Garran, ACT 2505, Australia
* Correspondence: christopher.nolan@anu.edu.au; Tel.: +61-2-5124-4224

Citation: Waters, M.F.; Delghingaro-Augusto, V.; Javed, K.; Dahlstrom, J.E.; Burgio, G.; Bröer, S.; Nolan, C.J. Knockout of the Amino Acid Transporter SLC6A19 and Autoimmune Diabetes Incidence in Female Non-Obese Diabetic (NOD) Mice. *Metabolites* **2021**, *11*, 665. https://doi.org/10.3390/metabo11100665

Academic Editors: Melkam Kebede and Belinda Yau

Received: 6 September 2021
Accepted: 27 September 2021
Published: 29 September 2021

Abstract: High protein feeding has been shown to accelerate the development of type 1 diabetes in female non-obese diabetic (NOD) mice. Here, we investigated whether reducing systemic amino acid availability via knockout of the *Slc6a19* gene encoding the system B(0) neutral amino acid transporter AT1 would reduce the incidence or delay the onset of type 1 diabetes in female NOD mice. *Slc6a19* gene deficient NOD mice were generated using the CRISPR-Cas9 system which resulted in marked aminoaciduria. The incidence of diabetes by week 30 was 59.5% (22/37) and 69.0% (20/29) in NOD.*Slc6a19*$^{+/+}$ and NOD.*Slc6a19*$^{-/-}$ mice, respectively (hazard ratio 0.77, 95% confidence interval 0.41–1.42; Mantel-Cox log rank test: $p = 0.37$). The median survival time without diabetes was 28 and 25 weeks for NOD.*Slc6a19*$^{+/+}$ and NOD.*Slc6a19*$^{-/-}$ mice, respectively (ratio 1.1, 95% confidence interval 0.6–2.0). Histological analysis did not show differences in islet number or the degree of insulitis between wild type and Slc6a19 deficient NOD mice. We conclude that Slc6a19 deficiency does not prevent or delay the development of type 1 diabetes in female NOD mice.

Keywords: amino acid; aminoaciduria; diabetes incidence; insulitis; neutral amino acid transporter; non-obese diabetic mouse; pancreatic islet; Slc6a19 deficiency; type 1 diabetes

1. Introduction

Islet beta cell metabolic stress in response to nutrient overload has been proposed to increase the susceptibility of islet beta cells to immune destruction, accelerating the development of type 1 diabetes in genetically at risk individuals [1,2]. Islet beta cell stress leads to activation of beta cell neoantigen production and stimulation of HLA class I expression on the cell surface [3–5]. This could underlie, at least in part, the increasing incidence of type 1 diabetes within older children and adolescents [6,7]. Accordingly, approaches to limit islet beta cell metabolic stress through pharmacological means, such as by using metformin, or reducing availability of nutrient secretagogues, could prevent type 1 diabetes [8,9]. Amino acids augment glucose-stimulated insulin secretion and have the potential to increase beta cell nutrient-induced stress [8]. There is pre-clinical evidence in situations of chronic beta cell stress, including endoplasmic reticulum stress, which shows transport of amino acids into beta cells is increased and this contributes to beta cell loss through apoptosis [10].

In human studies, in which the metabolome has been interrogated for biomarkers predictive of type 1 diabetes, trajectories of various amino acid plasma concentrations have been associated with the timing of islet antibody appearance [11–13]. In one study, elevated branched-chain amino acids (BCAA) and glutamine were shown to precede the development of insulin and glutamic acid decarboxylase antibodies in children who later progressed on to develop diabetes [13]. However, these trajectories are complex, differing according to which autoantibody appears first and at which age [11–13].

The non-obese diabetic (NOD) mouse spontaneously develops autoimmune diabetes at a high rate (80–90% in females) by 4–6 months of age [14]. The feeding of a protein-enriched diet to NOD mice was shown to accelerate the development of type 1 diabetes [15]. Furthermore, metabolomic analysis of plasma samples from NOD mice revealed elevated levels of BCAA in mice that did, compared to mice that did not, develop diabetes [16].

The *SLC6A19* gene encodes the broad-spectrum neutral amino acid transporter AT1 (B^0AT1), the major epithelial transporter for neutral amino acids in the intestine and kidney [17]. Mutations in *SLC6A19* cause Hartnup disorder, which is mostly a benign condition in adults, characterised by aminoaciduria [17]. Global *Slc6a19* knockout on the C57BL/6J mouse caused marked aminoaciduria, reduced fed-state plasma amino acid levels, increased fibroblast growth factor 21 (FGF21), reduced glucose stimulated insulin secretion and improved glucose tolerance [18,19]. This suggests amino acid off-loading through *Slc6a19* deficiency reduces islet beta cell work, and consequently has the potential to reduce islet beta cell metabolic stress for prevention of both type 1 and type 2 diabetes [8,20].

The aim of this study was to determine if amino acid off-loading via *Slc6a19* knockout would reduce the incidence or delay the onset of type 1 diabetes in NOD mice.

2. Results

2.1. Female Slc6a19 Deficient NOD Mice Are Markedly Aminoaciduric

The CRISPR/Cas9 mouse construction of *Slc6a19* deficient NOD mice, as described in Materials and Methods, generated three mice with mutations in the *Slc6a19* gene, of which one (ASD707:Maggot:::F0#13) had a 7 bp deletion and a 2 bp insertion within exon 2 of a single allele. ASD707:Maggot:::F0#13 (NOD.*Slc6a19*$^{+/-}$ founder mouse) was chosen for this study. Female NOD.*Slc6a19*$^{+/+}$ and NOD.*Slc6a19*$^{-/-}$ littermates were studied. The abundance in urine of alanine, glycine, isoleucine, leucine, methionine, phenylalanine, serine, threonine and valine were all markedly increased in 6–8 week old female NOD.*Slc6a19*$^{-/-}$ compared to NOD.*Slc6a19*$^{+/+}$ mice (ranging from 90 fold increased for urinary leucine to 6000 fold increased for urinary methionine), consistent with successful construction of Slc6a19 deficient NOD mice (Figure 1a). While there was an overall trend for a reduction in plasma amino acid abundance in the NOD.*Slc6a19*$^{-/-}$ compared to NOD.*Slc6a19*$^{+/+}$, the differences were not significant (Figure 1b).

2.2. Diabetes Incidence Is Unchanged in Slc6a19 Deficient NOD Mice

The body weight of Slc6a19 deficient NOD mice was not different from that of the wild-type mice from baseline to study end (Figure 2a). NOD.*Slc6a19*$^{-/-}$ compared to NOD.*Slc6a19*$^{+/+}$ mice, however, had mildly higher fed-state plasma glucose levels at 6 weeks of age (NOD.*Slc6a19*$^{+/+}$, 5.6 $\pm$ 0.1 mmol/L; NOD.*Slc6a19*$^{-/-}$, 6.0 $\pm$ 0.2 mmol/L; mean $\pm$ SEM, $p < 0.05$), however, this difference did not track with increasing age (Figure 2b). Diabetes incidence by 30 weeks of age was 59.5% (22/37) and 69.0% (20/29) in NOD.*Slc6a19*$^{+/+}$ and NOD.*Slc6a19*$^{-/-}$ mice, respectively (hazard ratio 0.77, 95% confidence interval 0.41–1.42; Mantel-Cox log rank test: $p = 0.37$) (Figure 2c). Diabetes developed as early as 14 weeks of age in two NOD.*Slc6a19*$^{-/-}$, however, the median survival time without diabetes was 28 and 25 weeks for NOD.*Slc6a19*$^{+/+}$ and NOD.*Slc6a19*$^{-/-}$ mice, respectively (ratio 1.1, 95% confidence interval 0.6–2.0) (Figure 2c).

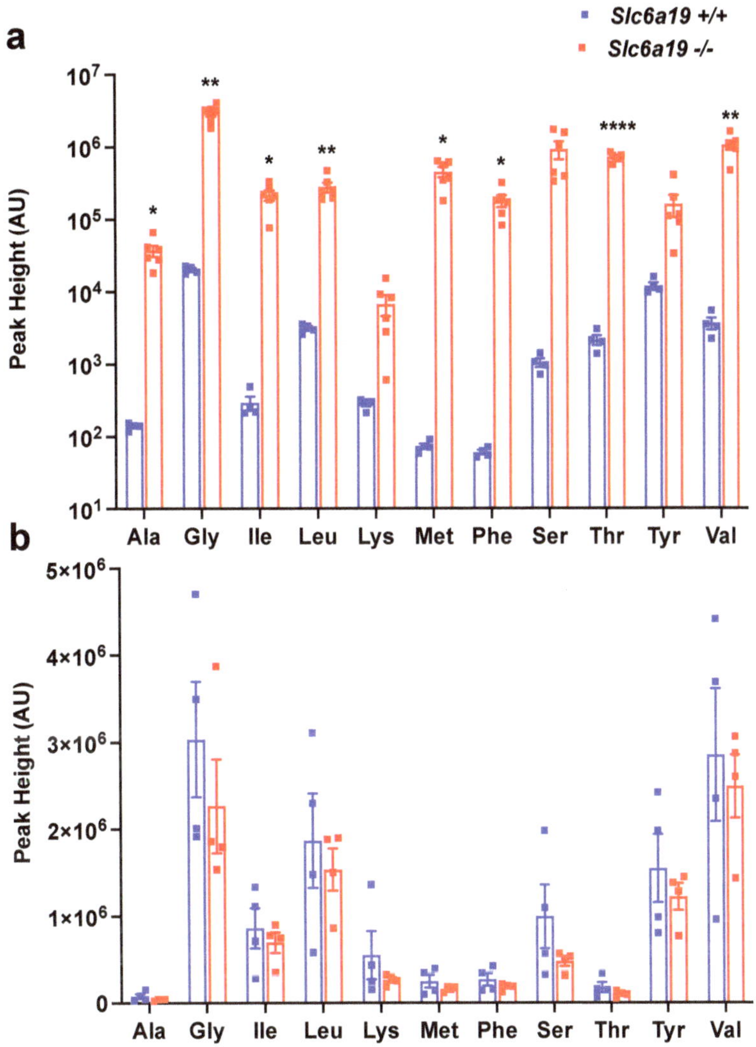

Figure 1. *Slc6a19* deficiency caused marked aminoaciduria in NOD mice. (**a**) Amino acid abundance measured in urine normalised to osmolality. (**b**) Plasma amino acid abundance. Samples were collected in the morning from mice 6–8 weeks of age, stored at −80 °C. Amino acid abundance is presented as peak height in arbitrary units (AU). All results are presented as mean ± SEM; * $p < 0.05$, ** $p < 0.01$, **** $p < 0.0001$ *Slc6a19*$^{+/+}$ versus *Slc6a19*$^{-/-}$; n = 4–6 mice/genotype.

2.3. Insulitis Severity Was Unaltered by Slc6a19 Deficiency in NOD Mice

All 37 pancreases of the NOD.*Slc6a19*$^{+/+}$ mice and 26 of 29 pancreases of NOD.*Slc6a19*$^{-/-}$ were available for analysis. All available pancreases were used for analysis of islet number per pancreas section. As 3 pancreases of the NOD.*Slc6a19*$^{+/+}$ had no identifiable islets on the analysed section, these pancreases were excluded from scoring of insulitis. The number of islets per pancreas section did not differ between the mouse genotypes (NOD.*Slc6a19*$^{+/+}$, 6 (4–9) islets/section; NOD.*Slc6a19*$^{-/-}$, 5 (4–9) islets per section; median (interquartile (IQ) range), $p = 0.94$) (Figure 2d). The severity of insulitis also did not differ with 19 of 34 (55.9%) and 11 of 26 (42.3%) of the NOD.*Slc6a19*$^{+/+}$ and NOD.*Slc6a19*$^{-/-}$ mice, respectively, having ≥50% of islets/pancreas section scoring at least grade 3 insulitis (chi-square test: $p = 0.30$).

The average percentage of islets within each insulitis score category is shown in Figure 2e. As expected, the mice that developed diabetes, irrespective of mouse genotype, had a reduced number of islets per pancreas section (non-diabetic, 8.5 (4.5–15) islets/section; diabetic, 5 (3–6) islets/section; median (IQ range), $p < 0.005$) (Figure 2d) and a higher percentage of islets/pancreas section with at least grade 3 insulitis (non-diabetic, 7 of 22 (31.8%); diabetic, 23 of 38 (60.5%); $p < 0.05$).

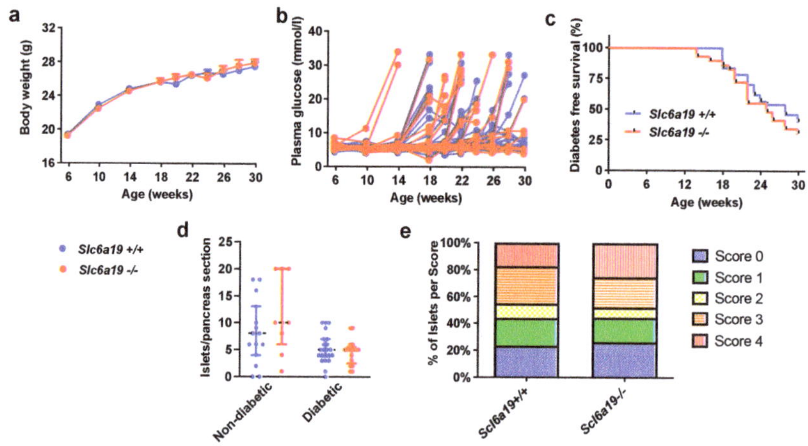

Figure 2. Slc6a19 deficiency in NOD mice does not alter body weight, incidence or timing of diabetes development, or severity of insulitis. (**a**) Body weight (g) and (**b**) fed-state plasma glucose (mmol/L) measured in the morning from weeks 6–30 of age, or until the time of diabetes diagnosis. Data are presented as mean ± SEM; *Slc6a19*$^{+/+}$, $n = 37$; *Slc6a19*$^{-/-}$, $n = 29$. (**c**) Diabetes free survival from weeks 6–30 of age. *Slc6a19*$^{+/+}$, $n = 37$; *Slc6a19*$^{-/-}$, $n = 29$; Mantel-Cox log rank test: $p = 0.37$. (**d**) Total number of islets per pancreas section according to genotype and diabetes status. *Slc6a19*$^{+/+}$, $n = 37$ ($n = 15$ non-diabetic, $n = 22$ diabetic); *Slc6a19*$^{-/-}$, $n = 26$ ($n = 9$ non-diabetic, $n = 17$ diabetic). (**e**) Number of islets at each grade of insulitis (grades 0–4) according to mouse genotype. *Slc6a19*$^{+/+}$, $n = 34$; *Slc6a19*$^{-/-}$, $n = 26$.

3. Discussion

The major finding is that neutral amino acid off-loading by knockout of the amino acid transporter gene *Slc6a19* in female NOD mice did not delay or reduce the incidence of autoimmune diabetes development. Approximately two-thirds of the female mice developed diabetes by 30 weeks of age in both the NOD.*Slc6a19*$^{+/+}$ and NOD.*Slc6a19*$^{-/-}$ mice groups. Furthermore, there were no differences in the severity of insulitis between the two genotypes. Considering that a protein enriched diet has previously been shown to accelerate the onset of diabetes in female NOD mice [14], it might have been expected that reducing the amino acid load would at least delay the onset of diabetes, but this was not seen. Untested is whether the NOD.*Slc6a19*$^{-/-}$ would have prevented the acceleration of autoimmune diabetes if they were fed a high protein diet.

The phenotype of NOD global *Slc6a19* knockout mice in this study had similarities and differences to C57Bl6J global *Slc6a19* knockout mice [18,19]. *Slc6a19* knockout induced marked aminoaciduria in mice of both backgrounds, but growth restriction, with reduced birth weight and subsequent weight gain, was seen only in C57Bl6J.*Slc6a19*$^{-/-}$ mice [18]. Interestingly, fasting glucose tended to be slightly higher in C57Bl6J.*Slc6a19*$^{-/-}$ mice at two months of age, which is similar to the mildly elevated fed-state glucose levels at six weeks of age in NOD.*Slc6a19*$^{-/-}$ mice. There was a trend for lower plasma amino acid abundance in the NOD.*Slc6a19*$^{-/-}$ compared to NOD.*Slc6a19*$^{+/+}$ mice, with two of four mice of the NOD.*Slc6a19*$^{+/+}$ having much higher levels than any of the other mice. Considering that B^0AT1 (SLC6A19) is the major transporter of neutral amino acids at the intestinal epithelium and absorbs the bulk of diet-derived neutral amino acids from the intestinal lumen [17], it is highly probable that post-prandial amino acid level rises were curtailed in the NOD.*Slc6a19*$^{-/-}$ mice, consistent with the absence of higher plasma amino acids in NOD.*Slc6a19*$^{-/-}$ mice sampled. In C57Bl6J.*Slc6a19*$^{-/-}$ compared to C57Bl6J.*Slc6a19*$^{+/+}$ mice, we previously showed that plasma levels of neutral amino acids, including the BCAAs valine, isoleucine, and leucine, were significantly reduced in Slc6a19 deficient mice after a high protein meal, indicating that amino acid malabsorption is a characteristic of Slc6a19 deficient mice [19].

While the NOD mouse is an established model for type 1 diabetes, it does not accurately reflect all the characteristics of human type 1 diabetes [21]. It is becoming clearer that human type 1 diabetes is a heterogenous condition, whether it be by the presence or absence of overweight/obesity, genetic susceptibility factors, order of appearance of islet antibodies, severity of insulitis progression, and age at diagnosis [22,23]. Thus, it remains possible that SLC6A19 could become a therapeutic target for certain subgroups at high risk of type 1 diabetes, such as those who are overweight or have overlapping genetic risk for type 1 and type 2 diabetes.

In conclusion, Slc6a19 deficiency did not reduce the high incidence of insulitis and type 1 diabetes development in female NOD mice. The research focus on SLC6A19 as a potential therapeutic target in diabetes should be for type 2 diabetes for now [17,18], with the possibility of reconsidering it as a target for type 1 diabetes as we learn more.

4. Materials and Methods

4.1. Mouse Model

NOD/ShiLtJArc mice (referred to as NOD mice herein) were sourced from the Animal Resource Centre (Canning Vale, WA, Australia). All mice were housed in a temperature and humidity controlled environment on a 12 h light/dark cycle. NOD mice globally deficient in a functional *Slc6a19* gene were generated using the CRISPR-Cas9 system and were genotyped using a PCR assay. Female NOD.*Slc6a19*$^{+/+}$ and NOD.*Slc6a19*$^{-/-}$ littermates were fed standard rodent chow diet (Gordon's Specialty Stockfeed, NSW, Australia) ad libitum from 6 to 30 weeks of age or until diabetes diagnosis (plasma glucose $\geq$ 20 mmol/L for 2 consecutive days) at which time the mice were ethically culled. All mice experiments were approved by the Animal and Experimentation Ethics Committee of the Australian National University (protocols: A2017/25 and A2017/44).

4.2. Generation of CRISPR/Cas9 Edited Slc6a19 Deficient Mouse Strain

For SpCas9 editing, two single guide RNAs (sgRNAs) were designed, guide 1 targeting exon 1, (5′- CCTGGCCTAGAGGAGCGGATTCC -3′) and guide 2 targeting exon 2, (5′- CATCGGTCAGAGGCTACGCAAGG -3′) of Slc6a19. The methodology for single guide RNA synthesis, as gBlock gene fragments (Intergrated DNA Techonologies, Baulkham Hills, NSW, Australia), is as previously described [24].

NOD studs and recipient mice (Animal Resource Centre, Perth, WA, Australia) were maintained at the Australian Phenomics Facility under specific pathogen–free conditions. Seven-wk-old NOD females were superovulated then mated with NOD males. Fertilized zygotes were collected from the oviduct and maintained under M16 medium (M7292; Sigma-Aldrich, St. Louis, MO, USA) overlaid with mineral oil. A mix of 50 ng/µL of SpCas9 protein (PNA Bio, Thousand Oaks, CA, USA), *Slc6a19*-purified gBlocks plasmids (guide 1, 5 ng/µL and guide 2, 5 ng/µL) were co-injected into the mouse zygotes. Pronuclear injections were performed as described [24], and microinjected zygotes were cultured overnight at 37 °C in a 5% CO_2 incubator and then surgically transferred at two-cell stage into the ampulla of NOD pseudo-pregnant females.

Eight mice were Sanger sequenced, of which three revealed *Slc6a19* gene allele indels as outlined in the Supplementary Material. The ASD707:Maggot:::F0#13 mouse was found to be wild type for both alleles targeted by guide 1 (exon 1), but had a 7bp deletion and 2bp insertion in one of two alleles targeted by guide 2 (exon 2) (wild type allele, 5′- CATCG-

GTCAGAGGCTACGCAAGG -3′; mutated allele 5′- CATCGGTCAG——**TG**GCAAGG -3′). The ASD707:Maggot:::F0#13 mouse was chosen as the NOD *Slc6a19*$^{+/-}$ founder mouse for this work. Following back-crossing of the founder mouse to NOD for 5 generations, keeping the compound 7 bp deletion and 2 bp insertion heterozygous *Slc6a19* mutation, marked aminoaciduria was confirmed in the phenotype of *Slc6a19*$^{-/-}$ offspring. Genotyping of mice was performed using DNA extracted from ear punch samples.

4.3. Body Weight, Glucose Measurement and Amino Acid Analyses

Serial measurements of body weight and 9 AM fed-state tail vein plasma glucose, using a glucose meter (StatStrip Xpress^{TM}, Nova Biomedical, Flintshire, United Kingdom), were performed four-weekly (weeks 6–18) and fortnightly (weeks 18–30), or up until diabetes diagnosis. Urine and plasma samples were collected from NOD.*Slc6a19*$^{+/+}$ and NOD.*Slc6a19*$^{-/-}$ mice between 6–8 weeks of age before the development of diabetes in any mice. Urine samples were standardised by osmolality. Amino acid analyses were performed using a gas chromatography mass spectroscopy (GC-MS) method, as previously described [19].

4.4. Pancreas Histology

All mice were euthanased by cervical dislocation at the end of the experimental period (upon the diagnosis of diabetes or at week 30). Pancreases were quickly excised and fixed in 10% neutral buffered formalin solution before embedding in paraffin using routine laboratory protocols. Histological analysis was conducted on 4 µm-thick sections stained with haematoxylin and eosin. Islets per whole pancreas section were counted and individually graded for insulitis using a method previously described with scores of 0 = no insulitis, 1 = peri-islet insulitis, 2 = intermediate insulitis, 3 = intra-islet insulitis and 4 = complete islet insulitis [25]. Analysis was performed by two blinded observers.

4.5. Statistical Analysis

Results are presented as mean ± SEM or median with interquartile range. Repeated measures two-way ANOVA and Bonferroni post hoc tests were used to compare differences in time course data between groups and replicates, respectively (unless otherwise stated). Comparisons between two groups were performed using the t-test for parametric analyses, with *p*-value Bonferroni adjustment for multiple comparisons, or the Mann–Whitney test for non-parametric analyses. Analyses were performed using GraphPad Prism (Version 9.2.0). $p < 0.05$ was considered statistically significant. Diabetes free survival was compared between NOD.*Slc6a19*$^{+/+}$ and NOD.*Slc6a19*$^{-/-}$ groups using the log-rank (Mantel-Cox) test. Thirty-two mice per group was estimated to provide 80% power to detect an increase in diabetes free survival from 50% to 85% with $\alpha = 5\%$.

Supplementary Materials: The following are available online at https://www.mdpi.com/article/10.3390/metabo11100665/s1, CRISPR/Cas9 edited Slc6a19 deficient mouse strains sequencing results.

Author Contributions: Conceptualization, C.J.N. and S.B.; methodology, M.F.W., V.D.-A., G.B., J.E.D., S.B., C.J.N.; investigation, M.F.W., V.D.-A., K.J., G.B., J.E.D., C.J.N.; formal analysis, M.F.W. and C.J.N.; resources, G.B., J.E.D., S.B., C.J.N.; data curation, M.F.W., C.J.N.; writing—original draft preparation, M.F.W.; writing—review and editing, M.F.W., V.D.-A., K.J., G.B., J.E.D., S.B., C.J.N.; visualization M.F.W. and C.J.N.; supervision, V.D.-A., S.B., C.J.N.; project administration, C.J.N.; funding acquisition, C.J.N. All authors have read and agreed to the published version of the manuscript.

Funding: This research was funded by a project grant (APP1128442, C.J.N.) from the National Health and Medical Research Council of Australia. G.B. receives funding from the National Collaborative Research Infrastructure (NCRIS) via the Australian Phenomics Network.

Institutional Review Board Statement: All mice experiments were performed in compliance with the Australian code of practice for care and use of laboratory animals for scientific purposes and were approved by the Animal and Experimentation Ethics Committee of the Australian National University (protocols: A2017/25 and A2017/44).

Informed Consent Statement: Not applicable.

Data Availability Statement: All data is available in the Supplementary Materials and on request to the corresponding author.

Acknowledgments: We acknowledge the technical assistance of Jenna Lowe in the generation of Slc6a19 deficient NOD mice and Elaine Bean in preparation of tissue sections for histological analysis.

Conflicts of Interest: The authors declare no conflict of interest.

References

1. Wilkin, T.J. The accelerator hypothesis: Weight gain as the missing link between Type I and Type II diabetes. *Diabetologia* **2001**, *44*, 914–922. [CrossRef]
2. Fourlanos, S.; Harrison, L.C.; Colman, P.G. The accelerator hypothesis and increasing incidence of type 1 diabetes. *Curr. Opin. Endocrinol. Diabetes Obes.* **2008**, *15*, 321–325. [CrossRef] [PubMed]
3. von Scholten, B.J.; Kreiner, F.F.; Gough, S.C.L.; von Herrath, M. Current and future therapies for type 1 diabetes. *Diabetologia* **2021**, *64*, 1037–1048. [CrossRef] [PubMed]
4. Kurien, B.T.; Scofield, R.H. Autoimmunity and oxidatively modified autoantigens. *Autoimmun. Rev.* **2008**, *7*, 567–573. [CrossRef] [PubMed]
5. Eizirik, D.L.; Szymczak, F.; Alvelos, M.I.; Martin, F. From Pancreatic beta-Cell Gene Networks to Novel Therapies for Type 1 Diabetes. *Diabetes* **2021**. [CrossRef] [PubMed]
6. Lawrence, J.M.; Divers, J.; Isom, S.; Saydah, S.; Imperatore, G.; Pihoker, C.; Marcovina, S.M.; Mayer-Davis, E.J.; Hamman, R.F.; Dolan, L.; et al. Trends in Prevalence of Type 1 and Type 2 Diabetes in Children and Adolescents in the US, 2001–2017. *JAMA* **2021**, *326*, 717–727. [CrossRef] [PubMed]
7. Ferrara-Cook, C.; Geyer, S.M.; Evans-Molina, C.; Libman, I.M.; Becker, D.J.; Gitelman, S.E.; Redondo, M.J.; Type 1 Diabetes TrialNet Study Group. Excess BMI Accelerates Islet Autoimmunity in Older Children and Adolescents. *Diabetes Care* **2020**, *43*, 580–587. [CrossRef] [PubMed]
8. Nolan, C.J.; Prentki, M. The islet beta-cell: Fuel responsive and vulnerable. *Trends Endocrinol. Metab.* **2008**, *19*, 285–291. [CrossRef]
9. Wilkin, T.; Greene, S.; McCrimmon, R. Testing the accelerator hypothesis: A new approach to type 1 diabetes prevention (adAPT 1). *Diabetes Obes. Metab.* **2016**, *18*, 3–5. [CrossRef]
10. Krokowski, D.; Han, J.; Saikia, M.; Majumder, M.; Yuan, C.L.; Guan, B.J.; Bevilacqua, E.; Bussolati, O.; Broer, S.; Arvan, P.; et al. A self-defeating anabolic program leads to beta-cell apoptosis in endoplasmic reticulum stress-induced diabetes via regulation of amino acid flux. *J. Biol. Chem.* **2013**, *288*, 17202–17213. [CrossRef]
11. Li, Q.; Parikh, H.; Butterworth, M.D.; Lernmark, A.; Hagopian, W.; Rewers, M.; She, J.X.; Toppari, J.; Ziegler, A.G.; Akolkar, B.; et al. Longitudinal Metabolome-Wide Signals Prior to the Appearance of a First Islet Autoantibody in Children Participating in the TEDDY Study. *Diabetes* **2020**, *69*, 465–476. [CrossRef]
12. Pflueger, M.; Seppanen-Laakso, T.; Suortti, T.; Hyotylainen, T.; Achenbach, P.; Bonifacio, E.; Oresic, M.; Ziegler, A.G. Age- and islet autoimmunity-associated differences in amino acid and lipid metabolites in children at risk for type 1 diabetes. *Diabetes* **2011**, *60*, 2740–2747. [CrossRef] [PubMed]
13. Oresic, M.; Simell, S.; Sysi-Aho, M.; Nanto-Salonen, K.; Seppanen-Laakso, T.; Parikka, V.; Katajamaa, M.; Hekkala, A.; Mattila, I.; Keskinen, P.; et al. Dysregulation of lipid and amino acid metabolism precedes islet autoimmunity in children who later progress to type 1 diabetes. *J. Exp. Med.* **2008**, *205*, 2975–2984. [CrossRef]
14. Delovitch, T.L.; Singh, B. The nonobese diabetic mouse as a model of autoimmune diabetes: Immune dysregulation gets the NOD. *Immunity* **1997**, *7*, 727–738. [CrossRef]
15. Schneider, K.; Laube, H.; Linn, T. A diet enriched in protein accelerates diabetes manifestation in NOD mice. *Acta Diabetol.* **1996**, *33*, 236–240. [CrossRef] [PubMed]
16. Grapov, D.; Fahrmann, J.; Hwang, J.; Poudel, A.; Jo, J.; Periwal, V.; Fiehn, O.; Hara, M. Diabetes Associated Metabolomic Perturbations in NOD Mice. *Metabolomics* **2015**, *11*, 425–437. [CrossRef]
17. Seow, H.F.; Broer, S.; Broer, A.; Bailey, C.G.; Potter, S.J.; Cavanaugh, J.A.; Rasko, J.E. Hartnup disorder is caused by mutations in the gene encoding the neutral amino acid transporter SLC6A19. *Nat. Genet.* **2004**, *36*, 1003–1007. [CrossRef]
18. Jiang, Y.; Rose, A.J.; Sijmonsma, T.P.; Broer, A.; Pfenninger, A.; Herzig, S.; Schmoll, D.; Broer, S. Mice lacking neutral amino acid transporter B(0)AT1 (Slc6a19) have elevated levels of FGF21 and GLP-1 and improved glycaemic control. *Mol. Metab.* **2015**, *4*, 406–417. [CrossRef]
19. Javed, K.; Broer, S. Mice Lacking the Intestinal and Renal Neutral Amino Acid Transporter SLC6A19 Demonstrate the Relationship between Dietary Protein Intake and Amino Acid Malabsorption. *Nutrients* **2019**, *11*, 2024. [CrossRef]
20. Yadav, A.; Shah, N.; Tiwari, P.K.; Javed, K.; Cheng, Q.; Aidhen, I.S.; Broer, S. Novel Chemical Scaffolds to Inhibit the Neutral Amino Acid Transporter B(0)AT1 (SLC6A19), a Potential Target to Treat Metabolic Diseases. *Front. Pharmacol.* **2020**, *11*, 140. [CrossRef]

21. Thayer, T.C.; Wilson, S.B.; Mathews, C.E. Use of nonobese diabetic mice to understand human type 1 diabetes. *Endocrinol. Metab. Clin. North. Am.* **2010**, *39*, 541–561. [CrossRef] [PubMed]
22. Battaglia, M.; Ahmed, S.; Anderson, M.S.; Atkinson, M.A.; Becker, D.; Bingley, P.J.; Bosi, E.; Brusko, T.M.; DiMeglio, L.A.; Evans-Molina, C.; et al. Introducing the Endotype Concept to Address the Challenge of Disease Heterogeneity in Type 1 Diabetes. *Diabetes Care* **2020**, *43*, 5–12. [CrossRef] [PubMed]
23. Tuomi, T.; Santoro, N.; Caprio, S.; Cai, M.; Weng, J.; Groop, L. The many faces of diabetes: A disease with increasing heterogeneity. *Lancet* **2014**, *383*, 1084–1094. [CrossRef]
24. Cardinez, C.; Miraghazadeh, B.; Tanita, K.; da Silva, E.; Hoshino, A.; Okada, S.; Chand, R.; Asano, T.; Tsumura, M.; Yoshida, K.; et al. Gain-of-function IKBKB mutation causes human combined immune deficiency. *J. Exp. Med.* **2018**, *215*, 2715–2724. [CrossRef] [PubMed]
25. Gutierrez, D.A.; Fu, W.; Schonefeldt, S.; Feyerabend, T.B.; Ortiz-Lopez, A.; Lampi, Y.; Liston, A.; Mathis, D.; Rodewald, H.R. Type 1 diabetes in NOD mice unaffected by mast cell deficiency. *Diabetes* **2014**, *63*, 3827–3834. [CrossRef]

metabolites

MDPI

Article

Short Duration Alagebrium Chloride Therapy Prediabetes Does Not Inhibit Progression to Autoimmune Diabetes in an Experimental Model

Danielle J. Borg [1,2], Pouya Faridi [3], Kai Lin Giam [3,†], Peta Reeves [4,†], Amelia K. Fotheringham [1], Domenica A. McCarthy [1], Sherman Leung [1], Micheal S. Ward [1], Brooke E. Harcourt [5], Rochelle Ayala [3], Jean L. Scheijen [6,7], David Briskey [8], Nadine L. Dudek [3], Casper G. Schalkwijk [6,7], Raymond Steptoe [4,‡], Anthony W. Purcell [3,‡] and Josephine M. Forbes [1,5,9,*]

1 Glycation and Diabetes Complications, Mater Research Institute, The University of Queensland, Translational Research Institute, Brisbane, QLD 4102, Australia; danielle.borg@mater.uq.edu.au (D.J.B.); amelia.fotheringham@mater.uq.edu.au (A.K.F.); domenica.mccarthy@mater.uq.edu.au (D.A.M.); sherman.leung@mater.uq.edu.au (S.L.); michael.ward5@msd.com (M.S.W.); brooke.harcourt@unimelb.edu.au (B.E.H.)
2 Pregnancy and Development, Mater Research Institute, The University of Queensland, South Brisbane, QLD 4101, Australia
3 Infection and Immunity Program, Department of Biochemistry and Molecular Biology, Biomedicine Discovery Institute, Monash University, Melbourne, VIC 3800, Australia; pouya.faridi@monash.edu (P.F.); kailin.giam@thermofisher.com (K.L.G.); rochelle.ayala.perez@monash.edu (R.A.); nadinedudekart@gmail.com (N.L.D.); anthony.purcell@monash.edu (A.W.P.)
4 Tolerance and Autoimmunity Group, The University of Queensland Diamantina Institute, Translational Research Institute, Brisbane, QLD 4102, Australia; peta.zorzetto@gmail.com (P.R.); r.steptoe@uq.edu.au (R.S.)
5 Murdoch Children's Research Institute, Royal Children's Hospital, Melbourne, VIC 3052, Australia
6 Laboratory for Metabolism and Vascular Medicine, Department of Internal Medicine, Maastricht University, 6211 Maastricht, The Netherlands; j.scheijen@maastrichtuniversity.nl (J.L.S.); c.schalkwijk@maastrichtuniversity.nl (C.G.S.)
7 Cardiovascular Research Institute Maastricht, 6211 Maastricht, The Netherlands
8 School of Human Movement and Nutrition Sciences, The University of Queensland, Brisbane, QLD 4067, Australia; d.briskey@uq.edu.au
9 Mater Clinical School, The University of Queensland, Brisbane, QLD 4101, Australia
* Correspondence: josephine.forbes@mater.uq.edu.au; Tel.: +61-7-3443-7642
† These authors contributed equally.
‡ These senior authors contributed equally.

check for updates

Citation: Borg, D.J.; Faridi, P.; Giam, K.L.; Reeves, P.; Fotheringham, A.K.; McCarthy, D.A.; Leung, S.; Ward, M.S.; Harcourt, B.E.; Ayala, R.; et al. Short Duration Alagebrium Chloride Therapy Prediabetes Does Not Inhibit Progression to Autoimmune Diabetes in an Experimental Model. *Metabolites* **2021**, *11*, 426. https://doi.org/10.3390/metabo11070426

Academic Editors: Melkam Kebede and Belinda Yau

Received: 24 May 2021
Accepted: 22 June 2021
Published: 28 June 2021

Abstract: Mechanisms by which advanced glycation end products (AGEs) contribute to type 1 diabetes (T1D) pathogenesis are poorly understood. Since life-long pharmacotherapy with alagebrium chloride (ALT) slows progression to experimental T1D, we hypothesized that acute ALT therapy delivered prediabetes, may be effective. However, in female, non-obese diabetic (NOD*ShiLt*) mice, ALT administered prediabetes (day 50–100) did not protect against experimental T1D. ALT did not decrease circulating AGEs or their precursors. Despite this, pancreatic β-cell function was improved, and insulitis and pancreatic CD45.1$^+$ cell infiltration was reduced. Lymphoid tissues were unaffected. ALT pre-treatment, prior to transfer of primed GC98 CD8$^+$ T cell receptor transgenic T cells, reduced blood glucose concentrations and delayed diabetes, suggesting islet effects rather than immune modulation by ALT. Indeed, ALT did not reduce interferon-γ production by leukocytes from ovalbumin-pre-immunised NOD*ShiLt* mice and NOD*scid* recipients given diabetogenic ALT treated NOD splenocytes were not protected against T1D. To elucidate β-cell effects, NOD-derived MIN6N8 β-cell major histocompatibility complex (MHC) Class Ia surface antigens were examined using immunopeptidomics. Overall, no major changes in the immunopeptidome were observed during the various treatments with all peptides exhibiting allele specific consensus binding motifs. As expected, longer MHC Class Ia peptides were captured bound to H-2D^b than H-2K^b under all conditions. Moreover, more 10–12 mer peptides were isolated from H-2D^b after AGE modified bovine serum albumin (AGE-BSA) treatment, compared with bovine serum albumin (BSA) or AGE-BSA+ALT treatment. Proteomics of MIN6N8 cells showed enrichment of processes associated with catabolism, the immune system, cell cycling and presynaptic endocytosis with AGE-BSA compared with BSA treatments. These data show that short-term ALT intervention, given prediabetes, does not arrest experimental T1D but transiently impacts β-cell function.

Keywords: advanced glycation end products; alagebrium chloride; cross-link breaker; immunopeptidome; MIN6N8 cell line; NOD mouse; autoimmune diabetes; type 1 diabetes

1. Introduction

Advanced glycation end products (AGEs) are formed by irreversible modifications of proteins or lipids with reducing sugars [1]. AGEs are synthesized endogenously in vivo and are ingested excessively in the western diet [2], although the availability of dietary AGEs and how they contribute to pathophysiological changes in the body remains to be fully understood [3]. AGEs are postulated to be regulated by multiple scavenger receptors expressed on a number of cell types, including the pattern recognition receptor for AGEs (RAGE), which can trigger events such as cellular adhesion and spreading [4,5], migration [6–8], and cytokine production [9] via nuclear factor kappa-light-chain-enhancer of activated B cells (NF-κB) activation [3].

Circulating AGEs are independent predictors of progression to type 1 diabetes (T1D) in islet cell autoantibody-positive children [10]. A smaller study in children and adolescents with pancreatic islet autoantibodies who were first-degree relatives of individuals with T1D, showed that progression to overt diabetes correlated with high circulating concentrations of AGEs [11]. In a population based nested case-control study, polymorphisms in the gene that encode for RAGE, advanced glycosylation end product-specific receptor (*AGER*), conferred increased T1D risk and reduced circulating soluble RAGE concentrations in young children with ketoacidosis carrying the HLA-DR3/DQ4 haplotype [12]. Additional studies have confirmed that in vitro excess AGE exposure results in pancreatic β-cell dysfunction reducing glucose stimulated insulin synthesis and secretion [11,13–15], increasing apoptosis [11,13], and mitochondrial abnormalities [11,16]. In vivo, acute AGE injections provided to healthy rodents [11,13], as well as chronic dietary AGE intake [17–19], each initiated β-cell dysfunction, resulting in insulin secretory deficiencies compared to control groups [11,13,16–19]. Temporal changes in circulating AGEs and RAGE expression are apparent pre-diabetes in the mouse model for T1D, non-obese diabetic (NOD) mice [20] when compared to non-diabetic NOD*scid* mice [21].

Thus, it is not surprising that anti-AGE therapies, such as diets low in AGEs [18], or prophylactic injections of soluble RAGE, a decoy receptor for cellular RAGE, reverse insulin secretion defects, dampen pancreatic islet immune cell infiltration and delay experimental autoimmune diabetes in NOD mice [22]. We have previously demonstrated that continuous, life-long administration of the therapeutic agent alagebrium (4,5-dimethyl-3-(2-oxo2-phenylethyl)-thiazolium chloride (ALT)), thought to cleave the intermediate α-dicarbonyls compounds which form AGEs [23,24], reduced diabetes progression in the NOD*ShiLt* mouse and protected β-cell function [11].

What is currently unknown is whether effects seen with AGE-lowering strategies, such as ALT on pancreatic β-cell function, are independent of improvements in the immune system in autoimmune diabetes. Secondly, it is unclear if a shorter duration of therapy prediabetes could delay autoimmune diabetes onset. To address this, we used the NOD mouse-derived β-cell MIN6N8 cell line and NOD strains with varying susceptibility to autoimmune diabetes and examined the efficacy of short-term ALT administration prior to diabetes onset, from day 50–100 of life on diabetes development.

2. Results

2.1. Short-Term AGE Lowering Therapy with ALT Prediabetes Does Not Impact Experimental Autoimmune Diabetes Development in NODShiLt Mice

Previous studies have shown benefits of AGE-lowering therapies in arresting autoimmune diabetes development in NOD*ShiLt* mice [11]. Fifty days of daily ALT therapy from 50 to 100 days of life modestly delayed the age of diabetes onset (control 141 ± 32 days vs. ALT 171 ± 38 days, $p = 0.05$; Figure 1A), but there were no differences in disease incidence when compared to control mice by the study end (control: 11/18; alagebrium: 15/20 mice, $p = 0.5$; Figure 1A).

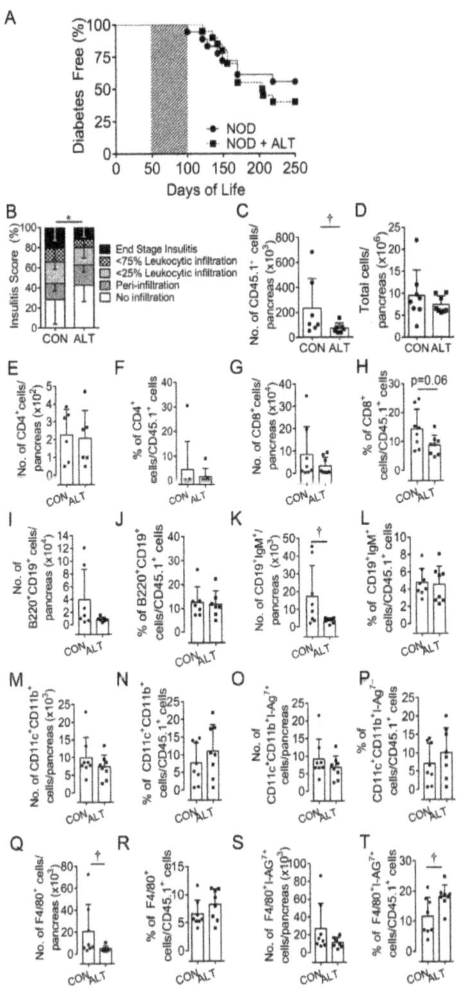

Figure 1. Short term AGE lowering therapy pre-diabetes in NOD mice does not protect against diabetes development despite attenuating β-cell infiltration. (**A**) Diabetes incidence of female NOD*ShiLt* (NOD) mice (n = 18–20/group) from day 50 to 100 of life (shaded grey) were untreated (NOD, •) or treated with the AGE lowering therapy alagebrium chloride (ALT, 1 mg/kg/day, s.c; ■). Mice were monitored daily, blood glucose tested weekly for diabetes incidence and mice were euthanised when either blood glucose concentrations exceeded 15 mmol/L for two consecutive days, day 250 of life was reached without diabetes diagnosis. (**B**) Islet immune cell infiltration at day 90 of life shown as

mean percentage of islets with no infiltrate (white), perivascular/periductal infiltrate (striped), <25% infiltrate (arrowhead), <75% infiltrate (checkerboard) or end stage insulitis (black) after no therapy (CON) or 40 days alagebrium chloride (ALT) treatment ($n = 4$ mice/group, 5 μm sections). After 30 days of ALT therapy, pancreata were digested and infiltrating immune cells quantified via surface staining for (**C**) total number of leukocyte common antigen positive ($CD45.1^+$) immune cells, (**D**) total number and proportions of (**E**,**F**) $CD4^+$ T cells, (**G**,**H**) $CD8^+$ T cells, (**I**,**J**) $CD19^+B220^+$, (**K**,**L**) $CD19^+IgM^+$ B cells, (**M**,**N**) $CD11b^+CD11c^+$, (**O**,**P**) $CD11b^+CD11c^+I\text{-}Ag7^+$, (**Q**,**R**) $F4/80^+$ and (**S**,**T**) $F4/80^+I\text{-}Ag7^+$ immune cells ($n = 7$–8 mice/group; $n = 2$ experiments). Data represented as mean ± SD. For analysis in (**C**,**D**) data was log transformed. * $p = 0.0005$ vs. NOD (Chi-square); † $p < 0.05$ vs. NOD (Student's *T* Test).

2.2. *Short-Term Therapy with ALT Increases Insulin Secretion Pre-Diabetes*

Beta cell function was examined in mice during and directly after ALT therapy. At approximately day 75 of life (after ~25 days of ALT therapy), intraperitoneal (IPGTT; Table 1; Figure S1A,B) and oral (OGTT; Table 1; Figure S1C,D) glucose tolerance tests were performed in two separate cohorts of mice. In both tests, insulin secretion across the test duration ($AUC_{INSULIN}$) was significantly greater in mice given ALT therapy (Table 1; Figure S1A–D). In addition, the first phase insulin response and action during the first 15 min of the IPGTT was higher in ALT treated mice (Table 1, $AUC_{INSULIN}$:$AUC_{GLUCOSE}$; Figure S1B). By day 107 of life, just after the completion of the ALT treatment period, circulating non-fasted insulin concentrations appeared higher in ALT treated mice when compared to controls but did not reach significance ($p = 0.06$; Table 2). During OGTT but not IPGTT, $AUC_{GLUCOSE}$ was greater in mice which were administered ALT (Table 1; Figure S1C).

Despite the slight elevation in circulating insulin concentrations over the course of ALT therapy (Table 2), blood glucose concentrations, body weight and glycated haemoglobin remained unchanged over the course of treatment (Table 2). When examined towards the end of the ALT treatment period (~40 days), pancreatic islets were a similar size to those seen in control mice (Figure S1E) and did not differ in insulin (Figure S1F,G) or proinsulin (Figure S3H,I) staining as compared to controls. Overall, these data imply that ALT may have a temporal effect on pancreatic islet function early in diabetes progression.

2.3. *AGEs and Precursor Molecules Were Unaffected by Short Term ALT Therapy*

AGE precursor molecules, the dicarbonyls methylglyoxal (MGO), glyoxal (GO), and 3-deoxyglucosone (3-DG) remained unchanged across the treatment period in NOD*ShiLt* mice receiving ALT therapy (Figure S2A–C). Although these dicarbonyls did not associate with changes in circulating insulin (data not shown), GO and 3-DG associated positively with glycated haemoglobin after 30 days of therapy (Figure S2D,E). Collectively, this data suggests that short-term, intensive alagebrium therapy does not reduce reactive dicarbonyl compounds or stop diabetes progression. There was a trend towards the AGE carboxyethyl-lysine (CEL) being reduced after ALT therapy, which was not observed with other AGE modifications such as carboxylmethyl-lysine (CML) and methylglyoxal-derived hydroimiadazolone (MG-H1) (Figure S2F).

Table 1. Metabolic parameters in pre-diabetic NOD mice following short term AGE lowering therapy. Female NOD*ShiLt* (NOD) mice from day 50 to 100 of life were untreated or treated with the AGE lowering therapy, alagebrium chloride (ALT, 1 mg/kg/day, s.c). Intraperitoneal (IP) and oral tolerance tests were performed after a 5–6 h fast in response to a 2 g/kg D-glucose bolus at ~day 75 of life (n = 5–10/group). Area under the glucose ($AUC_{GLUCOSE}$) or insulin curve ($AUC_{INSULIN}$) measurements were calculated from the IPGTT and OGTT curves (0 min–120 min; see curves presented in Figure S1). First phase $AUC_{GLUCOSE}$:$AUC_{INSULIN}$ has been taken as 0–15 min of the glucose tolerance test. Data are reported as mean ± SD. † $p = 0.002$ vs. NOD (Unpaired Student's t-test); ‡ $p < 0.0001$ vs. NOD (Mann-U Whitney); * $p = 0.02$ vs. NOD (Unpaired Student's t-test).

	IP Glucose Tolerance Test			Oral Glucose Tolerance Test			Fasting Insulin (p·mol/L)	Fasting Proinsulin (p·mol/L)
	$AUC_{GLUCOSE}$ (m·mol/L/min)	$AUC_{INSULIN}$ (n·mol/L/min)	$AUC_{INSULIN}$: $AUC_{GLUCOSE}$	$AUC_{GLUCOSE}$ (m·mol/L/min)	$AUC_{INSULIN}$ (n·mol/L/min)	$AUC_{INSULIN}$: $AUC_{GLUCOSE}$		
NOD	923.9 ± 81.4	37.3 ± 18.0	0.02 ± 0.008	903.5 ± 58.5	10.2 ± 0.6	0.03 ± 0.008	411.9 ± 391.3	14.0 ± 10.0
NOD + ALT	868.5 ± 45.5	75.8 ± 28.7 †	0.12 ± 0.04 ‡	1036.0 ± 78.9 *	11.8 ± 1.2 *	0.03 ± 0.004	398.4 ± 240.8	16.2 ± 7.6

Table 2. Biochemical parameters in pre-diabetic NOD mice following short term AGE lowering therapy. Female NOD*ShiLt* (NOD) mice from day 50 to 100 were untreated or treated with the AGE lowering therapy alagebrium chloride (ALT, 1 mg/kg/day, s.c). Non-fasting insulin concentrations and glycated haemoglobin (%) at baseline (day 50), after 30 days of treatment (day 80) and 7 days after cessation of therapy (day 107) are reported as median (IQR). Time courses were analysed using a mixed-effects model with Sidak post-hoc correction for multiple comparisons. [a] Non-fasting blood glucose concentrations and body weight are reported as slopes from linear regression (95% CI) across the study period until the end of the study period (day 250 of life).

	Non-Fasting Insulin (ng/mL)			Glycated Haemoglobin (%)			Non-Fasting Blood Glucose [a]	Body Weight [a]
	Day 50	80	107	Day 50	80	107		
NOD	0.34 (0.24–0.51)	0.31 (0.26–0.45)	0.30 (0.25–0.47)	3.12 (2.5–3.7)	4.4 (2.7–5.2)	3.8 (2.9–5.2)	0.0059 (0.002–0.010)	0.0429 (0.040–0.048)
NOD + ALT	0.37 (0.27–0.57)	0.32 (0.26–0.74)	0.44 (0.25–1.3)	2.9 (2.5–3.6)	3.4 (2.5–4.1)	3.2 (3.0–4.0)	0.0101 (0.005–0.016)	0.0393 (0.036–0.042)

2.4. Localised Pancreatic but Not Lymph Node or Splenic Immune Cells Are Modulated by Short-Term ALT Therapy

Based on the temporal effects on insulin secretion in the present study and our previous finding that long-term ALT administration reduced diabetes incidence in NOD mice [11], we hypothesized that local immunopathology within the pancreas may be altered. Semi-quantitative insulitis scoring of immune cell infiltrate after 40 days of ALT therapy confirmed reduced islet infiltration compared with control mice (Figure 1B). Whole pancreatic digests after 30 days of therapy showed a consistently lower number of CD45.1^{+} leukocytes per pancreas in ALT-treated mice as compared with untreated mice (Figure 1C). Total pancreatic cell numbers (Figure 1D) [25], did not differ between groups.

Within CD45.1^{+} cell populations in the pancreas (Figure S3A; gating strategy) the number and proportions of CD4^{+} T cells (Figure 1E,F), B220^{+}CD19^{+}B cells (Figure 1I,J), non-activated CD11b^{+}CD11c^{+} (Figure 1M,N) and activated CD11b^{+}CD11c^{+} I-Ag7^{+} conventional dendritic cells (cDC; Figure 1O,P) were similar between groups. However, the pancreatic CD45.1^{+} cell populations tended to contain a lower percentage of CD8^{+} T cells although numbers remained unchanged (Figure 1G,H) with ALT. There were fewer mature CD19^{+}IgM^{+} B cells and F4/80^{+} macrophages numbers (Figure 1K,Q) but not proportions and (Figure 1L,R), and higher proportions but not numbers of activated F4/80^{+}I-Ag7^{+} macrophages (Figure 1S,T) after ALT treatment compared to control.

Since islet antigen-specific T cells are commonly primed in the pancreatic lymph nodes by antigen-presenting cells [26,27], we next set out to verify whether the decreases in infiltrating pancreatic leukocytes with ALT were due to modulation of the local lymph node milieu. Flow cytometry analysis (Figure S3B; gating strategy), of pancreatic lymph node (pLN) cell suspensions isolated after 30–40 days of ALT therapy showed no change in the absolute numbers and proportions of CD4^{+} (Figure S4A,B) and CD8^{+} T cells (Figure S4I,J). The number and proportions of naïve (CD62^{+}CD44^{-}), effector/effector memory (CD62^{-}CD44^{+}) or central memory (CD62^{+}CD44^{+}) CD4^{+} (Figure S4C–H) or CD8^{+} (Supplementary Figure S4K–P) T cell subsets remained unchanged. Similarly, regulatory T cells (Treg; CD4^{+}CD25^{+}Foxp3^{+}), thought to play a key role in autoimmune diabetes in NOD mice [28] and T1D in humans [29], were not different between groups (Figure S4Q–R). Similar to the pancreas, CD11b^{+}CD11c^{+} cDC (Figure S5A,B), and CD11b^{-}CD11c^{+} plasmacytoid dendritic cells (Figure S5C,D) remained unchanged between groups. Unlike the pancreas, CD19^{+}B220^{+} B cells (Figure S4S,T) remained unchanged.

Flow cytometry analysis (Figure S3C; gating strategy) of cell suspensions prepared from spleens showed no change in numbers and proportions of dendritic cells (Figure S5E–H), CD4^{+} (Figure S6A–G), CD8^{+} (Figure S6I–P) T cells, Tregs (Figure S6Q,R) or B cells (Figure S6S,T) after ALT therapy, with the exception of a reduced proportion of central memory CD4^{+} T cells (Figure S6H). Collectively, these results suggest that mice exposed to short-term ALT therapy have reduced proportions of pancreatic CD8^{+} and splenic central memory CD4^{+} T cells that could be subsequent to changes in pancreatic antigen presenting cell proportions.

2.5. Pre-Treatment with ALT Delays Diabetes Development Following Adoptive Transfer of G9C8 CD8 T Cell Receptor (TCR) Transgenic Cytotoxic Lymphocytes

To further explore the effects of ALT on β-cells during diabetes development, we adoptively transferred activated, diabetogenic NOD G9C8 CD8 TCR transgenic cytotoxic T lymphocytes, that recognize aa15–23 of the insulin B chain [30], into non-diabetic NOD*ShiLt* recipient mice that were pre-treated for 30-days with or without ALT. ALT delayed diabetes onset by an average of 18 days compared to controls and halved the number of mice which developed diabetes (20% incidence with ALT pre-treatment and 40% incidence in untreated mice; Figure 2A). ALT pre-treatment decreased non-fasting blood glucose concentrations across the study duration ($p = 0.02$, Figure 2B). However, ALT pre-treatment exacerbated insulitis (Figure 2C), although this did not overtly reduce β-cell area (Figure 2D). Pancreatic expression of insulin (Figure 2E,F) remained unchanged.

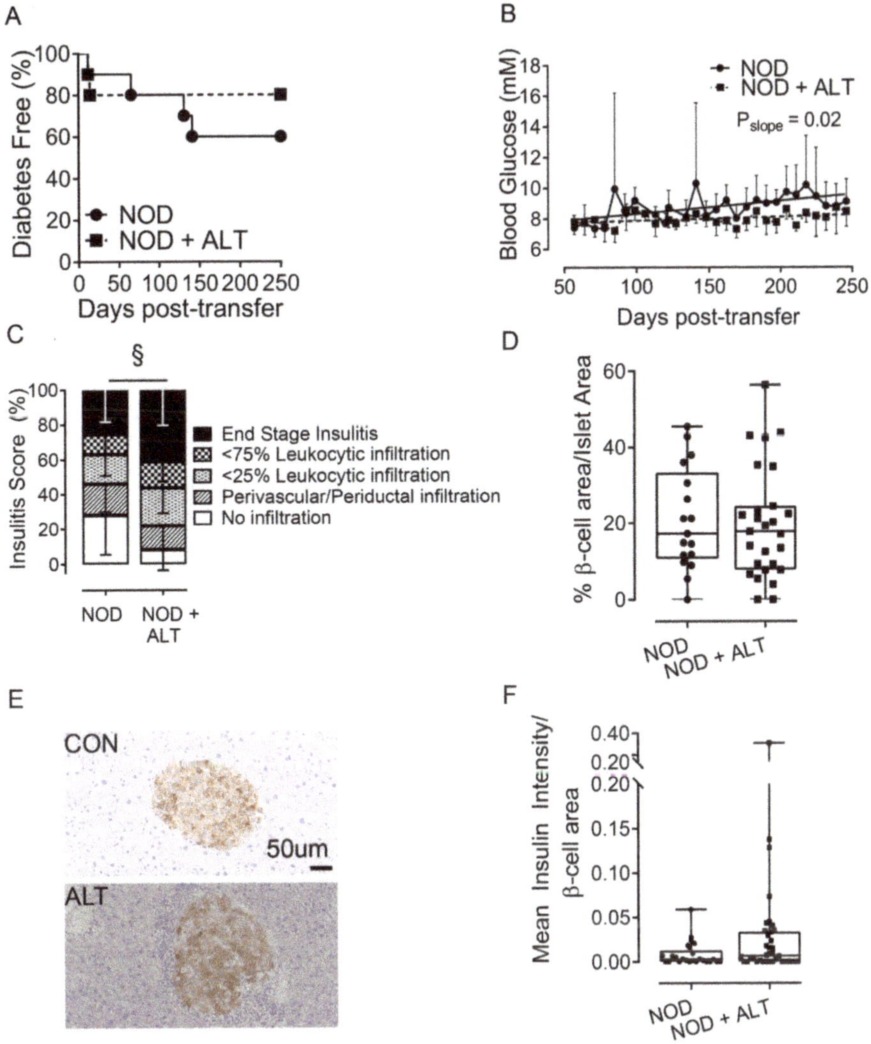

Figure 2. Alagebrium chloride delays diabetes development following adoptive transfer of G9C8 CD8 TCR transgenic cytotoxic lymphocytes. Diabetogenic splenocytes from NOD G9C8 CD8 TCR transgenic donors pre-primed with the aa15–23 of the insulin B chain (1×10^7 cells), were transferred into non-diabetic NOD*ShiLt* recipients (n = 10/group; day 80 of life). NOD*ShiLt* recipients were pre-treated for 30 days with (NOD + ALT, 1 mg/kg/day, s.c; ■, dashed line) or without alagebrium chloride (NOD, •, solid line). The recipient mice were monitored daily, tested for blood glucose and diabetes diagnosed by two consecutive blood glucose measurements > 15 m·mol/L and followed until diabetes diagnosis or until study end (day 250). (**A**) Diabetes incidence in NOD*ShiLt* recipients post-transfer of G9C8 cells. (**B**) Linear regression of non-fasting blood glucose concentrations over time, difference of slopes p = 0.02. (**C**) Pancreatic immune cell infiltration at day 250 (n = 6–8 mice/group), shown as mean percentage of islets with no infiltrate (white), perivascular/periductal infiltrate (striped), <25% infiltrate (arrowhead), <75% infiltrate (checkerboard) or end stage insulitis (black) of untreated splenocytes (CON; circles) or Alagebrium-treated splenocytes (ALT; squares) § p <0.0001, chi-square test. (**D**) β-cell area, determined from insulin IHC staining in (**E**) Representative images of islets detected with anti-insulin antibody in pancreatic sections taken from untreated (CON) or treated mice (ALT), scale bar = 50 μm. (**F**) Pancreatic expression of insulin (n = 26–40 sections/group, n = 2–39 islets/section, n = 6–8 mice/group, 5 μm sections, 96 μm apart. Data reported as either mean ± SD or box and whisker plots reporting median, interquartile ranges and min and max values.

2.6. *ALT Maintains Systemic Immune Function in the NODShiLt Mouse*

We next examined if ALT affected systemic cellular immune responses. After 28–32 days of ALT therapy, we determined the cytokine-response following an ovalbumin (OVA) challenge in splenocytes from pre-immunised NOD*ShiLt* mice one week after an ovalbumin immunisation (given at day 21–25 of treatment). Quantification of interferon-γ (IFN-γ) producing splenocytes by ELISpot revealed no differences between untreated mice and ALT-treated mice (Figure 3A). Consistent with this, the ability to adoptively transfer diabetes to NOD*scid* recipients by splenocytes from 28–32-day ALT-treated NOD*ShiLt* donors (9/13 diabetic; Figure 3B) was similar to that of control splenocytes (10/14 diabetic). The age at transfer and slight variances in cell number did not associate with diabetes transfer ($p = 0.06$ and $p = 1.0$, respectively; cox regression). This suggests short-term ALT therapy does not compromise systemic leukocyte function.

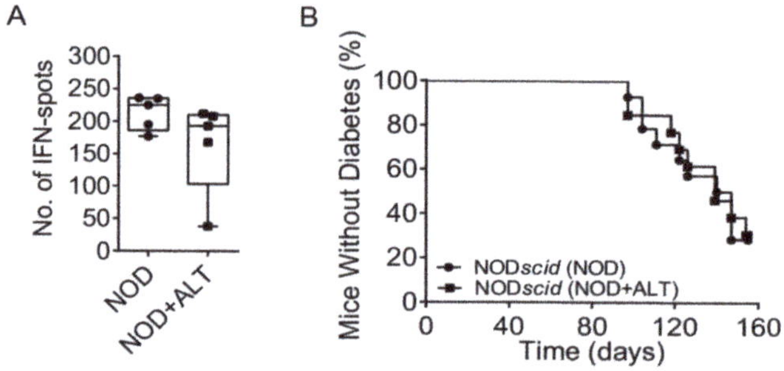

Figure 3. Systemic immune cells retain their function following short-term AGE lowering therapy with alagebrium chloride (ALT). (**A**) NOD*ShiLt* control mice (NOD, •) and NOD*ShiLt* treated with alagebrium (NOD + ALT, 1 mg/kg/day, s.c; ■) were immunised with OVA/QuilA and 1 week later spleens harvested and the number of cells producing IFN-γ in responses to OVA restimulation enumerated with ELISpot. ($n = 3$ replicates from $n = 5$ mice/group, $n = 2$ independent experiments). (**B**) Diabetes incidence after adoptive transfer of control or ALT treated (for ~30 days) diabetogenic NOD*ShiLt* splenocytes into NOD*scid* recipients aged 6–7 weeks old. N = 13–14/recipient group (1×10^6–2×10^7 cells/mouse). ■ NOD + ALT into NOD*scid* recipients; • NOD into NOD*scid* recipients.

2.7. *Immunopeptidomics and Proteomics of Pancreatic Beta Cells Following ALT Therapy*

We have previously demonstrated in vitro that advanced glycation end product modified BSA (AGE-BSA) impairs insulin secretion and adenosine triphosphate (ATP) production in MIN6N8 cells, which was reversed by ALT treatment [11]. To examine whether chronic exposure to a high AGE environment and ALT treatment resulted in changes in the peptide repertoire presented by MHC class Ia (immunopeptidome), bound peptides (pMHC-I; H-2D^b and H-2K^b) were immunoaffinity captured from treated MIN6N8 and sequenced via LC-MS/MS (Figure 4A). The complete dataset for both D^b and K^b are shown for untreated and treated MIN6N8 cells in Tables S1 and S2, respectively, and are accessible via the PRIDE data repository (see Data Availability Statement). Overall, no major changes in the immunopeptidome were observed during the various treatments with all peptides exhibiting allele specific consensus binding motifs. As expected, longer peptides were captured bound to H-2D^b than H-2K^b, with the H-2D^b peptides tending to be of longer chain length with AGE-BSA treatment (Figure 4C), when compared with either unmodified BSA (control) or AGE-BSA+ALT. H-2K^b peptides were more abundant but did not appear to differ among groups. H-2D^b and H-2K^b captured peptides were predominantly of 8–12 amino acids in length (Figure 4C) with similar bound motifs (Figure 4D), regardless of treatment (Figure S7A,B). Irrespective of the MHC-I class, the majority of H-2D^b and H-2K^b peptides that presented in each treatment, were found in at least one other condition (Figure 4E; Figure S7C), with differences likely reflecting subtle variation between experiments rather than significant changes to the bound peptide repertoire. The 8-mer

YQLENYCN peptide ($InsA_{14–21}$) where cysteine is modified by a glutathione disulfide, was elevated during AGE-BSA treatment compared to BSA control (Figure 4F). After the addition of ALT therapy, YQLENYC(glutathione disulfide)N decreased to a level similar to that of the BSA control (Figure 4F).

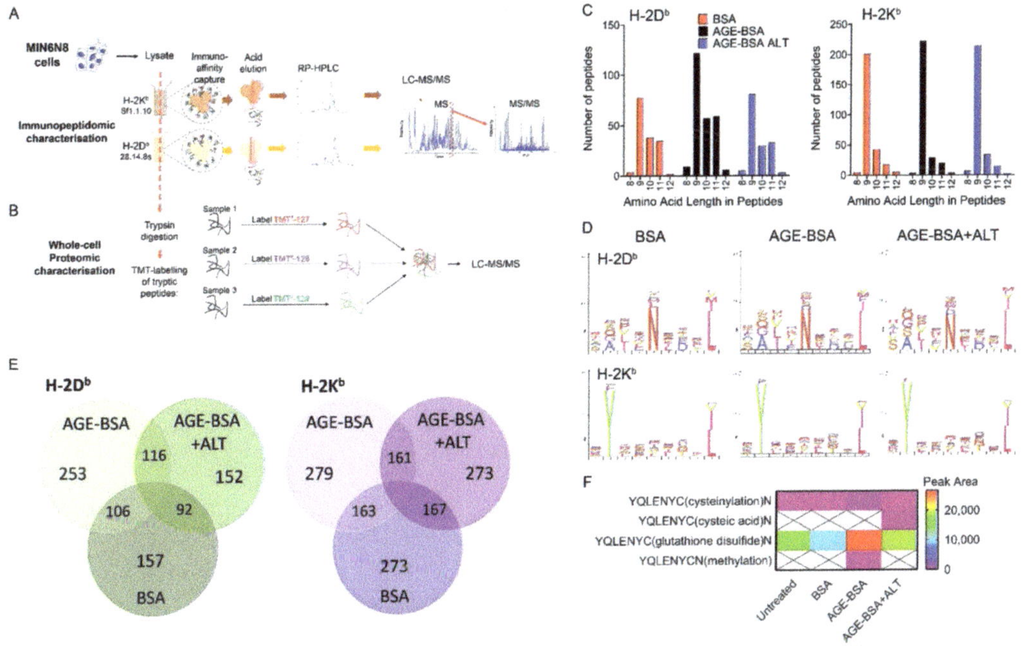

Figure 4. Immunopeptidomic and whole-cell proteomic characterisation of murine NOD β-cells. NOD*ShiLt* derived MIN6N8 β-cells were treated overnight with unmodified BSA (BSA; 100 μg/mL), AGE-modified BSA (AGE-BSA; 100 μg/mL), with or without alagebrium chloride (AGE-BSA+ALT; 40 μM) or untreated, with ALT vehicle control (PBS) or with BSA+ALT (Figures S7 and S8). Pooled cells (total 2×10^7 cells, pooled from $n = 4175$ cm^2 tissue culture flasks) from each treatment were lysed and antibodies against MHC Class 1a H-2K^b (clone sf1.1.10) and H-2D^b (clone 28.14.8 s) used to capture and identify presented peptides. Peptide complexes were run on RP-HPLC and identified via LC-MS/MS. Schematic of (**A**) immunopeptidomic and (**B**) whole-cell proteomic workflow used. (**C**) Number and length of MHC Class 1a H-2D^b (left) and H-2K^b (right) captured peptides isolated from BSA (red), AGE-BSA (black) and AGE-BSA+ALT (blue) treated MIN6N8 cells. (**D**) MHC Class Ia bound amino acid motifs of H-2D^b (top) and H-2K^b (bottom) peptides isolated. (**E**) Number of common H-2D^b (green; left) and H-2K^b (purple; right) immunocaptured MHC Class Ia associated peptides. (**F**) Modifications of insulin peptides precipitated with MHC Class Ia H-2D^b antibody from MIN6N8 cells. Scale bar shows peak area.

Next, we assessed if ALT affected the proteome of the same MIN6N8 cells from which cell surface pMHC-I complexes were isolated (Figure 4B). Gene Ontology analyses of MIN6N8 β-cells cultured in a high AGE environment alone, demonstrated an enrichment in proteins associated with catabolic processes, the immune system and cell cycling compared with BSA (Figure 5A). A high AGE environment enriched for predominately Carboxypeptidase N (*Cpn1*; Figure 5B; left) and Tubulin isoforms (*Tubb*; Figure 5B; middle). With the addition of ALT, Tubulin isoforms (*Tubb*; Figure 5B; middle) and proteins involved in catabolic processes in particular Multivesicular body subunit (*Mvb12a*), Cysteine protease (*Atg4b*), Calcium-independent phospholipase (*Pnpla8*), Carboxypeptidase N (*Cpn1*) and DNA damage binding protein 1 (DDB1)- and Cullin 4 (CUL4)-associated factor 11 (*Dcaf11*) were reduced, similar to that of the BSA control (Figure 5B; left). Conversely endocytotic proteins (Figure S8) Calnexin (*Cnx*) and Progranulin (*Grn*) were enriched (Figure 5B; right) after ALT therapy.

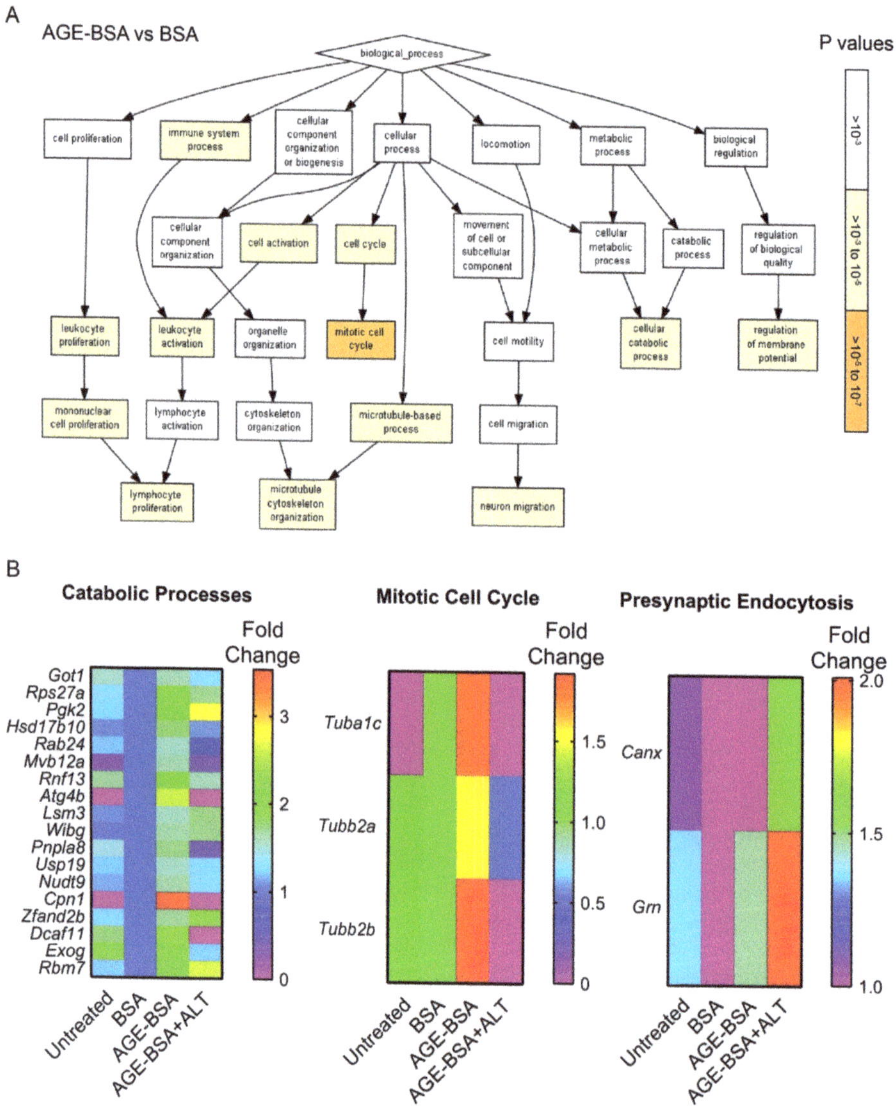

Figure 5. Gene Ontology enrichment analysis of biological processes in peptides isolated from MIN6N8 NOD derived beta cells. NOD*ShiLt* derived MIN6N8 beta cells were treated overnight with unmodified (BSA; 100 µg/mL), AGE-modified BSA (AGE-BSA; 100 µg/mL) without or with alagebrium chloride (AGE-BSA+ALT; 40 µM). The whole-cell proteome was performed on the cell lysate flowthrough that did not bind to MHC Class 1a antibodies. The effluents were trypsinised, peptides labelled with TMT using different isotypes and identified using LC-MS/MS. Workflow for proteomics is represented in Figure 4B. (**A**) Pathway analysis of peptides enriched following AGE-BSA vs. BSA treated MIN6N8 cells. *p* values (right; 10^{-3} to 10^{-7}) shown in white-orange scale. (**B**) Heatmaps of proteins associated with three major enriched Gene Ontology pathways after AGE-BSA and AGE-BSA+ALT vs. control (BSA) MIN6N8 treated cells. Scale shows fold change.

3. Discussion

The present study demonstrates that short-term ALT therapy given to NOD*ShitLt* mice prior to overt disease onset and then withdrawn, does not impact systemic immune cell function or diabetes incidence, but has transient effects on β-cell function. We observed an increase in insulin secretion during glucose tolerance tests performed halfway through the ALT therapeutic period (~day 75 of life). In addition, the first phase insulin response and action during the first 15 min of the IPGTT were higher in ALT treated mice. Increases in fasted insulin as compared with control mice also persisted until one week after ALT therapy cessation. It is clear however, that these effects on β-cell function were not sufficient to arrest the development of autoimmune diabetes in ALT treated mice after withdrawal of ALT, suggesting that these β-cell effects were transient. Blood glucose concentrations were increased during oral but not intraperitoneal glucose challenges at day 75 with ALT therapy. The reasons for this remain to be ascertained in the present study. Further supporting an effect on β-cells with ALT, were experiments where pre-treatment of NOD recipients delayed diabetes development and lowered blood glucose concentrations overtime following adoptive transfer of G9C8 CD8 TCR transgenic cytotoxic lymphocytes. Circulating glucose concentrations are impacted by various pathways which are impaired in NOD mice. These include changes in peripheral insulin sensitivity [31], hypoglycaemia-glucagon feedback loops, gastric emptying, glucose intestinal absorption and the incretin effect [32]. Certainly, the higher insulin concentrations and insulin:glucose ratios suggest that peripheral insulin sensitivity may be lower in NOD mice treated with ALT as compared with control at day 75 of life. Generally AGE lowering interventions have improved insulin sensitivity in humans who are obese [33] or with type 2 diabetes [34]. However, NOD mice prediabetes often show variable insulin sensitivity which is significantly impacted by the loss of insulin secretion and so it is difficult to ascertain in the present study whether the effects of ALT on insulin, are beneficial. This may be remedied by the determination of C-peptide during OGTT which would provide specific evidence for effects on insulin secretion *per se* without being confounded by changes in insulin sensitivity. Indeed, retention of circulating C-peptide concentrations is the most common primary end-point for clinical trials in T1D treatment and prevention [35].

Gastric emptying appears to change during disease development in NOD mice and is accelerated at diabetes onset [36], however in humans, blood glucose elevations in individuals with T1D can slow gastric emptying [37,38] impacting gastrointestinal glucose absorption [39]. Glucagon-like peptide-1 (GLP-1) and gastric inhibitory polypeptide expression in the gut of NOD mice appear unchanged by hyperglycaemia [40], and although combinatory therapies with GLP-1 receptor agonists in individuals with T1D have the potential to enhance β-cell function [41], recent Phase III trials (NCT01836523, NCT02098395) have observed elevations in hypoglycaemia and hyperglycaemia with ketosis [42,43]. One explanation for a rise in glucose concentrations after the OGTT may be that production of proinsulin is contributing to overall increases in insulin at day 75 and to the end of the ALT treatment period, but limited plasma volumes throughout GTTs, prevented proinsulin from being measured. Together with the relationship between insulin secretion and lack of efficacy on progression to diabetes, our study suggests that upon removal of ALT therapy prediabetes, benefits on β-cell function are unlikely to be maintained. This therapy if given prophylactically may be best suited during the neonatal period where priming of the immune system against β-cell antigens like proinsulin [44,45] and GAD65 [45] is key in NOD mice [44] and at-risk children [45].

The reduction in insulitis observed was confirmed using flow cytometry where CD45.1$^+$ cells were reduced in pancreatic digests. Here, lower proportions of CD8$^+$ T cells, B cells and F4/80$^+$ macrophages and higher proportions of activated F4/80$^+$I-Ag7$^+$ macrophages were seen. One major limitation of this work was the inability to track temporal changes to immune cells during and after ALT therapy and the inability to examine if ALT therapy impacted the expression of the cell surface receptor for AGEs, RAGE. It would be of interest to understand whether this therapy directly impacts RAGE expression on CD45.1$^+$ lymphocytes locally within the pancreas. Although there is disparity as to whether ALT can mediate effects directly through RAGE, this is suggested in diabetes complications [46,47] and RAGE knockout [48–50] mouse models. Further, studies by our team [22] and others [51] show that inhibiting RAGE by prophylactic soluble RAGE (sRAGE) therapy, a decoy receptor that clears circulating AGEs, elevates numbers of RAGE$^+$ CD4 and CD8 T cells, T_{reg} cells, DCs and macrophages in lymphoid tissue and improves Treg function [22] and prevents the transfer of diabetes by diabetogenic T cells in NOD*scid* mice [51]. With administration of ALT prediabetes, we saw no effect on IFN-γ production of OVA-stimulated splenocytes after 30 days of therapy. The lack of effect on systemic immune cell function within lymphocyte populations was further evidenced by the lack of effect of ALT-treatment of NOD donor splenocytes on diabetes incidence following adoptive transfer to NOD*scid* recipients.

Using an in vitro approach, we examined the MIN6N8 β-cell immunopeptidome; the peptide repertoire presented by MHC class Ia. The addition of ALT did not change the length of peptides recovered, which were a standard 8–12 amino acids in length as expected [52], but may have impacted their abundance, while motifs were shared amongst all treatment groups. To our knowledge, this is the first time that the immunopeptidome approach has been used to study MHC Class I peptide expression in the NOD derived MIN6N8 mouse β-cell line after AGE-lowering therapy. Although our work shows preliminary analysis and no major changes to the nature of MHC-I bound peptides, it was interesting to observe the increase of 8-mer YQLENYC(glutathione disulfide)N ($InsA_{14–21}$) after AGE-BSA which was reduced after ALT therapy to levels similar to similar to that of BSA control. Mannering et al., previously described post-translational modifications to cysteine residues of a human insulin A-chain T cell epitope [53], and using clustered regularly interspaced short palindromic repeats (CRISPR)/Cas9 to replace murine insulin 1 with human insulin in the NOD mouse, largely protected from diabetes development [54]. The modification of glutathionylation to insulin A should be investigated further beyond this study due the abundance of glutathionylation to cysteines under oxidative stress [55].

Proteomic analysis revealed an enrichment of selected proteins after the addition of AGE-modified BSA compared to unmodified BSA. From our preliminary experiment in MIN6N8 cells, we observed upregulation in tubulin proteins which was not seen in MIN6N8 cells treated with ALT. Microtubules are critical in glucose stimulated insulin secretion [56], regulating the availability of insulin [57] and allowing the travel of insulin granules along tubulin tracks [58]. Upon microtubule destabilisation, a subset of insulin granules are released [59]. High AGE environments consistently reduce the ability of β-cells to secrete insulin which has been attributed to an increase in oxidative stress [11,13,16]. What would be interesting to substantiate in the future, is if high AGE environments result in changes or depolymerisation [14] or hyper-stabilization of the dense microtubular structure in the β-cell, resulting in insulin secretory dysfunction [59,60].

Proteomic analysis further revealed that ALT therapy enriched for calnexin and progranulin, but reduced multivesicular body subunit and cysteine protease, proteins involved in secretory [61] and autophagy [62–64] pathways, respectively. Calnexin is a lectin chaperone protein in the endoplasmic reticulum (ER), responsible for the folding of *N*-glycosylated proteins destined for the plasma membrane [61]. ER stress is a major contributor of β-cell fragility in autoimmune diabetes [65], and while other ER chaperone proteins are reported in the removal of mis-folded proteins [66], to our knowledge, the importance of calnexin in the β-cell has yet to be determined. Interestingly, ER stress can increase in response to progranulin administration [67] and impair insulin sensitivity in rodent models [67,68]. Within the pancreas of *MEN1* transgenic mice, progranulin was found to be a potent stimulator of β-cell proliferation in pancreatic tumours [69] and in $Grn^{-/-}$mice, progranulin is vital for endolysosomal trafficking regulation [70]. Interestingly, while β-cells are high secretory cells like neurons, the role progranulins play in insulin secretion or β-cell survival is unknown.

Multivesicular body subunit forms a component of the regulatory complex that traffics ubiquinated cargos into multivesicular bodies [62,63], while cysteine protease is required for deconjugation where proteins are cleaved prior to the conjugation to phospholipids [64]. Receptor-mediated endocytosis is an important mechanism in both AGE uptake and removal [3], and in the recapture of exocytotic vesicles after insulin release in β-cells [71]. Although this is an interesting finding, conclusions on the mode of action within β-cells cannot be made. One exciting application that could further scrutinize the effect of ALT on islet cell secretion would be the rapidly developing methods that can monitor live, single vesicle fusion via two-photon microscopy [72].

Calcium-independent phospholipase hydrolyses ester bonds in phospholipids to release free fatty acids and lysophospholipids [73]. Arachidonic acid is a product generated from this reaction and can be further metabolised into proinflammatory lipid mediators. Within β-cells, calcium-independent phospholipase appears to be involved in glucose stimulated insulin secretion [74,75], proliferation and apoptosis in INS-1 cells [75] and human islets [76] and has thus been suggested as a target to improve β-cell longevity [76].

Carboxypeptidase N (*Cpn1*) is a circulating zinc metalloprotease which cleaves carboxy-terminal lysines and arginines from peptides found in the bloodstream [77]. Specific roles of carboxypeptidases have been difficult to ascertain due to the large number of protein members within the CPN family [77]. Within the islet, carboxypeptidase B1 has been shown to regulate rodent β-cell proliferation [78], carboxypeptidase E is regulated by insulin and is involved in proinsulin processing [79], carboxypeptidase H and E has been suggested as a T1D autoantibody in humans [80]. While the inhibition of carboxypeptidase M in rats reverses insulin resistance [81], and cross-talk between carboxypeptidase N and Carboxypeptidase B2 is observed during complement activation [82], the functional relevance of the downregulation of this protein with ALT is unknown.

DDB1- and CUL4-associated factors (DCAF) are proteins involved in substrate specificity for protein ubiquitination in the CUL4-DDB1 ubiquitin ligase [83]. While this ligase regulates cellular proliferation, survival and genomic integrity, DCAF11, which was down-regulated by ALT therapy, has an unknown function and cellular binding partner, like most identified DCAFs [83].

Given activation of catabolic processes can be an indication of nutrient scarcity [84], the reduction of these catabolic processes by ALT therapy appear strange without the reduction of circulating blood glucose. Still, it would be interesting to understand whether ALT, which reduced subunit proteins involved in endocytic [48,49] and autophagy vesicle formation [62–64], has the ability to confer stress-resistance via the change in metabolic pathways associated with catabolism; fatty acid oxidation and oxidative phosphorylation [84].

The major limitation of our *omic* work is the lack of technical replicates and the ability to perform dose escalation studies. This is due to the vast number of cells required for immunopeptide precipitation and capture and the slow growth of MIN6N8 cells. Our results were from pooled replicates using previously optimised concentrations and should be interpreted with as such, however, are consistent with the functional studies that suggest a β-cell effect of ALT treatment rather than an immune mediated effect. This further emphasizes the requirement for future *omic* experiments in treated preclinical and ex vivo human islets using escalating doses of ALT to validate the effect of AGEs on the protein pathways enriched. Future studies should focus on validation of differentially expressed proteins and pathways using orthogonal validation techniques, such as multiplexed Western blotting and targeted mass spectrometry. We suggest that our database of immunopeptides and proteins (PXD025998) start as a reference point for researchers interested in therapeutic agents of advanced glycation, and its associated receptors and pancreatic islets.

Another limitation of this work is that the specific mode of action of ALT continues to remain elusive. Although first thought as a chemical cross-link breaker of α-dicarbonyl-containing compounds, evidence supporting this is lacking and the most likely mechanisms of action include metal chelation, anti-oxidant activity and bypassing the generation of MGO, previously reviewed in [85–87]. Certainly, we observed that circulating AGE precursors, MGO, GO and 3-DG and AGEs CML, CEL and MG-H1 were not decreased directly after cessation of ALT therapy. This may point to the therapeutic dose and exposure time to ALT, the limited number of measurements performed in mice, or simply that the reduction of intermediates or AGEs is localised to the β-cell, islet, or pancreas. AGE precursors and AGEs were not measured within islets in our study and is a clear limitation which should be prioritised in future studies [86]. The time dependency of ALT is however supported by our previous study, in which long-term prophylactic ALT therapy delayed diabetes onset [11]. It has been highlighted by this experimental work and by others [41], that one therapy is unlikely to be effective for T1D prevention. Rather, future work should focus on innovative, combinatorial approaches that could slow β-cell demise and reduce, reset and/or correct immune system function to prolong diabetes remission [41].

Taken together, these studies present a case for future work to be performed to better understand the effects of AGE lowering therapies on β-cell function and how these may be temporally determined. Certainly, the AGE-RAGE axis is gaining increasing attention as a potential pathway which can be modulated to both preserve β-cell function and arrest the development of T1D. However, our present findings suggest that prophylactic therapy, with the AGE lowering therapy, alagebrium chloride prediabetes, is not sufficiently effective in this context and other approaches should be prioritised.

4. Materials and Methods

4.1. Rodent Studies

Animal experiments were approved by an animal ethics committee (University of Queensland) and adhered to national guidelines by the National Health and Medical Research Council, Australia. Sample size calculations were estimated using a power of 0.80 and $\alpha = 0.05$, where 80% of control mice were expected to develop diabetes by study end from previous studies in the animal facilities. Female NOD*ShiLt* mice were housed in the Biological Resource Facility in pathogen-free conditions at the Translational Research Institute, Brisbane Australia, and were randomly assigned to cages (n = 5/cage) and investigators were not blinded to treatments. Mice were given standard chow and water ad libitum, paper bedding and enrichment, and maintained on 12-h light-dark cycles, handled equally and allowed to acclimatise in the facility for at least 7 days prior to the start of procedures. Body weights and non-fasted blood glucose measurements by a glucometer (SensoCard) were taken weekly until day 100 of life, where measurements increased in frequency to at least twice weekly, when monitoring for diabetes incidence.

For diabetes incidence studies, groups of female NOD mice (n = 18–20/group) were either given no treatment or daily, subcutaneous injections of ALT chloride (1 mg/kg/day; Anthem Biosciences Ltd., Bangalore, Karnataka, India) for 50 days from day 50 to 100 of life. These mice were followed until diabetes was diagnosed (two consecutive, non-fasting blood glucose measurements of $\geq$15 m·mol/L) or until day 250 of life in the absence of diabetes. To determine the effect of ALT prediabetes, groups of female NOD mice (n = 10/group) were either given no treatment or daily, subcutaneous injections of ALT chloride (1 mg/kg/day; Anthem Biosciences Ltd.) for up to 40 days from day 50 of life, and were sacrificed from day 80–90 of life. To study effects on pancreatic islets, pancreatic infiltrate and splenocyte function, female NOD*ShiLt* mice were randomised (n = 8/group) to receive no treatment or subcutaneous injections of ALT (1 mg/kg/day; Anthem Biosciences Ltd.) for 28–32 days from day 50 of life.

Adoptive transfer studies involved either the treatment of female donor or recipient NOD*ShiLt* mice (n = 5–10/group) with or without subcutaneous injections of ALT (1 mg/kg/day; Anthem Biosciences Ltd.). Either treated or untreated NOD splenocytes were pooled and injected intravenously (1×10^6–2×10^7 cells/mouse) into 6-week old NOD.CB17-Prkdcscid/J recipient mice or NOD G9C8 cytotoxic T lymphocytes [30] were activated using recombinant insulin B15-23 peptide, and injected intravenously into 80- day-old-treated NOD*ShiLt* recipients [88]. Recipient mice were followed until diabetes diagnosis as described above or until day 250 of life in the absence of diabetes. At the end of the study or experiment end, mice were fasted and anesthetized as previously described [89].

4.2. MIN6N8 Cells

MIN6N8 cells (kindly provided by Professor Jun-ichi Miyazaki, Osaka University) were maintained in phenol-red free DMEM (ThermoFisher Scientific, Scoresby, VIC, Australia) containing 25 mM glucose (Sigma-Aldrich, Castle Hill, NSW, Australia), 10% *v*/*v* heat-inactivated FBS (ThermoFisher Scientific), 100 U/mL of penicillin, 100 µg/mL of streptomycin (ThermoFisher Scientific), 2 mM·L-glutamine and 71.5 µM beta mercaptoethanol (Sigma-Aldrich) [60]. Unmodified BSA (control) and AGE modified BSA were produced in-house and described elsewhere [11]. Prior to BSA (100 µg/mL), AGE-BSA (100 µg/mL) and ALT (40 µM) experimentation, MIN6N8 cells were cultured overnight in culture medium containing only 2% *v*/*v* heat-inactivated FBS.

4.3. Proteomic Extraction, Labelling, Detection and Quantification in MIN6N8 Cells

Refer to Supplementary Methods for full detail.

4.4. Metabolic and Biochemical Measurements

Glycated haemoglobin from whole blood was measured spectrophotometrically (Cobas Mira, Roche Diagnostics, Sydney, Australia) using a commercial ELISA (Crystal Chem Inc., Downers Grove, IL, USA). From fasting plasma, total insulin (Merck Millipore, Bayswater, VIC, Australia) and proinsulin (Mercodia AB, Uppsala, Sweden) were determined using commercial ELISAs following the manufacturer's instructions. From non-fasting plasma, circulating AGEs and α-dicarbonyls were measured using an LC-MS/MS technique as described previously [33].

4.5. Histology and Immunohistochemistry

The tail/body of pancreata from NOD mice were fixed (10% neutral buffered formalin), dehydrated and embedded in paraffin using standard techniques. For insulitis, serial sections (n = 4–8/mouse, 96 µm apart) were stained using haematoxylin and eosin and quantified in a blinded fashion (26–88 islets/mouse, n = 4–8 mice/group), as previously described [90]. Serial sections (n = 2–4/mouse, 96 µm apart) were stained using anti-insulin (R&D systems, Noble Park North, VIC, Australia; clone 182410) or anti-proinsulin (R&D systems; clone 253627) as previously described [17]. Sections were imaged using an automatic slide scanner (Olympus VS120, Olympus Australia Pty Ltd., Macquarie Park, NSW, Australia). Islet area and number were quantified in a blinded fashion using Visiopharm image analysis software v4.5.6.440 (Olympus Australia Pty Ltd.). Antigen quantification from DAB^+ areas was calculated as a reciprocal DAB intensity as detailed previously [91].

4.6. Flow Cytometry

Pancreata were isolated and digested to obtain single cell suspensions as previously described [92]. Antibodies against CD45.1 (A20), CD4 (GK1.5), CD8α (53-6.7), CD11b (M1/70), Cd11c (N418) and IA^{g7} (10.2.16) were purchased from Biolegend (San Diego, CA, USA). Antibodies against CD19 (ID3), CD45R (RA3-6B2) and IgM (II/41) were purchased from BD Biosciences (North Ryde, NSW, Australia). Antibodies against F4/80 (CI:A3-1) were purchased from AbD Serotec (Raleigh, NC, USA). Pancreatic cells were stained as previously detailed [92] and cytometric data were acquired in an unblinded fashion on the BD LSR II (BD Biosciences) and analysed using FlowJo software v8.

4.7. ELISpot Assay

Isolated splenocytes (5×10^5 cells/well) were loaded into polyvinylidene difluoride ELISpot plates (Merck Millipore) which were pre-absorbed with IFN-γ antibody. Cells were stimulated with anti-CD3 (0.1 µg/mL) and maintained for 48 h in RPMI, 5% *v*/*v* FBS, 2 mM of penicillin/streptomycin/glutamine (ThermoFisher Scientific), 50 nM beta mercaptoethanol and 1 mM sodium pyruvate (ThermoFisher Scientific). ELISpot assays were developed as previously described [25] using an immune-spot plate reader (AID GmbH, Strassburg, Germany).

4.8. Statistical Analyses

Data was analysed for normality by either the D'Agostino-Pearson test, Shapiro-Wilk test or by viewing histograms. Gaussian distributed data are expressed as mean ± S.D and analysed using the unpaired Student's t-test. Non-Gaussian distributed data are reported as medians with IQR and analysed by the Mann-Whitney U test, or mixed-effects model with Sidak's post-hoc test. Correlations were analysed via linear regression and Pearson's correlation test. Kaplan-Meier survival curves were compared by the Log-Rank test. Predicted variables associated with survival analysis were analysed via Cox regression. Frequency distributions were evaluated by chi-squared (χ^2). Calculations were performed using graph prism (v6.05 or v8.0.1). p values < 0.05 were considered statistically significant.

5. Conclusions

Overall, our data demonstrates that short duration of ALT therapy does not extend β-cell function, inhibit experimental autoimmune diabetes or effect systemic immune function. Future studies should dissect whether co-administration of immunomodulatory therapies with ALT during the neonatal period can have an additive protective effect and what cellular processes are changed in response to ALT in islets at the single cell level using sophisticated microscopic and *omic* approaches.

Supplementary Materials: The following are available online at https://www.mdpi.com/article/10.3390/metabo11070426/s1, Figure S1: ALT treatment increases insulin in response to glucose loads, Figure S2: Circulating advanced glycation end products and dicarbonyl compounds (AGE precursors) do not change throughout short-term ALT therapy but associate positively with glycated haemoglobin, Figure S3: Gating strategy for flow cytometry analysis, Figure S4: T and B cell numbers and proportions do not differ after ALT therapy in pLN, Figure S5: Conventional dendritic cell and plasmacytoid dendritic cell numbers and proportions do not differ after ALT in the pLN and spleen, Figure S6: T and B cell numbers and proportions do not differ after ALT therapy in the spleen, Figure S7: Immunopeptidomic characterisation of murine NOD beta cells, Figure S8: GO Pathway Mapping of BSA vs. AGE-BSA+ALT treated MIN6N8 cells, Table S1: K^b identified peptides list, Table S2: D^b identified peptides list, Table S3: D^b identified peptides list; proteins discovered in proteomic discovery, Supplementary Methods: Glucose and insulin tolerance tests, Flow Cytometry (spleen, pancreatic lymph node), Purification of MHC-peptide complexes, Identification of MHC-bound peptides using LC-MS/MS, Trypsin digestion of MIN6N8 lysates for proteomic characterization, TMT-labelling of tryptic peptides; Data Availability Statement.

Author Contributions: Conceptualization, D.J.B., A.W.P., R.S. and J.M.F.; methodology, D.J.B., D.B., N.L.D., C.G.S., R.S., A.W.P., J.M.F.; validation, D.J.B., P.R., C.G.S., R.S., A.W.P., J.M.F.; formal analysis, D.J.B., P.F., R.S., J.M.F.; investigation, D.J.B., P.F., K.L.G., P.R., A.K.F., D.A.M., S.L., M.S.W., B.E.H., R.A., J.L.S., N.L.D., D.B., R.S.; resources, C.G.S., R.S., A.W.P., J.M.F.; data curation, D.J.B., P.F., A.W.P., J.M.F.; writing—original draft preparation, D.J.B.; writing—review and editing, D.J.B., P.F., K.L.G., A.K.F., R.S., A.W.P., J.M.F.; visualization, D.J.B., P.F., K.L.G., P.R.; supervision, C.G.S., R.S., A.W.P., J.M.F.; project administration, D.J.B.; funding acquisition, C.G.S., R.S., A.W.P., J.M.F. All authors have read and agreed to the published version of the manuscript.

Funding: This was funded by the National Health and Medical Research Council of Australia (NHMRC; 1023664, 1165490, 1084283, 1043315 J.M.F., R.S., A.W.P.), the Victorian Government Infrastructure Program, and the Mater Foundation. Authors were specifically supported by: Mater Foundation (D.J.B.), Victorian Department of Health and Human Services acting through the Victorian Cancer Agency (P.F.), Australian Postgraduate Award Scholarship (P.R., A.K.F.), NHMRC Project Grant (1023664; D.M., J.M.F.), UQ/JDRF Postgraduate Scholarship (S.L.), JDRF Postdoctoral Fellowship (M.S.W.), NHMRC Peter Doherty Fellowship (B.E.H.), NHMRC Project Grant (1165490, 1084283; K.L.G., R.A.), NHMRC Principal Research Fellowship (1137739; A.W.P.), ARC Fellowship (FT110100372; R.S.) and NHMRC Senior Research Fellowship (1004503, 1102935; J.M.F.).

Institutional Review Board Statement: The study was conducted according to the guidelines of the NHMRC and approved by the Ethics Committees of the University of Queensland (TRI/MMRI/NHMRC: 169/12; 036/13; 210/15).

Informed Consent Statement: Not applicable.

Data Availability Statement: Availability of the immunopeptidome and proteomic datasets can be found in the PRIDE data repository, accession number: PXD025998, doi:10.6019/PXD025998.

Acknowledgments: The MIN6N8 cell line used for in vitro experiments were kindly provided by Jun-ichi Miyazaki, Osaka University. The TCRα and TCRβ founder lines used to generate the G9C8 mice were kind gifts from Susan F. Wong, Cardiff University. We would like to acknowledge the staff at the UQ Biological Research Facility, Crystal Chang of the Histology Facility and Sandrine Roy and Ali Ju of the Microscopy Facility, located in the Translational Research Institute. We would like to acknowledge the provision of instrumentation, training, and technical support by the Monash Biomedical Proteomics Facility. Computational resources for *omic* analysis were supported by the R@CMon/Monash Node of the NeCTAR Research Cloud, an initiative of the Australian Government's Super Science Scheme and the Education Investment Fund. We would like to extend our sincere thanks to Ristan Greer for statistical advice.

Conflicts of Interest: The funders had no role in the design of the study; in the collection, analyses, or interpretation of data; in the writing of the manuscript, or in the decision to publish the results. JMF holds two patents for alagebrium chloride.

References

1. Monnier, V.M.; Cerami, A. Nonenzymatic browning in vivo: Possible process for aging of long-lived proteins. *Science* **1981**, *211*, 491–493. [CrossRef] [PubMed]
2. O'Brien, J.; Morrissey, P.A. Nutritional and toxicological aspects of the Maillard browning reaction in foods. *Crit. Rev. Food Sci. Nutr.* **1989**, *28*, 211–248. [CrossRef]
3. Ott, C.; Jacobs, K.; Haucke, E.; Navarrete Santos, A.; Grune, T.; Simm, A. Role of advanced glycation end products in cellular signaling. *Redox Biol.* **2014**, *2*, 411–429. [CrossRef]
4. Chavakis, T.; Bierhaus, A.; Al-Fakhri, N.; Schneider, D.; Witte, S.; Linn, T.; Nagashima, M.; Morser, J.; Arnold, B.; Preissner, K.T.; et al. The Pattern Recognition Receptor (RAGE) Is a Counterreceptor for Leukocyte Integrins. *Nov. Pathw. Inflamm. Cell Recruit.* **2003**, *198*, 1507–1515. [CrossRef] [PubMed]
5. Demling, N.; Ehrhardt, C.; Kasper, M.; Laue, M.; Knels, L.; Rieber, E.P. Promotion of cell adherence and spreading: A novel function of RAGE, the highly selective differentiation marker of human alveolar epithelial type I cells. *Cell Tissue Res.* **2006**, *323*, 475–488. [CrossRef]
6. Schmidt, A.M.; Yan, S.D.; Brett, J.; Mora, R.; Nowygrod, R.; Stern, D. Regulation of human mononuclear phagocyte migration by cell surface-binding proteins for advanced glycation end products. *J. Clin. Investig.* **1993**, *91*, 2155–2168. [CrossRef] [PubMed]
7. Manfredi, A.A.; Capobianco, A.; Esposito, A.; De Cobelli, F.; Canu, T.; Monno, A.; Raucci, A.; Sanvito, F.; Doglioni, C.; Nawroth, P.P.; et al. Maturing Dendritic Cells Depend on RAGE for *In Vivo* Homing to Lymph Nodes. *J. Immunol.* **2008**, *180*, 2270–2275. [CrossRef]
8. Dumitriu, I.E.; Bianchi, M.E.; Bacci, M.; Manfredi, A.A.; Rovere-Querini, P. The secretion of HMGB1 is required for the migration of maturing dendritic cells. *J. Leukoc. Biol.* **2007**, *81*, 84–91. [CrossRef]
9. Serban, A.I.; Stanca, L.; Geicu, O.I.; Munteanu, M.C.; Dinischiotu, A. RAGE and TGF-β1 Cross-Talk Regulate Extracellular Matrix Turnover and Cytokine Synthesis in AGEs Exposed Fibroblast Cells. *PLoS ONE* **2016**, *11*, e0152376. [CrossRef] [PubMed]
10. Beyan, H.; Riese, H.; Hawa, M.I.; Beretta, G.; Davidson, H.W.; Hutton, J.C.; Burger, H.; Schlosser, M.; Snieder, H.; Boehm, B.O.; et al. Glycotoxin and autoantibodies are additive environmentally determined predictors of type 1 diabetes: A twin and population study. *Diabetes* **2012**, *61*, 1192–1198. [CrossRef]
11. Coughlan, M.T.; Yap, F.Y.; Tong, D.C.; Andrikopoulos, S.; Gasser, A.; Thallas-Bonke, V.; Webster, D.E.; Miyazaki, J.; Kay, T.W.; Slattery, R.M.; et al. Advanced glycation end products are direct modulators of beta-cell function. *Diabetes* **2011**, *60*, 2523–2532. [CrossRef]
12. Salonen, K.M.; Ryhanen, S.J.; Forbes, J.M.; Harkonen, T.; Ilonen, J.; Laine, A.P.; Groop, P.H.; Knip, M.; The Finnish Pediatric Diabetes Register. Circulating concentrations of soluble receptor for AGE are associated with age and AGER gene polymorphisms in children with newly diagnosed type 1 diabetes. *Diabetes Care* **2014**, *37*, 1975–1981. [CrossRef]
13. Kong, X.; Wang, G.-D.; Ma, M.-Z.; Deng, R.-Y.; Guo, L.-Q.; Zhang, J.-X.; Yang, J.-R.; Su, Q. Sesamin Ameliorates Advanced Glycation End Products-Induced Pancreatic β-Cell Dysfunction and Apoptosis. *Nutrients* **2015**, *7*, 4689–4704. [CrossRef]
14. You, J.; Wang, Z.; Xu, S.; Zhang, W.; Fang, Q.; Liu, H.; Peng, L.; Deng, T.; Lou, J. Advanced Glycation End Products Impair Glucose-Stimulated Insulin Secretion of a Pancreatic Beta-Cell Line INS-1-3 by Disturbance of Microtubule Cytoskeleton via p38/MAPK Activation. *J. Diabetes Res.* **2016**, *2016*, 9073037. [CrossRef] [PubMed]

15. Shu, T.; Zhu, Y.; Wang, H.; Lin, Y.; Ma, Z.; Han, X. AGEs Decrease Insulin Synthesis in Pancreatic β-Cell by Repressing Pdx-1 Protein Expression at the Post-Translational Level. *PLoS ONE* **2011**, *6*, e18782. [CrossRef] [PubMed]
16. Zhao, Z.; Zhao, C.; Zhang, X.H.; Zheng, F.; Cai, W.; Vlassara, H.; Ma, Z.A. Advanced Glycation End Products Inhibit Glucose-Stimulated Insulin Secretion through Nitric Oxide-Dependent Inhibition of Cytochrome c Oxidase and Adenosine Triphosphate Synthesis. *Endocrinology* **2009**, *150*, 2569–2576. [CrossRef] [PubMed]
17. Borg, D.J.; Yap, F.Y.T.; Keshvari, S.; Simmons, D.G.; Gallo, L.A.; Fotheringham, A.K.; Zhuang, A.; Slattery, R.M.; Hasnain, S.Z.; Coughlan, M.T.; et al. Perinatal exposure to high dietary advanced glycation end products in transgenic NOD8.3 mice leads to pancreatic beta cell dysfunction. *Islets* **2018**, *10*, 10–24. [CrossRef]
18. Peppa, M.; He, C.; Hattori, M.; McEvoy, R.; Zheng, F.; Vlassara, H. Fetal or Neonatal Low-Glycotoxin Environment Prevents Autoimmune Diabetes in NOD Mice. *Diabetes* **2003**, *52*, 1441–1448. [CrossRef] [PubMed]
19. Forbes, J.M.; Cowan, S.P.; Andrikopoulos, S.; Morley, A.L.; Ward, L.C.; Walker, K.Z.; Cooper, M.E.; Coughlan, M.T. Glucose homeostasis can be differentially modulated by varying individual components of a western diet. *J. Nutr. Biochem.* **2013**, *24*, 1251–1257. [CrossRef] [PubMed]
20. Pearson, J.A.; Wong, F.S.; Wen, L. The importance of the Non Obese Diabetic (NOD) mouse model in autoimmune diabetes. *J. Autoimmun.* **2016**, *66*, 76–88. [CrossRef]
21. Forbes, J.M.; Soderlund, J.; Yap, F.Y.; Knip, M.; Andrikopoulos, S.; Ilonen, J.; Simell, O.; Veijola, R.; Sourris, K.C.; Coughlan, M.T.; et al. Receptor for advanced glycation end-products (RAGE) provides a link between genetic susceptibility and environmental factors in type 1 diabetes. *Diabetologia* **2011**, *54*, 1032–1042. [CrossRef]
22. Leung, S.S.; Borg, D.J.; McCarthy, D.A.; Boursalian, T.E.; Cracraft, J.; Zhuang, A.; Fotheringham, A.K.; Flemming, N.; Watkins, T.; Miles, J.J.; et al. Expansion of Functional Regulatory T Cells Using Soluble RAGE Prevents Type 1 Diabetes. *bioRxiv* **2020**. [CrossRef]
23. Ulrich, P.; Cerami, A. Protein glycation, diabetes, and aging. *Recent Prog. Horm. Res.* **2001**, *56*, 1–21. [CrossRef] [PubMed]
24. Kim, T.; Spiegel, D.A. The unique reactivity of N-phenacyl-derived thiazolium salts toward α-dicarbonyl compounds. *Rejuvenation Res.* **2013**, *16*, 43–50. [CrossRef]
25. Steptoe, R.J.; Ritchie, J.M.; Wilson, N.S.; Villadangos, J.A.; Lew, A.M.; Harrison, L.C. Cognate CD4+ Help Elicited by Resting Dendritic Cells Does Not Impair the Induction of Peripheral Tolerance in CD8+ T Cells. *J. Immunol.* **2007**, *178*, 2094–2103. [CrossRef]
26. Gagnerault, M.-C.; Luan, J.J.; Lotton, C.; Lepault, F. Pancreatic Lymph Nodes Are Required for Priming of β Cell Reactive T Cells in NOD Mice. *J. Exp. Med.* **2002**, *196*, 369–377. [CrossRef] [PubMed]
27. Jaakkola, I.; Jalkanen, S.; Hänninen, A. Diabetogenic T cells are primed both in pancreatic and gut-associated lymph nodes in NOD mice. *Eur. J. Immunol.* **2003**, *33*, 3255–3264. [CrossRef] [PubMed]
28. Tang, Q.; Henriksen, K.J.; Bi, M.; Finger, E.B.; Szot, G.; Ye, J.; Masteller, E.L.; McDevitt, H.; Bonyhadi, M.; Bluestone, J.A. In Vitro–expanded Antigen-specific Regulatory T Cells Suppress Autoimmune Diabetes. *J. Exp. Med.* **2004**, *199*, 1455–1465. [CrossRef]
29. Viisanen, T.; Gazali, A.M.; Ihantola, E.-L.; Ekman, I.; Näntö-Salonen, K.; Veijola, R.; Toppari, J.; Knip, M.; Ilonen, J.; Kinnunen, T. FOXP3+ Regulatory T Cell Compartment Is Altered in Children With Newly Diagnosed Type 1 Diabetes but Not in Autoantibody-Positive at-Risk Children. *Front. Immunol.* **2019**, *10*, 19. [CrossRef]
30. Wong, F.S.; Khai Siew, L.; Scott, G.; Thomas, I.J.; Chapman, S.; Viret, C.; Wen, L. Activation of Insulin-Reactive CD8 T-Cells for Development of Autoimmune Diabetes. *Diabetes* **2009**, *58*, 1156. [CrossRef]
31. Chaparro, R.J.; Konigshofer, Y.; Beilhack, G.F.; Shizuru, J.A.; McDevitt, H.O.; Chien, Y.-h. Nonobese diabetic mice express aspects of both type 1 and type 2 diabetes. *Proc. Natl. Acad. Sci. USA* **2006**, *103*, 12475. [CrossRef]
32. Alquier, T.; Poitout, V. Considerations and guidelines for mouse metabolic phenotyping in diabetes research. *Diabetologia* **2018**, *61*, 526–538. [CrossRef]
33. De Courten, B.; de Courten, M.P.; Soldatos, G.; Dougherty, S.L.; Straznicky, N.; Schlaich, M.; Sourris, K.C.; Chand, V.; Scheijen, J.L.; Kingwell, B.A.; et al. Diet low in advanced glycation end products increases insulin sensitivity in healthy overweight individuals: A double-blind, randomized, crossover trial. *Am. J. Clin. Nutr.* **2016**, *103*, 1426–1433. [CrossRef] [PubMed]
34. Uribarri, J.; Cai, W.; Ramdas, M.; Goodman, S.; Pyzik, R.; Chen, X.; Zhu, L.; Striker, G.E.; Vlassara, H. Restriction of Advanced Glycation End Products Improves Insulin Resistance in Human Type 2 Diabetes. *Potential Role AGER1 SIRT1* **2011**, *34*, 1610–1616. [CrossRef]
35. Ehlers, M.R. Strategies for clinical trials in type 1 diabetes. *J. Autoimmun.* **2016**, *71*, 88–96. [CrossRef]
36. Choi, K.M.; Zhu, J.; Stoltz, G.J.; Vernino, S.; Camilleri, M.; Szurszewski, J.H.; Gibbons, S.J.; Farrugia, G. Determination of gastric emptying in nonobese diabetic mice. *Am. J. Physiol. Gastrointest. Liver Physiol.* **2007**, *293*, G1039–G1045. [CrossRef] [PubMed]
37. Fraser, R.J.; Horowitz, M.; Maddox, A.F.; Harding, P.E.; Chatterton, B.E.; Dent, J. Hyperglycaemia slows gastric emptying in type 1 (insulin-dependent) diabetes mellitus. *Diabetologia* **1990**, *33*, 675–680. [CrossRef] [PubMed]
38. Schvarcz, E.; Palmer, M.; Aman, J.; Horowitz, M.; Stridsberg, M.; Berne, C. Physiological hyperglycemia slows gastric emptying in normal subjects and patients with insulin-dependent diabetes mellitus. *Gastroenterology* **1997**, *113*, 60–66. [CrossRef]
39. Yki-Järvinen, H.; Helve, E.; Koivisto, V.A. Hyperglycemia Decreases Glucose Uptake in Type I Diabetes. *Diabetes* **1987**, *36*, 892. [CrossRef] [PubMed]

40. Berghöfer, P.; Peterson, R.G.; Schneider, K.; Fehmann, H.C.; Göke, B. Incretin hormone expression in the gut of diabetic mice and rats. *Metabolism* **1997**, *46*, 261–267. [CrossRef]
41. Von Scholten, B.J.; Kreiner, F.F.; Gough, S.C.L.; von Herrath, M. Current and future therapies for type 1 diabetes. *Diabetologia* **2021**, *64*, 1037–1048. [CrossRef] [PubMed]
42. Ahrén, B.; Hirsch, I.B.; Pieber, T.R.; Mathieu, C.; Gómez-Peralta, F.; Hansen, T.K.; Philotheou, A.; Birch, S.; Christiansen, E.; Jensen, T.J.; et al. Efficacy and Safety of Liraglutide Added to Capped Insulin Treatment in Subjects With Type 1 Diabetes: The ADJUNCT TWO Randomized Trial. *Diabetes Care* **2016**, *39*, 1693. [CrossRef]
43. Mathieu, C.; Zinman, B.; Hemmingsson, J.U.; Woo, V.; Colman, P.; Christiansen, E.; Linder, M.; Bode, B. Efficacy and Safety of Liraglutide Added to Insulin Treatment in Type 1 Diabetes: The ADJUNCT ONE Treat-To-Target Randomized Trial. *Diabetes Care* **2016**, dc160691. [CrossRef] [PubMed]
44. Jhala, G.; Chee, J.; Trivedi, P.M.; Selck, C.; Gurzov, E.N.; Graham, K.L.; Thomas, H.E.; Kay, T.W.H.; Krishnamurthy, B. Perinatal tolerance to proinsulin is sufficient to prevent autoimmune diabetes. *JCI Insight* **2016**, *1*, e86065. [CrossRef]
45. Heninger, A.-K.; Eugster, A.; Kuehn, D.; Buettner, F.; Kuhn, M.; Lindner, A.; Dietz, S.; Jergens, S.; Wilhelm, C.; Beyerlein, A.; et al. A divergent population of autoantigen-responsive CD4+ T cells in infants prior to β cell autoimmunity. *Sci. Transl. Med.* **2017**, *9*. [CrossRef]
46. Candido, R.; Forbes, J.M.; Thomas, M.C.; Thallas, V.; Dean, R.G. A breaker of advanced glycation end products attenuates diabetes-induced myocardial structural changes. *Circ. Res.* **2003**, *92*, 785–792. [CrossRef]
47. Forbes, J.M.; Yee, L.T.L.; Thallas, V.; Lassila, M.; Candido, R. Advanced glycation end product interventions reduce diabetes-accelerated atherosclerosis. *Diabetes* **2004**, *53*, 1813–1823. [CrossRef]
48. Harcourt, B.E.; Sourris, K.C.; Coughlan, M.T.; Walker, K.Z.; Dougherty, S.L.; Andrikopoulos, S.; Morley, A.L.; Thallas-Bonke, V.; Chand, V.; Penfold, S.A.; et al. Targeted reduction of advanced glycation improves renal function in obesity. *Kidney Int.* **2011**, *80*, 190–198. [CrossRef]
49. Tan, A.L.Y.; Sourris, K.C.; Harcourt, B.E.; Thallas-Bonke, V.; Penfold, S.; Andrikopoulos, S.; Thomas, M.C.; O'Brien, R.C.; Bierhaus, A.; Cooper, M.E.; et al. Disparate effects on renal and oxidative parameters following RAGE deletion, AGE accumulation inhibition, or dietary AGE control in experimental diabetic nephropathy. *Am. J. Physiol. Ren. Physiol.* **2010**, *298*, F763–F770. [CrossRef] [PubMed]
50. Tikellis, C.; Thomas, M.C.; Harcourt, B.E.; Coughlan, M.T.; Pete, J.; Bialkowski, K.; Tan, A.; Bierhaus, A.; Cooper, M.E.; Forbes, J.M. Cardiac inflammation associated with a Western diet is mediated via activation of RAGE by AGEs. *Am. J. Physiol. Endocrinol. Metab.* **2008**, *295*, E323–E330. [CrossRef]
51. Chen, Y.; Yan, S.S.; Colgan, J.; Zhang, H.-P.; Luban, J.; Schmidt, A.M.; Stern, D.; Herold, K.C. Blockade of Late Stages of Autoimmune Diabetes by Inhibition of the Receptor for Advanced Glycation End Products. *J. Immunol.* **2004**, *173*, 1399. [CrossRef]
52. Schuster, H.; Shao, W.; Weiss, T.; Pedrioli, P.G.A.; Roth, P.; Weller, M.; Campbell, D.S.; Deutsch, E.W.; Moritz, R.L.; Planz, O.; et al. A tissue-based draft map of the murine MHC class I immunopeptidome. *Sci. Data* **2018**, *5*, 180157. [CrossRef] [PubMed]
53. Mannering, S.I.; Harrison, L.C.; Williamson, N.A.; Morris, J.S.; Thearle, D.J.; Jensen, K.P.; Kay, T.W.H.; Rossjohn, J.; Falk, B.A.; Nepom, G.T.; et al. The insulin A-chain epitope recognized by human T cells is posttranslationally modified. *J. Exp. Med.* **2005**, *202*, 1191–1197. [CrossRef]
54. Elso, C.M.; Scott, N.A.; Mariana, L.; Masterman, E.I.; Sutherland, A.P.R.; Thomas, H.E.; Mannering, S.I. Replacing murine insulin 1 with human insulin protects NOD mice from diabetes. *PLoS ONE* **2019**, *14*, e0225021. [CrossRef] [PubMed]
55. Forman, H.J. Glutathione—From antioxidant to post-translational modifier. *Arch. Biochem. Biophys.* **2016**, *595*, 64–67. [CrossRef]
56. Ho, K.-H.; Yang, X.; Osipovich, A.B.; Cabrera, O.; Hayashi, M.L.; Magnuson, M.A.; Gu, G.; Kaverina, I. Glucose Regulates Microtubule Disassembly and the Dose of Insulin Secretion via Tau Phosphorylation. *Diabetes* **2020**, *69*, 1936. [CrossRef] [PubMed]
57. Zhu, X.; Hu, R.; Brissova, M.; Stein, R.W.; Powers, A.C.; Gu, G.; Kaverina, I. Microtubules Negatively Regulate Insulin Secretion in Pancreatic β Cells. *Dev. Cell* **2015**, *34*, 656–668. [CrossRef]
58. Fu, Z.; Gilbert, E.R.; Liu, D. Regulation of insulin synthesis and secretion and pancreatic Beta-cell dysfunction in diabetes. *Curr. Diabetes Rev.* **2013**, *9*, 25–53. [CrossRef]
59. Trogden, K.P.; McKinney, H.; Zhu, X.; Arpag, G.; Folland, T.G.; Osipovich, A.B.; Magnuson, M.A.; Zanic, M.; Gu, G.; Holmes, W.R.; et al. Microtubules regulate pancreatic beta cell heterogeneity via spatiotemporal control of insulin secretion hot spots. *bioRxiv* **2020**. [CrossRef]
60. Miyazaki, J.I.; Araki, K.; Yamato, E.; Ikegami, H.; Asano, T.; Shibasaki, Y.; Oka, Y.; Yamamura, K.I. Establishment of a Pancreatic Beta-Cell Line That Retains Glucose-Inducible Insulin-Secretion—Special Reference to Expression of Glucose Transporter Isoforms. *Endocrinology* **1990**, *127*, 126–132. [CrossRef] [PubMed]
61. Kozlov, G.; Gehring, K. Calnexin cycle—Structural features of the ER chaperone system. *FEBS J.* **2020**, *287*, 4322–4340. [CrossRef] [PubMed]
62. Filimonenko, M.; Stuffers, S.; Raiborg, C.; Yamamoto, A.; Malerød, L.; Fisher, E.M.C.; Isaacs, A.; Brech, A.; Stenmark, H.; Simonsen, A. Functional multivesicular bodies are required for autophagic clearance of protein aggregates associated with neurodegenerative disease. *J. Cell Biol.* **2007**, *179*, 485–500. [CrossRef] [PubMed]
63. Piper, R.C.; Katzmann, D.J. Biogenesis and function of multivesicular bodies. *Annu. Rev. Cell Dev. Biol.* **2007**, *23*, 519–547. [CrossRef] [PubMed]

64. Li, M.; Fu, Y.; Yang, Z.; Yin, X.M. Measurement of the Activity of the Atg4 Cysteine Proteases. *Methods Enzymol.* **2017**, *587*, 207–225. [CrossRef]
65. Engin, F.; Yermalovich, A.; Nguyen, T.; Hummasti, S.; Fu, W.; Eizirik, D.L.; Mathis, D.; Hotamisligil, G.S. Restoration of the unfolded protein response in pancreatic beta cells protects mice against type 1 diabetes. *Sci. Transl. Med.* **2013**, *5*, 211ra156. [CrossRef]
66. Oyadomari, S.; Takeda, K.; Takiguchi, M.; Gotoh, T.; Matsumoto, M.; Wada, I.; Akira, S.; Araki, E.; Mori, M. Nitric oxide-induced apoptosis in pancreatic beta cells is mediated by the endoplasmic reticulum stress pathway. *Proc. Natl. Acad. Sci. USA* **2001**, *98*, 10845–10850. [CrossRef] [PubMed]
67. Li, H.; Zhou, B.; Liu, J.; Li, F.; Li, Y.; Kang, X.; Sun, H.; Wu, S. Administration of progranulin (PGRN) triggers ER stress and impairs insulin sensitivity via PERK-eIF2α-dependent manner. *Cell Cycle* **2015**, *14*, 1893–1907. [CrossRef]
68. Matsubara, T.; Mita, A.; Minami, K.; Hosooka, T.; Kitazawa, S.; Takahashi, K.; Tamori, Y.; Yokoi, N.; Watanabe, M.; Matsuo, E.-i.; et al. PGRN is a Key Adipokine Mediating High Fat Diet-Induced Insulin Resistance and Obesity through IL-6 in Adipose Tissue. *Cell Metab.* **2012**, *15*, 38–50. [CrossRef]
69. Barbu, A.; Lejonklou, M.H.; Skogseid, B. Progranulin Stimulates Proliferation of Mouse Pancreatic Islet Cells and Is Overexpressed in the Endocrine Pancreatic Tissue of an MEN1 Mouse Model. *Pancreas* **2016**, *45*, 533–540. [CrossRef]
70. Arrant, A.E.; Davis, S.E.; Vollmer, R.M.; Murchison, C.F.; Mobley, J.A.; Nana, A.L.; Spina, S.; Grinberg, L.T.; Karydas, A.M.; Miller, B.L.; et al. Elevated levels of extracellular vesicles in progranulin-deficient mice and FTD-GRN Patients. *Ann. Clin. Transl. Neurol.* **2020**, *7*, 2433–2449. [CrossRef]
71. MacDonald, P.E.; Rorsman, P. The Ins and Outs of Secretion from Pancreatic β-Cells: Control of Single-Vesicle Exo- and Endocytosis. *Physiology* **2007**, *22*, 113–121. [CrossRef] [PubMed]
72. Tarasov, A.I.; Galvanovskis, J.; Rorsman, O.; Hamilton, A.; Vergari, E.; Johnson, P.R.V.; Reimann, F.; Ashcroft, F.M.; Rorsman, P. Monitoring real-time hormone release kinetics via high-content 3-D imaging of compensatory endocytosis. *Lab Chip* **2018**, *18*, 2838–2848. [CrossRef]
73. Akiba, S.; Sato, T. Cellular Function of Calcium-Independent Phospholipase A_2. *Biol. Pharm. Bull.* **2004**, *27*, 1174–1178. [CrossRef]
74. Caporarello, N.; Salmeri, M.; Scalia, M.; Motta, C.; Parrino, C.; Frittitta, L.; Olivieri, M.; Cristaldi, M.; Avola, R.; Bramanti, V.; et al. Cytosolic and Calcium-Independent Phospholipases A2 Activation and Prostaglandins E2 Are Associated with Escherichia coli-Induced Reduction of Insulin Secretion in INS-1E Cells. *PLoS ONE* **2016**, *11*, e0159874. [CrossRef]
75. Ramanadham, S.; Song, H.; Bao, S.; Hsu, F.-F.; Zhang, S.; Ma, Z.; Jin, C.; Turk, J. Islet Complex Lipids. *Diabetes* **2004**, *53*, S179. [CrossRef]
76. Lei, X.; Zhang, S.; Bohrer, A.; Barbour, S.E.; Ramanadham, S. Role of calcium-independent phospholipase A(2)β in human pancreatic islet β-cell apoptosis. *Am. J. Physiol. Endocrinol. Metab.* **2012**, *303*, E1386–E1395. [CrossRef] [PubMed]
77. Matthews, K.W.; Mueller-Ortiz, S.L.; Wetsel, R.A. Carboxypeptidase N: A pleiotropic regulator of inflammation. *Mol. Immunol.* **2004**, *40*, 785–793. [CrossRef] [PubMed]
78. Yu, S.-L.; Han, S.; Kim, H.R.; Park, J.W.; Jin, D.I.; Kang, J. Phosphorylation of carboxypeptidase B1 protein regulates β-cell proliferation. *Int. J. Mol. Med.* **2017**, *40*, 1397–1404. [CrossRef]
79. Liew, C.W.; Assmann, A.; Templin, A.T.; Raum, J.C.; Lipson, K.L.; Rajan, S.; Qiang, G.; Hu, J.; Kawamori, D.; Lindberg, I.; et al. Insulin regulates carboxypeptidase E by modulating translation initiation scaffolding protein eIF4G1 in pancreatic beta cells. *Proc. Natl. Acad. Sci. USA* **2014**, *111*, E2319–E2328. [CrossRef]
80. Roep, B.O.; Peakman, M. Antigen targets of type 1 diabetes autoimmunity. *Cold Spring Harb. Perspect. Med.* **2012**, *2*, a007781. [CrossRef] [PubMed]
81. Haddad, Y.; Couture, R. Kininase 1 As a Preclinical Therapeutic Target for Kinin B1 Receptor in Insulin Resistance. *Front. Pharmacol.* **2017**, *8*, 509. [CrossRef]
82. Leung, L.L.K.; Morser, J. Carboxypeptidase B2 and carboxypeptidase N in the crosstalk between coagulation, thrombosis, inflammation, and innate immunity. *J. Thromb. Haemost.* **2018**, *16*, 1474–1486. [CrossRef]
83. Lee, J.; Zhou, P. DCAFs, the Missing Link of the CUL4-DDB1 Ubiquitin Ligase. *Mol. Cell* **2007**, *26*, 775–780. [CrossRef] [PubMed]
84. Wang, A.; Luan, H.H.; Medzhitov, R. An evolutionary perspective on immunometabolism. *Science* **2019**, *363*, eaar3932. [CrossRef] [PubMed]
85. Borg, D.J.; Forbes, J.M. Targeting advanced glycation with pharmaceutical agents: Where are we now? *Glycoconj. J.* **2016**, *33*, 653–670. [CrossRef]
86. Schalkwijk, C.G.; Stehouwer, C.D.A. Methylglyoxal, a Highly Reactive Dicarbonyl Compound, in Diabetes, Its Vascular Complications, and Other Age-Related Diseases. *Physiol. Rev.* **2019**, *100*, 407–461. [CrossRef]
87. Sell, D.R.; Monnier, V.M. Molecular Basis of Arterial Stiffening: Role of Glycation—A Mini-Review. *Gerontology* **2012**, *58*, 227–237. [CrossRef] [PubMed]
88. Reeves, P.L.S.; Rudraraju, R.; Liu, X.; Wong, F.S.; Hamilton-Williams, E.E.; Steptoe, R.J. APC-targeted proinsulin expression inactivates insulin-specific memory CD8+ T cells in NOD mice. *Immunol. Cell Biol.* **2017**, *95*, 765–774. [CrossRef]
89. Gallo, L.A.; Ward, M.S.; Fotheringham, A.K.; Zhuang, A.; Borg, D.J.; Flemming, N.B.; Harvie, B.M.; Kinneally, T.L.; Yeh, S.-M.; McCarthy, D.A.; et al. Once daily administration of the SGLT2 inhibitor, empagliflozin, attenuates markers of renal fibrosis without improving albuminuria in diabetic db/db mice. *Sci. Rep.* **2016**, *6*, 26428. [CrossRef]

90. Leiter, E.H. The NOD Mouse: A Model for Insulin-Dependent Diabetes Mellitus. Available online: https://currentprotocols.onlinelibrary.wiley.com/doi/10.1002/0471142735.im1509s24 (accessed on 28 June 2021).
91. Nguyen, D.; Zhou, T.; Shu, J.; Mao, J. Quantifying chromogen intensity in immunohistochemistry via reciprocal intensity. *Cancer InCytes* **2013**, *2*. [CrossRef]
92. James, C.R.; Buckle, I.; Muscate, F.; Otsuka, M.; Nakao, M.; Oon, J.S.H.; Steptoe, R.J.; Thomas, R.; Hamilton-Williams, E.E. Reduced interleukin-2 responsiveness impairs the ability of Treg cells to compete for IL-2 in nonobese diabetic mice. *Immunol. Cell Biol.* **2016**, *94*, 509–519. [CrossRef] [PubMed]

MDPI
St. Alban-Anlage 66
4052 Basel
Switzerland
Tel. +41 61 683 77 34
Fax +41 61 302 89 18
www.mdpi.com

Metabolites Editorial Office
E-mail: metabolites@mdpi.com
www.mdpi.com/journal/metabolites

www.ingramcontent.com/pod-product-compliance
Lightning Source LLC
Chambersburg PA
CBHW061504210926
51850CB00008B/1812